# READING NATURE IN THE EARLY MIDDLE AGES

## PREMODERN ECOSYSTEMS — CLIMATE, ENVIRONMENT, PEOPLE

**Further Information and Publications**

www.arc-humanities.org/series/book-series/

# READING NATURE IN THE EARLY MIDDLE AGES

## WRITING, LANGUAGE, AND CREATION IN THE LATIN *PHYSIOLOGUS*, ca. 700–1000

by
ANNA DOROFEEVA

ARC HUMANITIES PRESS

This book is the winner of the ANZAMEMS–
Arc Humanities Award for Original Research 2021.
Its Open Access publication was made possible by funds
provided by ANZAMEMS (Australian and New Zealand

**British Library Cataloguing in Publication Data**
A catalogue record for this book is available from the British Library.



ISBN (hardback): 9781802700022
ISBN (paperback): 9781802701661
e-ISBN: 9781802701654

**www.arc-humanities.org**
Printed and bound in the UK (by CPI Group [UK] Ltd), USA (by Bookmasters), and elsewhere using print-on-demand technology.

# CONTENTS

# LIST OF ILLUSTRATIONS

# ABBREVIATIONS

| | |
|---|---|
| *CCSL* | *Corpus Christianorum Series Latina*. Turnhout: Brepols, 1953– |
| *CLA* | *Codices Latini Antiquiores: A Palaeographical Guide to Latin Manuscripts Prior to the Ninth Century*. Edited by Elias A. Lowe. 11 vols. Oxford: Clarendon Press, 1934–66. |
| *MBK* | *Mittelalterliche Bibliothekskataloge Deutschlands und der Schweiz*. Edited by Paul Lehmann et al. 4 vols. Munich: Beck, 1918–2009. |
| *MGH Capit.* | Monumenta Germaniae Historica Legum sectio III. Capitularia regum francorum 1. Edited by Alfred Boretius. Berlin: Weidmann, 1883. |
| *MGH Conc.* | *Monumenta Germaniae Historica Concilia*. Edited by Friedrich Maassen, Albert Werminghoff et al. 13 parts in 8. vols. Hannover: Hahnsche Buchhandlung, 1893–2010. |
| *MGH Epp.* | *Monumenta Germaniae Historica Epistolae sectio IV. Epistolae karolini aevi* 2. Edited by Ernst Dümmler. Berlin: Weidmann, 1895. |
| *MGH Form.* | *Monumenta Germaniae Historica Legum sectio V. Formulae merowingici et karolini aevi. Accedunt Ordines iudiciorum Dei*. Edited by Karl Zeumer. Hannover: Hahnsche Buchhandlung, 1886. |
| *MGH Poet.* | *Monumenta Germaniae Historica Poetae latini aevi carolini*. Edited by Ernst Dümmler, Ludwig Traube, Paul von Winterfeld and Karl Strecker. 4 parts in 5 vols. Berlin: Weidmann, 1881–99. |
| *MGH SRG* | *Monumenta Germaniae Historica Scriptores rerum germanicarum in usum scholarum separatim editi*. 83 vols. Hanover: Hahnsche Buchhandlung, 1871–1987. |
| *MGH SRM* | *Monumenta Germaniae Historica Scriptores rerum merovingicarum*. Edited by Bruno Krusch and William Levison. 8 parts in 7 vols. Hannover: Hahnsche Buchhandlung, 1885–1920. |
| *PL* | *Patrologiae cursus completus. Series Latina*. Edited by Jacques P. Migne. 221 vols. Paris: Migne, 1844–1864. |
| *PL Suppl.* | *Patrologiae cursus completus. Series Latina. Supplementum* Edited by Adalberto Hamman. 5 vols. Paris: Garnier, 1958–74. |

# PREFACE

THIS MONOGRAPH HAS been over a decade in the making and is therefore in many ways a collaborative project. The contribution of a widely flung network of friends and colleagues is vital to any project as large as this. To them all I would like to express my thanks.

I am very grateful to my editor Anna Henderson and managing editor Becky Straple-Sovers for their expert support, and to Amanda Power, the editor of the series in which it appears, for improving my writing immeasurably. All remaining faults are of course my own. For the opportunity to publish this monograph in full open access, and for the early career rewards associated with my membership, I am indebted to the Australian and New Zealand Association for Medieval and Early Modern Studies (ANZAMEMS) and the committee of the Arc Humanities Award for Original Research. My thanks also go to the reviewers and editors for their constructive criticism at various stages of this book's production.

Many friends and colleagues have read chapters, or contributed to my progress with advice, help, and debate. They include Tina Bawden, Anna Boreczky, Claire Burridge, Ainoa Castro Correa, Pierre Chambert-Protat, David Defries, Ilya Dines, Carlotta Dionisotti Mateusz Fafinski, Roy Flechner, Sarah Gilbert, Thom Gobbitt, Zachary Guiliano, Matthew Holford, Rutger Kramer, Emmanuelle Kuhry, Meg Leja, Ana de Oliveira Dias, James Palmer, Carine van Rhijn, Evina Steinová, Mariken Teeuwen, Baudouin Van den Abeele, Graeme Ward, and Charles West.

I owe a great deal to mentors and friends, whose advice and conversations through the long and precarious postdoctoral lifetime of this book were invaluable. I am particularly thankful to Jo Story and Erik Kwakkel for their mentorship and generosity with their time. In addition, I am grateful to Sinéad O'Sullivan and Kristina Odenweller for their support.

The research for this book was based on my doctoral work, which was supervised by Rosamond McKitterick. I am deeply grateful for her encouragement and support, faith in my ability, and infallible guidance. I am likewise grateful to Teresa Webber, whose training taught me compassion, integrity, and rigour, as well as to David Ganz, for his continuing intellectual generosity and friendship.

My original research was made possible by Arts and Humanities Research Council UK funding. In addition, I would like to thank Trinity Hall, the University of Cambridge Faculty of History Trust Funds, the London Bibliographical Society, and the Monumenta Germania Historica Institute in Munich for major grants towards archival research and travel.

No manuscript-based work is possible without the librarians and archivists who curate the collections. For their expert assistance and friendly welcome, I owe thanks to Florian Mittenhuber of the Bern Burgerbibliothek; Charlotte Denoël of the Bibliothèque nationale de France; Gerlinde Strauß of the Herzog August Bibliothek in Wolfenbüttel; Don Faustino Avagliano of the Archivio dell'Abbazia in Montecassino; Sabine Bachofner of the Stiftsbibliothek St. Gallen; and the librarians of the Biblioteca

Apostolica Vaticana, the Bibliothèque municipale in Orléans, the Bibliothèque royale de Belgique, the Bodleian Library in Oxford and the Parker Library of Corpus Christi College in Cambridge. Finally, these acknowledgements would not be complete without a note of thanks to Jason Jewell and Anne Simon, who originally fostered my love of reading and the past.

Over the years, a significant contribution to this book has been made by the Twitter medievalist community, which has frequently found images, participated in inspiring and productive discussions, and even supplied hard-to-find literature. It has been a joy to participate in this community.

However, this book would still never have been finished if not for the unfailing and unselfish support, encouragement even during the most challenging times, unstinting love and managerial talent of my partner, Nik Myers. I dedicate it to him. He has to read it now.

Introduction

# THE *PHYSIOLOGUS* BETWEEN EAST AND WEST

CONCEPTIONS OF THE natural world have often resisted definition and categorization. This is particularly the case for the period from ca. 700 to ca. 1000 in the Latin West, when Roman culture ceased to dominate, but before the rise of scholastic natural philosophy and the proliferation of encyclopedias about the natural world. There was no single clear statement about nature during this time, and written sources from this period largely rely on the classical tradition. Many histories have therefore glossed over the early Middle Ages as contributing little to the medieval understanding of nature. But the problem with this interpretation, as this book will show, is that the early Middle Ages were in fact an essential period of re-contextualisation, using complex textual approaches, of an inherited relationship between the physical world and the human imagination of it. As Walter Pohl put it, referring to historiography, "fact continually threatens to subvert fiction and to break up its coherence, so that all sorts of textual strategies are required to impose significance and narrative unity upon resistant material."[1] Early medieval texts, both newly composed and copied from older sources, are in the process of trying to impose narrative unity on a daunting subject: the mental, physical, and spiritual worlds represented in the concept of "nature." At the same time, the whole idea of what constitutes a text is in flux, blurring the lines between language, the thing that it describes, and the description itself. The *Physiologus* and the manuscripts in which it was copied serve as a lens on this early medieval semiotics, and on the processes by which it re-shaped an inherited discourse about nature along new lines.

The *Physiologus* and its manuscript context also highlights the distinctiveness of early medieval intellectual practices. The unrecorded lived experience of nature in the early Middle Ages—across diverse ecological zones, changing modes of governance and agricultural regimes—was quite different from the practice of abstract debate of the concept of nature and the ideas about it. It's vitally important to understand the cultural choices that were made by early medieval thinkers over a long period of knowledge acquisition and network-building, and through the very slow process called "Christianisation," which profoundly changed political, economic, agrarian, and intellectual worlds. At the same time, the manuscripts themselves represent an obvious transitional point between nature and thinking about nature, containing as they do texts copied on animal membrane. The sheets of membrane, in turn, represent hundreds—thousands—of animals, raised, fed, and slaughtered. This is not a new point, but surely, for literate early medieval people, the physical world was encapsulated in the physical manuscript.

1 Pohl, "History in Fragments," 347.

Manuscript context is, then, the crucial component of an early medieval natural history that has until now remained absent from most studies on the subject. The early Middle Ages saw a proliferation of books that contained multiple texts—a continuation of a late antique trend. The lack of standardisation in manuscript descriptions makes it difficult to assess numbers, but multi-text manuscripts were almost certainly the norm rather than the exception.[2] In these books, excerpts were copied alongside full texts; unrelated works, both full and partial, were linked by being copied in groups, or within the same quire or set of quires; features of layout were used either to homogenise or to distinguish between various texts, sometimes in surprising ways; and new cultural contexts were created for both old and new works, simply by copying them together in a single volume whose unique early medieval uses can be traced from its script, materiality, and paratext. Although early medieval literary and intellectual activity was substantially dependent on inherited knowledge—and has therefore often been dismissed as unoriginal—the selection, organization and re-structuring of that knowledge, as evident in multi-text books, was both intelligent and innovative. The resulting re-contextualization of existing knowledge, together with a substantial amount of new (if anonymous) writing, set the course of post-Roman learning. Even more fundamentally, the production of multi-text manuscripts enacted a belief, deeply rooted in centuries of Christian and Neoplatonic thought, that there was an essential relationship between words and God's Creation.

This book explores that imagined relationship, using the manuscripts of the *Physiologus* as a case-study. The *Physiologus* is an important text within this framework because it quickly became part of the medieval knowledge corpus, at first through its connections to works by well-known authors like Ambrose and Isidore of Seville, and then as part of the gradual development of the bestiary, which played an important role in the later medieval understanding of nature, from schoolrooms to royal courts. In addition, the miscellanies in which the *Physiologus* is exclusively found are a useful lens through which to examine the impact of early medieval written culture on perceptions of nature, as well as to assess how such perceptions fed into contemporary social and political discourse. Most importantly, the *Physiologus* was the most widely read text on nature that cannot be relegated to a separate, clearly defined field of early medieval learning—astronomy, computus, geography and medicine among them. The *Physiologus* was not entangled with other, specific preoccupations of early medieval intellectuals and it was therefore readily altered, as well as copied with a diverse selection of texts. This makes it particularly well suited for a broad investigation of "nature."

2 Dorofeeva, "Reading Early Medieval Miscellanies."

## The Stories of the *Physiologus*

The *Physiologus* consists of short chapters on various beasts, stones, and plants, real and imagined, that explain their "naturae" or particular characteristics, which are given a moral Christian interpretation.[3] The whale, for example, is described as follows:

> Physiologus spoke of a certain whale in the sea called the aspidoceleon that is exceedingly large like an island, heavier than sand, and is a figure of the devil. Ignorant sailors tie their ships to the beast as to an island and plant their anchors and stakes in it. They light their cooking fires on the whale but, when he feels the heat, he urinates and plunges into the depths, sinking all the ships. You also, O man, if you fix and bind yourself to the hope of the devil, he will plunge you along with himself into the hell-fire.[4]

The structure of the chapters generally follows a formula, which, at its most complete, consists of a Biblical citation; a phrase beginning "Physiologus dixit..." followed by a description of the subject; the meaning of this in Christian doctrine with another supporting biblical citation; and a concluding phrase which begins "bene Physiologus dixit..."[5] The stories of the *Physiologus* were very popular in the ancient and medieval world, and some of them are still familiar to us as echoes of anecdotes or symbols. It is from the *Physiologus* that we have heard about the hedgehog which rolls on the ground to spear fruit on its spines, which it then carries to its young; the phoenix, which bursts into flame at the end of its life and is reborn from its own ashes; the unicorn, which can only be tamed by a maiden; and the elephant, which carries houses and even castles on its back. These and other tales are also found in other, earlier texts; but it is through the *Physiologus* that they were popularized and embedded in Latin culture.

For a long time, the simplicity of the *Physiologus* stories impeded any serious attempt to understand its function. They evoke the fables of Aesop and other "stories with a moral" that are often read to children. Such stories seem to have didactic but otherwise no real intellectual value and little historical significance or influence. The editor of the facsimile of the *Physiologus* in Bern, Burgerbibliothek, MS 318, for example, was sure of the text's "kunstlose Naivität" and "Volksbuchcharakter."[6] But this kind of assessment is unjust to the *Physiologus* and especially to fables. None of the variety of forms taken by fables is found in the *Physiologus*.[7] The Greeks had used fables as useful material for the practice of rhetoric: that is, as elements of speech or thought. Writers such as Babrius, Aelian, and Phaedrus, active between the third century BCE and the third century CE, who took up and helped transmit Aesop's material, aimed to entertain and engage audiences as well as to educate them.[8] Priscian, the

3 See appendix II for a list of *Physiologus* chapters by manuscript and textual family.

4 Curley, trans., *Physiologus*, 45–46.

5 "Physiologus said ... well spoke the Physiologus." Vidal, Álvarez, and Osende, "La versión C," 30.

6 "Artless naïveté," "chapbook character". Steiger and Homburger, *Physiologus Bernensis*, 9.

7 Perry, *Babrius and Phaedrus*, xi and xxii–xxiii.

8 Perry, *Babrius and Phaedrus*, xxiv. See also Perry, *Studies in the Text History of the Life and Fables of Aesop*.

sixth-century grammarian (widely read in the early medieval Latin West), said that fable teaches and delights.[9] Historically, fables were much more than simple morality tales. Dismissive views of such stories are in fact grounded in a false dichotomy between "great" and "folk" literature, which until not very long ago affected even the most excellent philological scholarship. It ultimately derives from the antiquated and racist notion that "classical" Graeco-Roman literature is somehow superior to the written output of other cultures.

Setting these views aside, it becomes clear that the *Physiologus* was a new and important kind of text. It made use of the flexibility and interest of its natural material, but combined this with the emerging moral and spiritual character of a relatively new religion—Christianity. The text was copied and translated all over the European and Mediterranean worlds. It was enormously successful.[10] But the Latin manuscripts have never been studied as a corpus. We do not really know, despite extensive research on many aspects of the Latin text, who read it in the Middle Ages, or why it was copied. The period before the eleventh century, when the *Physiologus* gave way to the bestiary, has been especially neglected. Yet this was perhaps one of the most significant periods in the evolution of this text, with wide-ranging implications. What can the *Physiologus* tell us about early medieval intellectual culture? How did it affect the way in which nature was understood in the post-Roman Latin West? What does its hitherto ignored manuscript context reveal about the connections between literacy, faith, and the physical world? The evidence suggests that there are many interesting answers to these questions, which are addressed throughout this book.

In order to situate the manuscript study presented in the following chapters, we need to see how past scholarship has received the Latin *Physiologus*—not only the groundwork it has laid, but also the complications it has introduced.

## The Greek *Physiologus*

The term "Physiologus" translates literally as "the naturalist." Rather than referring to an enthusiast for natural history, it invokes a Christian authority on natural philosophy. The Greek *Physiologus* first appeared within this context, between ca. 150 and 200 CE.[11] This dating can be established with relative precision. The Greek text appears to have been known to Clement of Alexandria (ca. 150–215), Tertullian (ca.

---

**9** Delhaye, "'Grammatica' et 'Ethica'," 67–78; Priscian, "Praeexercitamina," 551–2.

**10** Among the few in-depth studies that address the function and cultural context of the *Physiologus* in its vernacular medieval forms are Baxter, *Bestiaries and Their Users*; Belli, *Il Physiologus*; Bitterling, "Physiologus und Bestiarien," 153–70; Corrigan, "The Smyrna Physiologos"; Glendinning, *A Critical Study*; Guglielmi, ed., *El Fisiólogo*; Henkel, *Studien zum Physiologus*; Hoek, "Anglo-Saxon Innovation"; Lazic and Kotarcic, *Fisiolog*; Rossi-Reder, "The Physiologus"; Wegera, "Zur Rezeption des Physiologus."

**11** McCulloch, *Medieval Latin and French Bestiaries*, 15. For a detailed discussion of the scholarship on the date of the *Physiologus*, see Alpers, "Physiologus," 598–99.

160–220), Hippolytus of Rome (ca. 170–235), and Origen (184/5–253/4).[12] Since it contains a quote from the apocryphal Gospel of James, which dates to ca. 150, the Greek *Physiologus* can be no older.[13] Some scholars have argued that the text's date of composition was ca. 254 or ca. 370, based on the use of the *Physiologus* by other authors, but these dates are much too late in light of the other evidence.[14] Alexandria was almost certainly the place where the text was compiled, to judge from the Coptic names of the month and Egyptian animals present in the Greek text, as well as a possible association with Pantaenus, the teacher of Clement of Alexandria.[15]

The *Physiologus* derives from a number of different antique sources and traditions, which probably included Indian and Hebrew material. These sources were brought to Alexandria, a centre for exchanges of various kinds in the ancient Greek world.[16] The tales derive from texts by authors including Aristotle, Ctesias, Hermes Trismegistos, Herodotus, and Plutarch, as well as Pliny; many of these texts are likely to have entered oral tradition before being brought to Egypt by merchants and other travellers.[17] The *Physiologus* therefore probably has no single author, although different versions of it have at various times been ascribed, in whole or in part, to Ambrose, Athanasius of Alexandria, Basil of Caesarea, Bolos of Mendes, Epiphanius, Jerome, John Chrysostom, John IV Nesteutes Bishop of Constantinople, and Pope Peter of Alexandria.[18] None of these attributions have been proven.

Much work remains to be done on the sources of the Greek *Physiologus*, particularly because the foundational scholarship is now fifty or more years old. Since Francesco Sbordone's work on the seventy-seven medieval Greek copies of the *Physiologus* in 1936, more manuscripts have been discovered. Sbordone was not aware of the ear-

**12** There is some debate about Origen's use of the *Physiologus*, since the Seventeenth Homily on Genesis, which has been attributed to him and which contains the phrase "nam physiologus de catulo leonis haec scribit" ("For Physiologus writes thus about the lion's cub"), is in fact an original work by his Latin translator, Rufinus. See Rowland, "The Relationship," 492; Wellmann, *Der Physiologos*. The list of locations in Origen's text that may derive from the *Physiologus* are listed in Alpers, "Physiologus," 598. See also McCulloch, *Medieval Latin and French Bestiaries*, 19 and Allport, "Three Early Christian Interpretations."

**13** See Scott, "The Date of the Physiologus," Lauchert, *Geschichte des Physiologus*, 69, Alpers, "Physiologus," 598, and Alpers, *Untersuchungen zum Griechischen Physiologus*, 53.

**14** Wellmann, *Der Physiologus*, 11 and 13; Scott, "The Date of the Physiologus." Wellmann's study is generally problematic and does not take into account some of the evidence, according to Alpers, *Untersuchungen zum Griechischen Physiologus*, 15. See also Treu, "Zur Datierung des Physiologus," 101–4.

**15** Not, as Wellmann thought, Syria. See Alpers, "Physiologus," 598.

**16** Steiger and Homburger, *Physiologus Bernensis*, 10.

**17** Woodruff, "The *Physiologus* of Bern," 234. Wüstefeld, "Catalogue," 190. The exact sources of the *Physiologus* are still not fully documented, though much of the scholarship on the Greek, Armenian, and Coptic texts, and on the animal turn, investigates these sources in a general way. See, for example, Muradyan, *Physiologus. The Greek and Armenian Versions*; Porcier, Ikram, and Pasquali, ed., *Creatures of Earth, Water and Sky*; and Suciu, "Quotations from the Physiologus."

**18** Rowland has shown that Basil wrote independently of the compiler of the *Physiologus*; see "The Relationship".

liest known witness of the Greek *Physiologus*, the late tenth-century codex New York, Morgan Library and Museum, MS M.397, from southern Italy.[19] But although Sbordone's study depended on even older (both nineteenth and early-twentieth century) research, particularly the study of the Greek and German versions of the *Physiologus* made by Friedrich Lauchert in 1889, it has still not been superseded.[20] More recent scholarship on the Greek text has included critical editions of three of the text's families and useful historical essays, but this work is essentially a supplement to Sbordone's.[21] The classic studies of the Latin text have also examined its Greek and Alexandrian roots, but these are also either very old or limited in scope.[22] Despite the high quality of much of this scholarship, it needs to be updated and re-evaluated in the light of our current understanding of cultural exchanges around the Mediterranean. This is essential in order to situate the Greek *Physiologus* in its proper place as a new textual product of vibrant multicultural learning and Christian belief in antiquity.

Further work on the Greek transmission would also shed more light on the translation of the *Physiologus* into Latin, which is the language of the largest extant group of *Physiologus* manuscripts (taking into account all copies to ca. 1500). The translation is likely to have taken place in the fourth century. A plausible *terminus ante quem* is based on Ambrose's *Hexaemeron*, composed between 386 and 388, as this text incorporates some of the description of the partridge from the Latin *Physiologus*.[23] However, this resemblance between the two texts may have come about because Ambrose, who knew Greek, independently used the same Greek sources that went into the *Physiologus*.[24] This may also have been the case with the African bishop Verecundus of Junca,

---

**19** Sbordone, *Physiologus*. The manuscript has been partially digitized at http://ica.themorgan.org/manuscript/thumbs/112333. The first published editions of the *Physiologus* by Ponce de Leon in Rome (1587) and Antwerp (1588), and Denis Pétau (1622), were based on Paris, Bibliothèque nationale de France, MS gr. 835. All the editions attribute the authorship of the text to Epiphanius. See Perry, "Physiologus," cols. 1103 and 1111, and Côté, "Un manuscrit oublié."

**20** Lauchert, *Geschichte des Physiologus*; Heider, *Der Physiologus*; Cahier and Martin, eds., *Mélanges*; Ahrens, *Zur Geschichte des sogenannten Physiologus*; Goldstaub and Wendriner, *Ein tosco-venezianischer Bestiarius*; and Goldstaub, "Der Physiologus."

**21** Offermanns, ed., *Der Physiologus nach den Handschriften G und M*; Kaimakis, ed. *Der Physiologus nach der ersten Redaktion*; Muradyan, *Physiologus: The Greek and Armenian Versions*; Curley, "Physiologus"; Curley, trans., *Physiologus*; Lazaris, *Le Physiologus grec*, 1 and 2; Kindschi Garský and Hirsch-Luipold, eds., *Christus in natura*. The historiography of the Greek *Physiologus* beyond the key works already mentioned is too extensive to be given in detail here. For more information, see Henkel, *Studien zum Physiologus*, and the detailed summary of this work by Guglielmi, "Review: Nikolaus Henkel. Studien zum *Physiologus* im Mittelalter."

**22** Wellmann, *Der Physiologus*; Carmody, *Physiologus latinus: éditions préliminaires, versio B*; Sbordone, "La tradizione manoscritta"; Woodruff, "The *Physiologus* of Bern"; Carmody, "De bestiis et aliis rebus," *Physiologus latinus versio Y*, and *Quotations in the Latin Physiologus*. On the *Physiologus* as a Christian text, see also Cox, "The Physiologus"; Evdokimova, "Deux traductions"; Gerlach, "Physiologus"; Seel, *Der Physiologus*; and Treu, "Zur biblischen Überlieferung im Physiologus."

**23** McCulloch, *Mediaeval Latin and French Bestiaries*, 21; Carmody, "De bestiis et aliis rebus," p. 153, n. 3.

**24** Benjamin, "Review: Florence McCulloch. *Mediaeval Latin and French Bestiaries*."

who used the *Physiologus* in the mid-sixth century.[25] Currently, two convincing arguments exist for ascribing the translation to before the fourth century. The first was made by the art historian Helen Woodruff, who pointed out that most of the Latin versions contain a list of heretics condemned at the First Council of Constantinople in 381 in the story of the ant, but do not mention the Nestorians, against whom the First Council of Ephesus was called in 431. She believed that the translation into Latin must therefore have been made before this date.[26] The second argument for a fourth-century date was made by Max Wellmann, who noted that the commentary on the *Hexameron* of Pseudo-Eustathius of Antioch (d. ca. 337) quotes directly from the *Physiologus.*[27] This internal textual evidence is useful, but the lack of a sustained comparison between the Greek and Latin versions, or indeed a proper text critical edition of either, means that we still know very little about the origins of the *Physiologus.*[28] How was it used by early Eastern Christian writers, and what do the circumstances of its origin and translation tell us about late antique Christian culture and intellectual exchange? The popularity of the *Physiologus* across this region is evidenced by its translation into Arabic, Armenian, Coptic, Ge'ez (an Ethiopic language), and Syriac, as well as Latin, but none of these traditions have been extensively compared either.[29] Since no contextual assessment of the function of the *Physiologus* in late antiquity exists, the conclusions of the textual scholarship to date sketch out a broad history that is incomplete and has little to say about the roots of the Latin text.

## The Bestiary

By contrast, a great deal of information is available about the later medieval development of the *Physiologus*. Between the tenth and twelfth centuries, the Latin *Physiologus* underwent several transformations. Firstly, it became one of the texts used in the developing cathedral schools. This is suggested by the appearance of a *Physiologus* on an eleventh-century book list from the school of the Cathedral of Le Puy (Haute-Loire).[30] Secondly, different recensions began to appear. These include a metrical ver-

**25** Manitius, *Geschichte der lateinischen Literatur*, 1:117.

**26** Woodruff, "The *Physiologus* of Bern," 237.

**27** Wellmann, *Der Physiologus*, 15.

**28** The most recent publication on the Greek *Physiologus* contains comparatively little new research: Lazaris, *Le Physiologus grec*, 2.

**29** One exception is Muradyan, *Physiologus: The Greek and Armenian Versions*. A study of the Ge'ez tradition has recently been made by the project "Fra Alessandria e Aksum. La tradizione greco-etiopica del Fisiologo (secoli III–VI)" (2018–20), directed by Prof. Gianfrancesco Lusini at the Università degli Studi di Napoli L'Orientale, supported by the Associazione Internazionale di Studi sul Mediterraneo e l'Oriente (IsMEO). See also Macé and Gippert, *The Multilingual Physiologus*.

**30** The book catalogue was copied onto the final bifolium of Paris, Bibliothèque nationale de France, MS lat. 7501 (archivesetmanuscrits.bnf.fr/ark:/12148/cc126140q). The *Physiologus* entry reads "Tunc sequitur phisialogus" (fol. 59r). Clark, *A Medieval Book of Beasts*, 9n20, based on Glauche, *Schullektüre im Mittelalter*, 70. Unlike Glauche, Henkel, *Studien zum Physiologus*, 56, thought this was the *Physiologus Theobaldi*.

sion of twelve chapters, the *Physiologus Theobaldi* attributed to the eleventh-century Abbot Theobaldus of Montecassino—also used in cathedral schools—and the prose *Dicta Chrysostomi*, which formed the basis of the German vernacular families of the text.[31] These versions did not exist before the eleventh century and are therefore not discussed in this book. The later bestiary appears to be based on the B recension of the *Physiologus*, the largest of all the recensions, which had reached England by the twelfth century.[32] There, it received expansions and additions from a large number of different sources, principally the third-century *Collectanea rerum memorabilium* of Solinus (itself an adaptation of Pliny's *Historia naturalis*) and the seventh-century *Etymologiae* of Isidore of Seville. From the twelfth century, its major sources included Hugh of Fouilloy's *Aviarium* and Peter of Cornwall's *Pantheologus*.[33]

The bestiary was a significant central and late-medieval cultural phenomenon, and it has been extensively studied as a result. Its impact, and its gradual development from the Latin *Physiologus*, have meant that the latter has been somewhat marginalised. The difficulty of expressing the exact difference between the bestiary and the *Physiologus* has also contributed to a view of the *Physiologus* as a mere bestiary source text. A precise itemisation of the variant characteristics of the two texts would require a separate study, both because the texts in question vary greatly in their sources and contents across all the manuscripts, and because the line between the two cannot easily be drawn. For instance, the *Physiologus* in the tenth-century manuscript Vatican City, Biblioteca Apostolica Vaticana, MS Pal. lat. 1074 has been classified as a B-Is bestiary—that is, a B-recension copy of the *Physiologus* which had been expanded from Isidore of Seville's *Etymologiae*. However, the historical context of this codex indicates that the manuscript was read in the same ways as other early medieval miscellanies, and not as a bestiary (see Chapter 5). Variations to the form and structure of a text, on their own, are not enough to make a bestiary or a *Physiologus* in historical, if not in text critical terms.

For this reason, the bestiary cannot be seen as an improvement on the *Physiologus*, although that is how it is represented in some histories of the bestiary. The thorough study by Willene Clark is one example: "[The] new bestiary had much greater breadth than *Physiologus*, a far more rational organization, a lack of monastic ideas and language, and forward looking art...the moral/ethical didacticism and the ancient authority of the text's sources was unquestionably old fashioned."[34] Yet to view the bestiary as more rational, and as having greater breadth, than the *Physiologus*, is to do them both a disservice, as these are relative concepts that depend entirely on cultural context. The bestiary had more content, and was arranged by category—but it does not then follow that the *Physiologus* was its poorer version. The extensive scholarship on

---

**31** Henkel, *Studien zum Physiologus*.

**32** On the various textual families of the bestiary, see James, *The Bestiary*; McCulloch, *Mediaeval Latin and French Bestiaries*; Baxter, *Bestiaries and Their Users*; Clark, *A Medieval Book of Beasts*. See also Yapp, "A New Look"; and Dines, "A Critical Edition" and "The Problem."

**33** Clason, "Animals, Birds and Fish," p. 22.

**34** Clark, *A Medieval Book of Beasts*, 9.

the bestiary has greatly contributed to the history of the *Physiologus*, but also overshadowed its function before the bestiary's rise. As we shall see in later chapters, the success of the *Physiologus* in the eighth, ninth, and tenth centuries indicates that it had important uses of its own.

## The Latin *Physiologus* Text

The content and order of the *Physiologus* chapters was frequently modified or expanded, which meant that the early Middle Ages inherited a complex text that lacked a single fixed form. This continued throughout the medieval period: of the approximately 500 extant manuscripts of the *Physiologus* and bestiary, both Western and Eastern, no two are identical in terms of text, chapter order or illustration.[35] Although this seems to make it impossible to assess the *Physiologus* as a single coherent text, each addition or modification actually highlights a different aspect of its reception. The early medieval Latin copies are particularly interesting in this respect, since they not only shaped early medieval ideas about the natural world, but also preserve the evidence of how textual difference and variety, which defined learning at a local level, was a key element of early medieval intellectual culture.

As a consequence of the early scholarly interest in the *Physiologus*, however, this contextual history has been ignored in favour of more traditional philological approaches. Perhaps the greatest part of the scholarship has concentrated on certain aspects of its textual development, manuscript survival and transmission. This was at first essential to establishing the connection between the Greek and Latin texts: although the *Physiologus* was translated into Latin soon after its composition in the second century, the earliest extant copies (in any language) date only from the eighth and ninth centuries. There exists a long historiography of attempts to define the various recensions of the text, beginning with the simple A, B, and C determined by Charles Cahier in his nineteenth-century edition.[36] Subsequent work on the Latin recensions showed that Cahier's divisions were not comprehensive, as some manuscripts fell outside his classification. This led scholars to adopt their own systems: for example, certain of the manuscripts that could not be classified using Cahier's system were referred to by Sbordone as M, N, and E, and by Carmody as Y, Y$^2$, and Y$^3$.[37] The systems used by Cahier and Carmody have survived: the Latin *Physiologus* manuscripts are now divided into four main recensions designated A, B, C, and Y (see Appendix II), with several sub-branches (AB, B-Is).[38] But this system is out of date. More manuscripts are now known, and other kinds of evidence are available that suggest a much more complex transmission history. The *Physiologus* used to help compile the *Liber*

---

**35** Muratova, "Problèmes de l'origine," 303. See also Pakis, "A Note in Defense," 732.

**36** Cahier and Martin, eds., *Mélanges*, vol. 1. A list of the editions and translations of the Latin *Physiologus* is provided in Schönberger et al., eds., *Repertorium*, 452–53.

**37** Sbordone, *Physiologus*; Carmody, *Physiologus latinus versio Y*.

**38** There are a few studies of individual versions, but they are rare. For an example, see Villar Vidal and Álvarez, "El Fisiólogo latino."

*glossarum*, for example, must have represented a different textual version compared to any still extant, and was perhaps older as well.[39] The recensions of the Latin *Physiologus* also shed light on the bestiary, but their initial classification is now generally cited uncritically in bestiary studies. There exists no authoritative study of the manuscript dissemination, nor, until now, a comprehensive handlist of manuscripts. Much more work remains to be done on the history of the text.

Historical scholarship, including the summaries provided in handbooks and encyclopedias, tends to rely upon these earlier, pioneering studies, which appear to have identified the text's applications and traced its development. They often assume that the authority of the antique sources and moral teachings of the *Physiologus* meant that it was a schoolroom textbook, which set out moral Christian principles in a simple, readily accessible format for both the teacher and the pupil. One scholar suggested that it is in fact this very use that accounts for the relative scarcity of extant manuscripts in the eighth to tenth centuries, as they would have been subjected to greater wear and tear than non-didactic texts.[40] Yet, apart from the use of the *Physiologus* in cathedral schools from the eleventh century, and the simplicity of its stories, no evidence has yet been provided that the *Physiologus* was a schoolbook, or that it was used to teach at all in classrooms, during the eighth, ninth, and tenth centuries. That this assessment of the text persists may be due to a propensity on the part of some modern *Physiologus* scholars to defer to the judgement of those who wrote during the nineteenth and early twentieth centuries, the era of definitive scholarship on the Greek and Latin texts.[41] The textual studies of these and other authors remain authoritative, but their judgement of the *Physiologus* is not entirely free of that bias against the fourth to tenth centuries—the period which saw the appearance and widespread use of the Latin *Physiologus*—which viewed them as part of an intellectual "Dark Age." This label is never explicitly used, but its judgement is expressed nonetheless. Writing about the *Physiologus* in Bern, Burgerbibliothek, MS 318 in 1851, Cahier remarked that the scribe was "un copiste patient et attentif, mais de courte intelligence; une espèce *d'homme de peine* dévoué à sa tâche avec une obéissance véritablement aveugle."[42] This statement reflects a tendency in nineteenth-century scholarship to regard early medieval scholars essentially as preservers and copyists, whose work lacked originality and was therefore of limited interest, though instrumental in the transmission of antique learning.

Another important reason for the neglect of the early medieval *Physiologus* is that the text appears to have been regarded by modern scholars as less valuable in itself than as a witness to the transmission of a number of antique intellectual traditions and textual influences. In this respect, the Greek versions have naturally been seen

---

**39** Gorla, "Some Remarks."

**40** Orlandi, "La tradizione del Physiologus," 1104.

**41** Lauchert, *Geschichte des Physiologus*; Goldstaub, "Der Physiologus"; Wellmann, *Der Physiologus*; Carmody, *Physiologus latinus versio Y*.

**42** "A patient and attentive copier, but of limited intelligence; a kind of handyman, devoted to his task with a truly blind obedience." Cahier and Martin, *Mélanges*, 2:98.

as more relevant. Moreover, because the Latin *Physiologus* served as the basis for the expansion of the bestiary, it has been included in most studies of the bestiary's development, where it is of secondary importance. For this reason, among the most important works on the Latin *Physiologus* are those whose primary focus is in fact on the bestiary.[43]

The impression gained from the published scholarship is that interest in the pre-bestiary *Physiologus* as an independent text has been sporadic. The first study to bring together the results of earlier research on the Latin *Physiologus* families, and the problems of their transmission and classification, was published by Henkel in 1976.[44] He made clear how much work still remains to be done: the family groupings are complex and there is still disagreement over the classification of individual manuscripts. This is partly due to contamination, where groups of chapters overlap across the different versions.[45] No *stemma* of any Latin family exists, despite the historiographical focus on the textual history of the *Physiologus*. Henkel himself dedicated half of his study to the German vernacular families, which derive from the Latin. Perhaps the only scholar to focus exclusively on the Latin *Physiologus* has been Giovanni Orlandi, whose 1984 article is notable for its thorough reference to earlier research and attempt to draw some conclusions about the applications of the Latin text.[46]

There is, however, little incentive for textual scholars, taking into account the current state of knowledge about the function of the *Physiologus*, to publish a critical edition of the text or to work any further on its sources and textual history. It continues to be considered a *Volksbuch* or chapbook. Yet the complexity and interest of the Latin *Physiologus* lie not only in the text itself but also in the use that was made of it by the Latin West in the early Middle Ages. Each translation and change in the text meant an adaptation of its use for the needs of a new audience. In the early medieval period the *Physiologus* began to be copied more frequently, to judge from the pattern of survival: fourteen Latin manuscripts and fragments from between the eighth and tenth or early eleventh centuries are extant. This suggests that the Latin *Physiologus* was considered to be of some significance, in a period characterized by a cultural and intellectual renewal.

The manuscripts of the *Physiologus* themselves corroborate this. The *Physiologus* was collocated with a broad range of other widely read texts: passages from works by Isidore and Augustine, glossaries, riddles, Fredegar's *Chronicle*, Solinus' *Collectanea rerum memorabilium*, the *Psychomachia* of Prudentius, fragments and extracts from Cicero, Boethius, Eucherius, and Jerome, and the *Liber monstrorum*. Twenty-two entries in the great early medieval glossary compilation, the *Liber glossarum*, are taken from the *Physiologus*. It was perhaps used by Ambrose, Gregory of Tours, Isidore of

43 McCulloch, *Mediaeval Latin and French Bestiaries*; Clark and McMunn, *Beasts and Birds*. See also Diekstra, "The Physiologus."

44 Henkel, *Studien zum Physiologus*.

45 Henkel, *Studien zum Physiologus*. See also Kuhry, "Zoological Inconsistency."

46 Orlandi, "La tradizione del Physiologus."

Seville, and Jerome in passages about the meaning of animals and constellations.[47] That it had applications outside the schoolroom is implied by the will of Eberhard, count of Friuli, and his wife Gisela, made in 863 or 864, which lists a *Liber bestiarum* among the private chapel books that were bequeathed to their eldest son Unruoch (see Chapter 2). Such a bequest suggests that the material of the *Physiologus* was of some interest to the educated layman in the Carolingian period. Together these links reveal the *Physiologus* to have been part of an early medieval intellectual tradition that had its roots in the inherited, encyclopaedic knowledge of the late antique and early medieval Christian spheres, and in knowledge about the created world.

The art historical context of the Latin *Physiologus* is equally important. It is a witness to the continuity of Byzantine iconography in medieval Europe.[48] Its earliest extant illustrated copy in the manuscript Bern, Burgerbibliothek, MS 318, created in the second third of the ninth century, contains luxurious pictures that point both to the survival of the illusionistic style of painting found in late antique Greek manuscripts and to the development and use of this style by the Carolingians. More specifically, the manuscript has been studied as an important witness to the "Rheims School": a ninth-century creative and intellectual revival, probably initiated by Archbishop Ebo upon his appointment to the see of Rheims in 816.[49] This manuscript not only provides a rich source of evidence for some of the ways in which antique art and knowledge was received and adapted in the early Middle Ages, but also reveals some of the innovative ways in which pictorial sources could be used for teaching natural allegory (see Chapter 4).[50]

It is clear, therefore, that the uses, recipients, and context of the early medieval Latin *Physiologus* require re-assessment. This is one of the goals of the present book. Through an examination of the manuscripts and their texts, it will consider the relationship between the *Physiologus* and the early medieval understanding of nature. In addition, this book will situate this process in its early medieval political and cultural context. Although the general evidence for the *Physiologus* in the early medieval West is not restricted to continental Europe, all the manuscripts were demonstrably made

**47** Ambrose of Milan, *Hexaemeron*, 6.3.13; Gregory of Tours, *De cursu stellarum ratio*, 12; Isidore of Seville, *Etymologiae* 11.3.36; Jerome, *In Hieremiam prophetam* 17:11. Little serious study has been done on the textual evidence for the use of the *Physiologus* in Latin-speaking Europe before the eighth century, however, and the possibility remains that other sources were used by the authors cited here. On Gregory of Tours and the *Physiologus* see Van den Broek, *The Myth of the Phoenix*, 203.

**48** Tselos, "A Greco-Italian School," 1; Muratova, "La production des manuscrits"; Leclercq, "De l'art antique à l'art médiéval." See also Wittkower, "'Physiologus' in Beatus Manuscripts," 253–54, for a more specific iconographical discussion. On art in the Greek *Physiologus*, see the overview volumes Lazaris, *Le Physiologus grec*, 1 and 2.

**49** Swarzenski, "Die karolingische Malerei"; Koehler and Mütherich, *Die karolingischen Miniaturen*; Mütherich, "Carolingian Manuscript Illumination"; Nees, *Frankish Manuscripts*.

**50** Swarzenski has shown that the *Physiologus* in Bern, Burgerbibliothek, MS 318, influenced some later psalter illustrations; Swarzenski, "Die karolingische Malerei," 88.

in continental centres. They were therefore shaped by the Carolingians and their political, cultural, and intellectual program.

## Nature and Reform

The impact of the Carolingian rulers and their activities on early medieval society, culture and economy has been much debated. The eighth and ninth centuries saw a program of changes, many initiated by Charlemagne, which have been described variously as "reform," "renaissance," or "*correctio*."[51] Its earliest (but by no means only) decisive statements are usually considered to be the Capitulary of Herstal in 779, the *Epistola de litteris colendis* in the late 780s and the *Admonitio generalis* in 789. The latter was an extensive piece of legislation in eighty-two clauses that restated previously instituted regulations for the clergy, updated for the Frankish church at the end of the eighth century. Clause 72 of the *Admonitio generalis* is often cited as a summary of the reform aims:

> Let us establish schools that many may be drawn to God's service by their upright way of life and they may gather and associate to themselves not only children of servile condition but also the sons of freemen. And let schools for teaching boys the psalms, the *notae*, singing, computation and grammar be created in every monastery and episcopal residence. And correct catholic books properly, for often, while people want to pray to God in the proper fashion, they yet pray improperly because of uncorrected books. And do not allow your boys to corrupt them, either in reading or in copying; and if there is need to copy the gospel, Psalter, or missal, let men of full age do the writing, with all diligence.[52]

This clause commanded monastery and cathedral schools to teach psalms, *notae*, singing, computation, and grammar, and to correct books.[53] In the list of these subjects we also see the emphasis of learning as a means to an end: monks were to be taught in

---

**51** A small selection of the most important studies includes *I problemi della civiltà carolingia*; Borst, *Die karolingische Kalenderreform*; Contreni, "The Carolingian Renaissance"; Fried, *Die Formierung Europas*; Kramer, *Rethinking Authority*; McKitterick, *The Carolingians and the Written Word*; Nelson, "On the Limits of the Carolingian Renaissance"; Reuter, "'Kirchenreform' und 'Kirchenpolitik'"; Schramm, "Karl der Große"; Smith, "Emending Evil Ways"; Ullmann, *The Carolingian Renaissance*; and Wormald and Nelson, ed., *Lay Intellectuals*. For a more extensive bibliography, see Noble, "Carolingian Era."

**52** "Et ut scolae legentium puerorum fiant. Psalmos, notas, cantus, compotum, grammaticam per singula monasteria vel episcopia et libros catholicos bene emendate; qui saepe, dum bene aliqui Deum rogare cupiunt, sed per inemendatos libros male rogant. Et pueros vestros non sinite eos vel legendo vel scribendo corrumpere; et si opus est uangelium, psalterium et missale scribere, perfectae aetatis homines scribant cum omni diligentia." *Admonitio generalis*, chap. 72, MGH Capit. I, 60. Translated in King, ed., *Charlemagne: Translated Sources*, 217.

**53** The word *notae* can be interpreted broadly as "signs in writing"; suggested translations have included "Tironian notes" and "musical notation" (though this is less likely). Critical and abbreviation signs may also have been meant. Given the amount of supra-alphabetic signs involved in early medieval writing, there seems no reason to interrogate the translation "signs in writing" any further. The term "notae" may indeed have been left deliberately open to the interpretation of individual bishops and abbots, depending on their resources. The *Admonitio generalis* itself presents difficulties, as several variants of its Latin text exist and their translation and interpretation are not

order to improve their Latin, ability to participate in liturgy, and time-keeping, so as to understand the calculation of the date of Easter. Obedience to canon law, hierarchy, and order within the Church were set up as essential, together with the education and literacy of the clergy. The ultimate purpose of these changes was to guide populations towards salvation through the Church. Charlemagne presented himself in the tradition of the biblical king Josiah: "For we read in the Books of Kings how the holy Josiah strove to recall the kingdom given to him by God to the worship of true religion, by visitation, by correction, by admonition."[54]

The changes proposed in the *Admonitio generalis* and other texts over the following two hundred years were often put into practice, but it is anachronistic to call them either a "reform" or a "renaissance." Recently, Carine van Rhijn and Rutger Kramer have set out the range of problems associated with the historiography of this "reform history."[55] They have pointed out that, to a large extent, the vocabulary of "reform," "correction," and "renaissance" is rooted in the interests of twentieth-century scholars, and is not especially well reflected in the sources. Wider problems with a discourse of reform include implicit standardisation, which is not at all evident in the early Middle Ages, and an over-emphasis on elite power. A better way to describe the drivers of changes that occurred under the Carolingians is weight given to a Christian idea of betterment for both society and the self, "the centrality of education and learning, and sustained generous patronage." In addition, van Rhijn has called for a new look at the Carolingian "reform" that re-centres the agency of writers who are anonymous to us.[56] Such writers constituted the majority of early medieval scholars producing new books and texts, but their works have rarely been edited or studied, in sharp contrast to named authority figures such as Theodulf of Orléans. The extant sources indicate that the ability of early medieval minority elites like Theodulf to effect top-down transformative change was limited, and that there was significant horizontal knowledge exchange as well.

This view of the transformation that occurred from the late eighth century places books and their contents firmly centre stage, as perhaps the most significant product of the Carolingian period. That is not to say that there was no intellectual activity in previous centuries: the Carolingian scribes selecting and copying texts for the day-to-day work of teaching and securing the Christian faith were continuing a tradition stretching back at least to Vivarium in the sixth century. The difference was, firstly, in extent—there was an explosion in book production under Charlemagne and his successors—and secondly, in a conscious awareness of book-making as a political state-

straightforward. See Steinová, *Notam superponere studui*. For a discussion of musical notation, see Atkinson, *The Critical Nexus*, pp. 49–50, n. 3; and Steinová, "*Psalmos, notas, cantus.*"

**54** "Nam legimus in regnorum libris, quomodo sanctus Iosias regnum sibi a Deo datum circumeundo, corrigendo, ammonendo ad cultum veri Dei studuit revocare." Prologue, *Admonitio generalis*, 54.

**55** Kramer, *Rethinking Authority*.

**56** Van Rhijn, "Introduction," *Rethinking the Carolingian Reforms*. I am grateful to the author for sharing this chapter with me before publication.

ment. Books and script became oriented as the products of centres that projected particular identities, through the cultivation of house or regional styles. There was a great deal of variation and complexity here, as with anything else in the early Middle Ages. In many places, however, books, and script became one of the means through which it was possible to exercise power. This extended to texts as well. They now had a purpose that went beyond practical utility: they were part of a wider statement about faith, within the context of the whole books in which they appeared and as part of the intellectual networks such books represented.

For scholars of early medieval manuscripts, this is not a new idea.[57] But this context is still absent from research on the natural world as an intellectual concept during this period, although it, too, was politicised and made to serve as a tool for achieving the aims of those who controlled the production of writing. This book attempts to set aside the ideas of "correction" and "reform," therefore, focusing instead on education, language, and the exchange of Christian knowledge as the key elements of a conscious cultural shift in which the *Physiologus*, and its statement about nature, participated.[58] It is a history of how the natural world was made to serve the aims of early medieval intellectuals. As such, it takes up a different challenge from that posed by the conceptualization of the Carolingian effort. Instead, it attempts to bring the sources into clear focus, and to accommodate their full range of diverse and problematic readings, as a way to understand the fundamental connection between nature and writing in the early Middle Ages. Each chapter therefore represents a semi-independent study of the relationships between the Church, ruling elites and monastic communities, and the institutions and institutional needs, that formed and directed changing attitudes to the natural world.

## The Manuscripts

The early medieval *Physiologus* cannot simply be seen as an indifferent replication of an antique text during a period of such wide-ranging change, and particularly not in the light of scholarship from the past fifty years that has shown how the Carolingians not only inherited but also transformed knowledge. But in precisely what ways did the *Physiologus* participate in this transformation? This question will be answered by an examination of its manuscript context: the alterations and additions to the text, the script and layout of its copies, its textual collocations and production circumstances. This study will particularly focus on manuscripts as evidence for the living human traditions that they embodied, with all their limitations. This includes when these manuscripts were produced and for what discernible purpose. Where relevant, each chapter considers how they were made and using what materials, human resources, and effort. Each chapter looks at what their decoration, marginalia, dimensions, and

57 The research group consisting of Steffen Patzold, Carine van Rhijn and Bastiaan Waagmeester in particular has done groundbreaking research on this subject; see Patzold and van Rhijn, eds., *Men in the Middle*; Patzold, *Presbyter*; Waagmeester, "Pastoral Works."

58 On education in this context, see Contreni, "John Scottus."

other characteristics reveal about the use to which they may have been put—considering that use may not have coincided with purpose of production. At the same time, this study investigates how the texts were put together, what their script may tell us about their origin and status, and what they can reveal about the interests of the compilers and intellectual context of each codex.

The full list of the contents and features of the early medieval manuscripts of the pre-bestiary Latin *Physiologus* is provided in Appendix I. It is one of the new contributions of this study, and underpins the analysis presented here. The historiographical focus on the textual families of the Latin *Physiologus* has, as noted above, contributed to the scholarly neglect of its physical copies. They have never been considered as a group; some have not been examined at all in *Physiologus* scholarship. This is the case for the Montecassino and Chartres manuscripts. Some are not described in any detail in available catalogues. This has meant that many of the texts contained within these books have remained unidentified, beyond a Latin description or *incipit*. Yet the fourteen manuscripts identified in Appendix I represent a definable period in the history of the *Physiologus*, which facilitates their comparison and evaluation. All the codices can be dated to between ca. 700 and ca. 1000. Although the *Physiologus* continued to be copied until at least the sixteenth century, by the end of the eleventh it had already begun to be expanded into the bestiary, while variant versions such as the metrical *Physiologus Theobaldi* had begun to appear by the middle of the eleventh century. Its use appears to have been evolving as the bestiary gradually emerged. Copies from the mid-eleventh century onwards cannot therefore be used as evidence of earlier practice and attitudes alongside the text used by the Carolingians and Ottonians. For this reason, Appendix I does not extend beyond ca. 1000 CE. One of its important functions is to correct many catalogue errors and omissions. In some cases, the origin or palaeography of a codex has been revised, with a fuller discussion provided in the corresponding analysis of that codex elsewhere in the book.

Chapter 1 examines our understanding of nature and science, including the different terms used to describe it both today and in the early Middle Ages. The chapter uses a range of texts to explore the complex ideas about the visible and invisible world, the natural and the supernatural, God and humanity's place in Creation, that informed early medieval thought. It concludes that the early medieval Latin West did not understand the natural world in the modern sense as something that can be investigated by the senses, but rather as a canvas of portents and mysteries to search for order, truth, and signs of God's plan for Creation. This conclusion rejects the adversarial and apocalyptic views of nature in late antique and early medieval Europe that have prevailed in much of the scholarship since being proposed by David Herlihy in 1980. It also questions the prevailing opinion that before the twelfth century, with its attendant view of nature as a logical, scientifically observable physical realm, the incoherent natural world was imbued with symbols whose intelligibility depended only on a higher order. Such a teleological, "Dark Age" dismissal ignores the historical context of early medieval attempts to understand the natural world. This chapter shows that the relationship between humanity and nature in this period was instead shaped by a long discourse rooted in the past which, though sometimes heterogeneous, nevertheless emphasized a single idea: the hope of human salvation through divine Creation. Humanity, the physical world and the divine were considered as part of a whole, and this view of "nature," while very different from later medieval and modern conceptions, was nevertheless complex, well-articulated and unifying.

Chapter 2 considers the geographical spread of the *Physiologus* manuscripts and other known mentions of the *Physiologus* from early medieval library catalogues and booklists. Based on this, it considers who the audiences and readers of the *Physiologus* might have been. It then examines the complex contents of the collection that contains the earliest known copy of the *Physiologus* in any language: Bern, Burgerbibliothek, MS 611 + Paris, Bibliothèque nationale de France, MS lat. 10756. This collection dates to the year 727 and situates the *Physiologus* in the context of a dynamic early medieval culture led by monasteries. Some of the folia that make up the manuscript were brought to an eastern French monastery from Italy, where they were cleaned of text and collated with other booklets; the whole was then written over with a very large number of text extracts, among them poetry, administrative templates, and astronomical tables, as well as medical, grammatical and patristic works. Tironian shorthand was used throughout. This important miscellany shows that the *Physiologus* was transmitted very early on in a compilation that sought to collect practically useful texts for a wide variety of purposes, which included the calculation of Easter, public chancery and administrative activity, liturgy, and teaching, among others. The compilatory effort represented here reflects the beginning of early medieval monastery-led attempts to integrate a variety of far-flung intellectual traditions. This process is examined using the historical context of seventh- and eighth-century monastic book production. The *Physiologus* is, in this period, one of a range of practical works required for both the internal and external running of the monastery and its community, but it is not yet used as part of a coherent system of thought on nature. Rather, its function in this manuscript appears to be restricted to its general utility—itself very diverse.

Chapter 3 explores a shift in the way the *Physiologus* was used from the second half of the eighth century onwards. This shift is linked to the eighth- and ninth-century legislative, educational, and ecclesiastical reforms initiated by Pippin the Short and his son Charlemagne, and to the emergence of a new type of book in the early medieval period: the miscellany. This chapter explores the structure and functions of the miscellany using the example of a variety of *Physiologus* manuscripts. In contrast with the Merovingian manuscript discussed in Chapter 2, these Carolingian manuscripts are internally coherent compilations that responded to a contemporary need to understand the Bible better, particularly through language. The definition and role of such miscellanies in early medieval learning reflect the importance of the written word in this period. One result of the Carolingian reform is that texts were brought together to respond to the demand for better education of priests. But priests also needed to teach each other, as well as lay children and adults in a variety of settings. Information on and around core learning subjects therefore appears in the *Physiologus* manuscripts in a variety of forms that could be adapted or used in different contexts. Education and compilation processes are richly represented in St. Gallen, Stiftsbibliothek, Cod. Sang. 230, which is a very long collection of works used for the education of priests. The range of textual techniques used in this manuscript are examined to explore how the natural material of the *Physiologus* supplemented Christian learning. The material in this manuscript is compared with the similarly large collection represented by Bern, Burgerbibliothek, MS 225 + Bern, Burgerbibliothek, MS 233 + Orléans, Médiathèque, MS 313. These miscellanies demonstrate that from the late eighth century, the *Physiologus* was used to promote a view of the physical, created world as a narrative tapestry—a text that could be read—with language as a means of doing this; and for teaching the mysteries of Scripture, and eternal salvation as the ultimate reward, as a logical consequence of this kind of reading.

The later ninth- and early tenth-century development of the *Physiologus* is examined in Chapter 4. It explores allegory as a moral tool in several other ninth-century manuscripts of the *Physiologus*, and allegoresis as an approach to reading texts in this period. The four-fold method of Christian exegesis, which includes the allegorical, has already been extensively studied, but this chapter explores in more depth the link between spiritual allegory, morality, and nature in the early Middle Ages. Nature can be regarded as the basis of allegorical interpretation, which leads to the moral God and therefore to human salvation. This has implications for good living. The practical way these ideas were disseminated and implemented is evident from three collections containing the *Physiologus*. Montecassino, Archivio dell'Abbazia, MS 316 + MS 323 is one of these books. It is very heavily focused on understanding the Bible, with nearly 150 folios dedicated to glossaries; but it also explains baptism, and natural occurrences (winds, the sun, and thunder). The inclusion of the *Physiologus* in such a focused collection of important introductory texts can only be explained by a strong perceived link between spiritual and physical life. In addition, etymology is shown to have linked the written word with nature and with God: as part of the theory of natural signification, used both before and after the early Middle Ages by the Stoics and William Ockham, sounds imitate the true natures of the things they represent, and, through etymology, can lead humanity to the Creator. Words therefore slot into a spiritually rich and intelligible conception of the created world, as evident in Munich, Bayerische Staatsbibliothek, Clm 14388 and Munich, Bayerische Staatsbibliothek, Clm 19417.

Chapter 5 tracks the developments that may have led to the later medieval modifications in the content and function of the *Physiologus* from the last quarter of the ninth century. Firstly, the manuscripts reflect a growing tendency to assemble texts on natural topics and link them with moral material—a thematic collocation not observable in the earlier ninth-century manuscripts. This is especially clear in Wolfenbüttel, Herzog August Bibliothek, MS Gud. lat. 148. Around half of this manuscript's texts are allegorical and moral interpretations of animal behaviour and appearance, stones, plants, and geographical features. This thematic selection is clearly no accident. Most interestingly, however, the remaining texts—which discuss Paradise, monsters, and morality—give this focus on the natural world an extra dimension by exploring spiritual salvation. A similar dimension is evident in Brussels, Bibliothèque Royale, MS lat. 10066–77. The pictorial cycle in this manuscript, which links the *Physiologus* to the late antique allegorical poem *Psychomachia* (a title that translates as "battle of the spirit"), emphasizes the importance of living a moral Christian life by resisting vice and embracing virtue. Finally, in Vatican City, Biblioteca Apostolica Vaticana, MS Pal. lat. 1074, the *Physiologus* is presented alongside two other texts on animals and is followed by texts dealing with fundamental credal statements. In all these manuscripts, orthodoxy and salvation are the logical result of observing the natural world. The Carolingian program of Christian and moral education continues to be reflected in this period as well—as shown by the tenth-century Oxford, Bodleian Library, MS Auct. T.2.23—but there is a growing awareness of "nature" as a guiding principle for the collocation of texts. This, in turn, may reflect a new perception of the natural world as a distinct subject.

Secondly, we see that the *Physiologus* was being increasingly used in schools. In the late tenth or early eleventh-century manuscript Paris, Bibliothèque nationale de France, Nouv. acq. lat. 455 (previously only cursorily mentioned in *Physiologus* scholarship), I have found evidence of three inexperienced scribes trained in Visigothic minuscule using the *Physiologus* to practice Caroline minuscule. This manuscript reveals the movement of scribes between France and Spain, and suggests how the *Physiologus* may have reached Catalonia, where the Vatican manuscript Pal. lat. 1074 was produced. This is also the first copy of the *Physiologus* which was demonstrably used in a schoolroom setting. Since, with this manuscript, all the evidence for the use of the Latin *Physiologus* in schools dates from the tenth century at the very earliest, and since this evidence is richest in the eleventh and twelfth centuries, it is clear that the cultural context underwent a major change during the period ca. 950–1100. Overall, the shift of the *Physiologus* to a school setting and the new emphasis on "nature" as a theme paved the way both for the expansion of the *Physiologus*, leading to the bestiary, and for the creation of new didactic versions of the text.

The conclusion briefly draws together the wider strands of the *Physiologus*' history. It unfolds the *Physiologus* as a work which, once it had reached early medieval Latin Europe, was inserted into manuscript collections that represented deliberate compilations of texts used both for a variety of local purposes and for promoting an allegorical understanding of nature as God's moral creation. The *Physiologus* has long deserved to be re-evaluated as an important text on the natural world, which, within its highly innovative early medieval miscellany context, played a significant role in re-shaping inherited thought along new lines. Its simple format, far from reducing it to a collection of childish stories, meant that it plugged easily into the encyclopaedic culture that characterized much early medieval intellectual activity from Isidore of Seville onwards. It must, therefore, be read within its early medieval manuscript and historical context to be fully understood.

In the early Middle Ages, the prestige accorded to the written word blurred the lines between the sacred text, the words on the page, and the real-life objects that these discussed. The whole world was intelligible, if only you knew enough language to read it. This link between the Bible and the natural world ultimately represents a fascination with the surrounding environment, and more particularly, with humanity's place in Creation. It is this fascination that underpins the success of the *Physiologus*, the "authority on nature," in the early Middle Ages.

Chapter 1

# THE NATURAL WORLD IN THE EARLY MIDDLE AGES

THE NATURAL WORLD is a human construct. Despite their seeming simplicity, the terms "natural world" and "nature" were used to mean a wide range of things in the pre-modern period. It is only relatively recently, following the rise of the physical sciences as a category, that they have come to refer to a widely accepted idea familiar to anyone who has attended a Western-style school or university.[1] But it is worth remembering that these ideas are still far from universal. There are continuing indigenous or non-European models of ecosystems in which the natural world looks very different, and which are not well-recognized by mainstream science.[2]

A very similar problem exists for the early Middle Ages because the modern idea of "nature" as shaped by "science" tends not to recognize early medieval models of the natural world. Compounding this issue is a persistent perception of the early Middle Ages as a period of compilation and preservation, which produced few works of real originality or lasting value. The concept of "nature" continues to be seen, even by twenty-first-century historians, as shaped entirely by the Classical past and advanced little or not at all before the twelfth century in the Latin West.[3] But recent scholarship has demonstrated that almost every aspect of early medieval intellectual culture purposely forged new contexts out of older ones.[4] Against this dynamic background, it would have been almost impossible for early medieval thinkers not to take up older ideas about nature and shape them to their own needs. As this chapter argues, that is just what they did.

Before turning to early medieval innovation, however, it is necessary to disentangle the somewhat confused narrative about medieval nature from the much more dominant narrative of a progress of science. The first part of this chapter first looks critically at the ways in which past scholarship has approached the concept of "nature" in the Middle Ages, and how it has mapped contemporary preoccupations onto past material. This includes the ideas inherited by the Middle Ages: the Neoplatonic understanding of a progenerative being, and the concept of *natura* in philosophy. This discussion situates the *Physiologus* within a broader debate while highlighting its poten

---

**1** See Buc, *The Dangers of Ritual*, 3, on the blind spots of writing in the same tradition as the object of study.

**2** Jessen et al., "Contributions of Indigenous Knowledge."

**3** Astronomy, as driven by computus, is perhaps the only exception.

**4** Two recent illustrations of this are Chambert-Protat, "Le manuscrit Montpellier 157"; and Westwell, "The Content and the Ideological Constructionof the Early Pontifical Manuscripts." See also the four excellent "Storehouses of Wholesome Learning" volumes: Bremmer and Dekker, eds., *Foundations of Learning, Practice in Learning*, and *Fruits of Learning*; and Giliberto and Teresi, *Limits to Learning*.

tial to draw out early medieval specificities. Then, in order to examine the cultural and intellectual reception of "nature" in the early Middle Ages, and to obtain a historically sensitive reading of the *Physiologus*, this chapter reviews the evidence for how the physical world was perceived. This evidence includes not only texts but also art and material objects. The early medieval context evidenced by these sources had a rich cultural and spiritual dimension that is generally missing from the modern scientific definition of the natural world as a precisely measurable entity.

The second and longer part of this chapter offers a closer reading of these sources and the different aspects of the early medieval worldview that they represent. From the seventh century onwards, attempts were made to integrate this inherited classical learning about the physical world with the particular understandings of Christianity that were becoming dominant among intellectuals of those centuries. Although there was no centralized oversight of these intellectual works—which meant that they were often disjointed and contradictory—they were nevertheless products of monastic learning, which advocated broadly similar teaching methods and reading material across the Latin West. By the late eighth century, those texts (old or recent) which dealt with the physical or created world had, under the particular influence of the liberal arts curriculum and especially Isidore of Seville's *Etymologiae*, been reframed into a new context. Central to this was an approach based on language, and the use of allegory and etymology not only to interpret texts but to link seemingly disparate things: the Bible, morality, human bodies, portents and omens, the position and movement of celestial bodies, plants, stones, and animals, among others.[5] This universalizing system was, in effect, a new cosmography, based on inherited ideas but using methodologies that were distinctive to this time and place, dealing with its particular challenges and ambitions. By the ninth century, the *Physiologus*, and the plurality of readings its material context represented, was a sophisticated tool for learning and applying these new concepts. Although far removed from our own ideas about nature, as we will see in subsequent chapters, this developing early medieval worldview was both intellectually complex and deeply innovative.

## Nature and Science

Medieval concepts of nature can be studied from a philosophical or literary perspective, but categories such as ethics, philosophy, and theology—as well as zoology and botany—have their own histories that have led to their current place in our system of knowledge. This system was not the same a hundred years ago and is a great deal less recognizable twelve hundred years ago. In order to be understood, it must be taken on its own terms. The same problem with the scientific approach, which asks the wrong questions of seventeenth-century naturalism by looking "for the roots of modern zoology and botany," was recognized by the historian William Ashworth in 1990.[6]

---

5 On portents, especially, see Foot, "Plenty, Portents and Plague."

6 Ashworth, "Natural History," 304.

Nevertheless, the history of science provides a very powerful narrative from which it is difficult to separate the history of nature, especially as it is recounted for the early Middle Ages. In this narrative, the concept of nature is traced back to science in ancient Greece. The Greek philosophical schools developed various models for understanding the universe, drawing on the wider intellectual inheritance of the ancient world. Traditional accounts of the history of Western science have credited the Greeks, especially though not exclusively Aristotle, with inventing the idea of "scientific" rigour.[7] There are good reasons to see the ancient Greek philosophers as important contributors to the beginnings of modern science, but they were not therefore dispassionate, objective observers of the world who separated moral and emotional concerns from intellectual ones. Scientific objectivity as we know it today is a nineteenth-century development, growing out of an "epistemic virtue" that historians of science Lorraine Daston and Peter Galison called truth-to-nature. By following truth-to-nature, naturalists examining natural objects aimed to represent their ideal or universal form or type. Both truth-to-nature and the other forms of objectivity that succeeded it were closely linked to the philosophical ideas of the eighteenth and nineteenth centuries.[8] Objectivity in science more generally was a product of the scientific revolution, which had begun around the sixteenth century, and which saw the rise of the scientific method. These ideas certainly had their roots in ancient and medieval knowledge both in the West and beyond—they did not arise out of nothing—but they are much more recent than we tend to think. Even when they seem to be recognizable in ancient Greek philosophy, their roots, purposes and uses often seem alien. An example of this is the science of acoustics, which in ancient Greece was inextricably associated with the art of poetry.[9] Similarly, Natalia Lozovsky has shown that in the early Middle Ages, geography was considered to be a sub-section of history, in particular sacred history.[10] For the vast majority of human history, objectivity in science has been far less important than its cultural and social functions. And the modern emphasis on objectively provable knowledge does not mean the scientific achievements of the pre-modern past were somehow of less value. As David D'Avray has put it, "it would be arrogant and culture-bound to assume that the modern West's attempt to understand the central problems of existence—e.g., whether there is a difference in kind as well as in degree between humans and other animals, or whether personal identity continues in any way after death—is any more "scientific" than that of medieval people."[11]

Histories of modern science have now also moved on beyond traditional narratives. They range from critiques of modern science based on environmental and health disasters; the anthropocentrism of modern science; the lack of women's and indig

7 For example, Agazzi, *Scientific Objectivity*; Graham, *Science Before Socrates; Harrison, The Bible, Protestantism, and the Rise of Natural Science*. See also Arnaldez and Beaujeau, eds., *La Science*.

8 Daston and Galison, *Objectivity*, 20–21.

9 Barker, "Words for Sounds."

10 Lozovsky, *The Earth is Our Book*, chap. 3.

11 D'Avray, "Symbolism and Medieval Religious Thought."

enous voices and knowledge; the far wider global roots of science before more narrow Enlightenment ideas came to dominate; and new "climate" narratives which see the Anthropocene as a time of regrowth and possibly better human health.[12] These lines of inquiry critically interrogate the way human beings interact with the world in which they live, while taking into account past contributions. Likewise, the conceptualization of the world in the early European Middle Ages in locally produced miscellaneous manuscripts, as investigated in this book, can and should be critically explored as a particular (historical) instance of human interaction with nature. What we can learn from this is both useful for early medieval history and transferable to the present, both from an evolutionary and a comparative perspective.

For the history of science, a view of knowledge that takes modern categories and terms as its starting principles demonstrates a particular kind of progress and permits historians to judge the success or failure of ideas or cultures within the context of that progress.[13] A consciously partisan position may in fact be necessary for any *longue durée* history if it is to be more than a descriptive chronology. But it makes no sense to do the same for the history of nature during the early Middle Ages. Seen from the point of view of scientific objectivity, the early medieval contribution to the Western understanding of the natural world was negligible, and it disappears altogether when compared with ideas put forward in antiquity or in the later medieval universities. Viewed on its own terms, the early medieval understanding of nature was so bound up with the very idea of what constituted knowledge and faith, and with the social means of exercising these—education, language, worship—that it serves as a window onto the landscape of early medieval society. Alone the strangeness of this landscape to our eyes makes it worth investigating, both as an important part of our shared past and as a challenge to our own inherited imperatives.

## Early Medieval *natura*

Viewed separately from the story of scientific progress, then, what was the natural world in the early Middle Ages? The available conceptual frameworks are strikingly rich in what they encompass. In antiquity and into the Middle Ages, the Latin word for nature, *natura* (which in turn derives from Latin *nascor*, to be born) was most frequently used to mean either an active, usually divine force that animates the universe or the human body, or the essential quality of a person or thing. This is the case in the *Instructiones*, a set of sermons composed by the sixth-century Irish saint Columbanus, in which the word *natura* occurs thirteen times but always refers to God or an inner quality rather than the natural world.[14] In the seventh century, Isidore of Seville affirmed this in the *Etymologiae*: "Nature [is] called *natura* because it causes something to be born, *nasci*, and is capable of producing and creating. Certain people have said nature is God, by whom all things are created and have

12 See for example Brooke, *Climate Change and the Course of Global History*.

13 The same argument is briefly made in Graham, *Science Before Socrates*, 28.

14 Stancliffe, "The Thirteen Sermons." 165.

their being."[15] The Earth, therefore, can be seen as acting under the impulse of this divine force, and while the effects of gravity, heat, light, and so on are felt, they are not forces in themselves, but rather agents of the primary force. God and the physical world are not strictly distinguished, an attitude inherited by the early Middle Ages from antiquity. Boethius expressed this clearly in his *Consolation of Philosophy*: "The generation of all things [*omnium generatio rerum*] and the development of things that change and move [*cunctus mutabilium naturarum progressus*] take their order and forms and causes [*ordinem, formas, causas*] from the unchangeable mind of God [*ex diuinae mentis stabilitate*]."[16]

In associating *natura* with a progenerative and divine God, Isidore and Boethius followed Neoplatonic tradition.[17] Neoplatonic thought was based on the ideas developed by Plato in the *Timaeus*, and by Aristotle (Plato's student) in the *Categories*. Among those Neoplatonists who exercised the most influence on early medieval ideas of nature were Plotinus (d. 270), and through him, Augustine (d. 430) and Boethius (d. 524/5). Boethius not only translated Aristotle's *Categories, On Interpretation* and *Prior Analytics*, and Porphyry's *Isagoge*—an introduction to the *Categories*—but also wrote several highly influential works on music (*De institutione musica*) and philosophy (*De consolatione philosophiae*). The translation and commentary on the *Timaeus* by the fourth-century philosopher Chalcidius, the fifth-century commentary on Cicero's *Dream of Scipio* (*Somnium Scipionis*) by Macrobius, and Martianus Capella's fifth-century poem *On the Marriage of Philology and Mercury* (*De nuptiis Philologiae et Mercurii*) also helped to transmit Neoplatonic ideas in the early Middle Ages. These ideas encompass a rich and diverse tapestry of meaning, function on a grand scale as a synthesis of almost the entire Hellenic intellectual tradition and are too extensive to be properly summarized here. (The word "Neoplatonic" itself, coined in the 1830s, was not used in antiquity.) We can, however, trace two key Neoplatonic ideas that had an impact on early medieval thought about nature: the idea of an unlimited single being, a divine "One," from whom reality proceeds; and the Aristotelian categories or divisions of knowledge, which are used to divide and classify reality.[18]

The Neoplatonic idea of the One was particularly influential. Although Plotinus' original description of the One was not the same as the Christian understanding of a triune God, it helped to articulate Christian belief, and had been thoroughly absorbed into Christian philosophy—particularly through the works of St. Augustine—by the early Middle Ages.[19] The Christian focus was, however, not on creation itself as divine, which was the Neoplatonist view, but on the divine Creator, from whom the natural

15 Isidore, *Etymologiae* 11.1.1.

16 Boethius, *The Consolation of Philosophy* 4:6. Trans. Slavitt, *The Consolation of Philosophy*, 131; Latin from Boethius, *Theological Tractates*, trans., Stewart, Rand, and Tester, 340.

17 Natural and extraordinary generation has been covered for the central Middle Ages in Lugt, *Le ver, le démon et la vierge*.

18 Marenbon, *Early Medieval Philosophy*; Wildberg, "Neoplatonism."

19 For an introduction, see O'Meara, ed., *Neoplatonism and Christian Thought*.

world derived its origin. In his commentary on Genesis, the eighth-century scholar and teacher Alcuin restated the position of the Church on this, which had been articulated early on by Augustine and others. Alcuin explained why God did not give mankind his laws to be written down in the beginning, as he later did to Moses:

> The first men kept the law of good nature [*lex bonae naturae*] for a long time; but when the natural law [*naturalis lex*] vanished, caused by the habit of sinning, the written law [*lex litterae*] was given by means of Moses, that man might have the authority of the good things to know, and that these, that had begun to be hidden, would be made manifest; and so that the fear of punishment would correct offenders, and restore the faithful to God.[20]

Here the natural was equated with virtue and described as something internal and instinctive, that nevertheless weakens over time and requires correction to be restored.[21] This is faith—upheld by the written word—by means of which the human soul finds God. Thus the Creator is the essence of both nature and virtue, as the ultimate law and the only progenerative being. Alcuin's emphasis on the need for correction bestowed a moral aspect on Creation, which was a reflection of the Creator and so of human salvation. The natural law was a philosophical concept that had wide-ranging implications for medieval jurisprudence, but its inclusion in Alcuin's commentary on Genesis also indicates that it was made part of an approach to nature that focused on its morality, as we shall see in Chapter 4.

## Categorizing Nature

Ultimately, the term "natura" wasn't really used to discuss the environment. It was a tool of logical and philosophical discourse, recognized by the educated elite as having its roots in ancient ideas. The same was true for the word *physica*, which had its roots in Ciceronian rhetoric. In the dialogue *Disputatio de rhetorica et de virtutibus*, written for Charlemagne, Alcuin applied the traditional division of philosophy into physics, ethics, and logic, and divided physics, in turn, into arithmetic, astronomy, astrology, mechanics, medicine, geometry, and music.

The philosophical definition of nature in the early Middle Ages was therefore both dominant and highly theoretical. It does, however, highlight the extent to which "nature" was associated with God: a seemingly simple idea which, as we shall see throughout this book, had complex consequences. In his major work *Periphyseon*, also known as *On the Division of Nature*, the ninth-century philosopher John Scottus Eriugena described *natura* as "the general term for all things that are and all things that

---

**20** "Quia in hominibus primis diu lex bonae naturae seruabatur; at ubi naturalis lex euanuit, oblata consuetudine peccandi, data est lex litterae per Moysen, ut bona quae sciebantur auctoritatem haberent, et quae latere coeperant, manifestarentur; et ut terror disciplinae corrigeret delinquentes, et fidem reformaret in Deum." Alcuin of York, *Interrogationes et responsiones in Genesin*, PL 100.518.

**21** Alcuin's sources here are unclear. The phrase *lex naturalis* may derive from Augustine; Chroust, "The Fundamental Ideas," 68–69. The phrase *lex litterae* may derive from an Irish source; Wright, *The Irish Tradition*, 14. Both phrases were used by Stoic philosophers such as Gaudentius of Brescia, but Stoic writings were not read in the early Middle Ages; Lapidge, "The Stoic Inheritance," 83.

are not" (*Periphyseon* I.441a). His four categories for nature were based on whether it created or was itself created, a division that echoes Augustine (*De civitate Dei* V.9), among others:[22]

1. That which creates and is not created.
2. That which is created and creates.
3. That which is created and does not create.
4. That which is neither created nor creates.

Nevertheless, *natura* was also a single entity: *universitas rerum*, "the sum of all things" or the one God, who is Beginning, middle and End (*Periphyseon* I.11). By categorizing nature in this way, Eriugena also followed the pre-Socratics, who understood nature as origin (ἀρχή), as a process of development, and as the final end. Eriugena's work is testament to a continuing and sophisticated philosophical tradition in the early Middle Ages, but it is also among the most advanced such texts in the West in this period. It was not much read by other Carolingian intellectuals, though there is evidence that Eriugena's fellow Irishmen Martin of Laon and Sedulius Scottus knew it.[23] The exact number of manuscripts in circulation is difficult to judge, as in 1225 Pope Honorius III ordered all extant copies to be brought to Rome and burned as a heretical work, due to a new hostility associated with Aristotelian ideas.[24] The influence of the *Periphyseon* and its ideas of nature are, however, discernible in glosses on the word *natura* in Bede's *De natura rerum*, which was a popular early medieval school text.[25]

Aristotle's *Categories* was also significant for the early medieval understanding of *natura*. After the sixth century, it was Boethius' translation of and commentary on Aristotle's *Categories*, and a summary of it called the *Ten Categories* (*Categoriae decem*), together with a composite translation, that were the principal texts for the study of logic. Together with grammar and rhetoric, logic was a component of the trivium and so a key part of early medieval education.[26] Alcuin's work is once more an apt illustration of this, as he wrote a poem for Charlemagne introducing the *Ten Categories*. This text provides a supplemented explanation of Aristotle's logical categories, which it lists as substance (*ousia*), quantity, relation, quality, action, passion, situation, place, time, and condition.[27] Like other Platonic texts and ideas, the *Ten Categories* informed early medieval theology and were debated in anonymous texts such as the *Dicta Albini*

---

**22** Moran, "John Scottus Eriugena"; Eriugena, *Periphyseon*; Augustine, *De civitate Dei. Book V*. See also Moran, *The Philosophy of John Scottus Eriugena*.

**23** Marenbon, *From the Circle of Alcuin*, 111.

**24** Otten, "Overshadowing or Foreshadowing," 211.

**25** Petrov, "Karolingskiye schkolniye teksty."

**26** Gracia and Newton, "Medieval Theories of the Categories." The *Categoriae decem* are also sometimes known as the *Paraphrasis Themistiana*, as the anonymous author cites Themistius, a fourth-century philosopher. On early medieval education, see Hildebrandt, *The External School*, and Sullivan, ed., *The Gentle Voices of Teachers*.

**27** For a fuller summary of the contents, see Gersh, *Concord in Discourse*, 74–75.

and the *Dicta Candidi Presbyteri de imagine Dei*, as well as in the first book of the *Periphyseon*. Alcuin's poem, which imbues the text with Christian authority by ascribing it to Augustine, was reproduced in many of the manuscripts of the *Ten Categories*. It begins by describing the proper application of Aristotle's categories for the logical understanding of reality:

> This little book contains ten words of nature
> amazing words which by their power represent every property of things,
> which can be perceived by our minds.
> Let him who reads it praise the marvelous learning of the ancients,
> and let him strive to exercise his own with the same diligence,
> adding praiseworthy honors to his allotted lifespan.[28]

These phrases make two important points: Firstly, in this poem *natura* is "every property of things which can be perceived by our minds"—that is, the physical world as we understand it with the bodily senses, and through them with the mind; and secondly, the study of nature is directly linked to education, and consequently to living a good life. Both of these points derive from antiquity, during which the study of philosophy—the knowledge of all things—was essential to good education and consequently to civic life.[29] These ideas became enshrined in Western monastic education in the early Middle Ages, and continued at least into the twelfth century, when the *studia naturalium*, also called the *studia philosophiae*, were a standard (if advanced) part of the schooling of Dominican friars.[30]

Following Augustine, the *Ten Categories* were read in the early Middle Ages as a commentary on God's substance or *ousia*.[31] Like many other Neoplatonic texts that circulated in Western Europe from the seventh century, they were used for the study of Christian theology, and the metaphysical philosophical principles that helped to shape and inform that theology. In large part, this was a program of reading that the early Middle Ages had inherited from late antique didactic practices grounded in the liberal arts and informed by the principles of monastic life and learning. These were enshrined in monastic rules and guides to education such as Cassiodorus' *Foundations of Divine and Secular Literature* (*Institutiones divinarum et saecularium litterarum*) and the Rule of St. Benedict, both composed in the sixth century. In so far as nature is mentioned in these traditions, it is never articulated as an independent concept, and relies on inherited ideas: those that derive from Greek philosophy on the one hand,

---

**28** "Continet iste decem naturae verba libellus / Quae iam verba tenent rerum ratione stupenda / Omne quod in nostrum poterit decurrere sensum / Qui legat, ingenium veterum mirabile laudet / Atque suum studeat tali exercere labore / Exornans titulis vitae data tempora honestis." Minio-Paluello, ed., *Aristoteles latinus*, lxxxvii. "He" has been used instead of "they," despite the ambiguous gender inherent in the Latin, to emphasise the individuality of the addressee and Alcuin's male-dominated intellectual context. My thanks to Evina Steinová for assistance with the translation.

**29** On Creation and goodness, see Crouse et al., ed., *Divine Creation*; and Mähl, *Quadriga Virtutum*.

**30** Teeuwen, *The Vocabulary of Intellectual Life*, 85n236, and 142.

**31** Marenbon, *Early Medieval Philosophy*, 25.

and those that rely on the biblical hierarchy (in which nature is subordinated to God and man) on the other.

In the early Middle Ages, then, the natural world, as represented by the concepts of *natura* and *physica*, seems to become fully amalgamated into the Christian philosophical tradition. Moreover, these concepts existed within the narrow boundaries of the liberal arts curriculum, which meant that they were learned only by a small proportion of the educated elite, and then in order to train the mind rather than for any broader purpose related to knowledge about the physical world. In this form, knowledge about nature was both theoretical, and recognized to be an inheritance of the past.

Since the early medieval natural world is not to be found through the term *natura*, it is set aside in this book. This "natural world"—also referred to here as the "physical world" and as "nature"—is not quite what we would call the environment, which implies ecological systems. Early medieval people engaged with these as well (see below), but in more pragmatic ways. The natural world discussed here is an intellectual concept, forged gradually and repeatedly out of inherited texts. It is, as stated at the start of this chapter, a human construct. I offer a preliminary set of features of this "natural" or "physical" world as it may have been imagined by those early medieval people who were responsible for the compilation, dissemination, and local use of *Physiologus* miscellanies in the Conclusion to this book.

## Manuscript Compilations and Early Medieval Innovation

Although the early Middle Ages and their contribution to our understanding of nature have been neglected, the medieval natural world nevertheless has a long historiography. Some historians have viewed the interaction of nature and humanity in the early Middle Ages as inherently antagonistic. This was the case with the influential article published in 1967 by Lynn White Jr., in which he argued that nature was a space of exploitative, violent human dominion until the advent of St. Francis of Assisi. Thirteen years later, David Herlihy rejected this view, instead outlining four attitudes to nature that chronologically succeeded each other in importance: the eschatological, adversarial, collaborative, and recreational.[32]

In this model, the adversarial attitude is particularly typical of the early medieval period, which is characterized by fear of the hostile monsters and beasts that populate the natural world. But the more nuanced view put forward by Herlihy did not challenge the statement made explicit by Lynn "that some essential feature of Western thinking created the precondition for an assault on the natural world."[33] Other scholarship of the late twentieth century also took up the idea of an opposition between humanity and nature in the early Middle Ages. One work claimed, for example, that "change in the external world and one's appreciation of it were separated by an unbridgeable gulf. The result was anxiety, lack of comprehension and a whole range of compensa-

32 White, "The Historical Roots"; Herlihy, "Attitudes Toward the Environment." See also the introduction in Arnold, *Negotiating the Landscape*, and White, *Medieval Religion and Technology*.

33 Bruce, "Introduction: Hoffmann in the Historiography of Environmental History," 15.

tory techniques. For, if the hard facts of life could not be altered, at least they could be fitted into a system of belief that made them understandable and acceptable."[34]

However, although the metaphor of an opposition between nature and culture, particularly in studies on medieval wilderness, continues to exert influence, more recent work has moved the study of nature in the Middle Ages away from the paradigm of conflict.[35] Scholarship across a range of disciplines has shown that attitudes to nature in early medieval sources were much more diverse and complex than they have been portrayed. They include subordination to humanity, but also resistance to anthropocentrism.[36] The medieval built environment (sometimes called the anthroposphere) has also been examined in terms of integration rather than conflict.[37] All of these studies together demonstrate conclusively that early medieval intellectual responses to the natural world were not delineated merely by the paradigm of nature versus culture.[38]

Nevertheless, the way in which modern historians have engaged with the natural world as a cultural and intellectual phenomenon in early medieval Western Europe has meant that it has been incorporated into histories of the environment or of a more general idea of "nature" in the ancient and medieval worlds.[39] The message in these histories has been—either implicitly or explicitly—that the early medieval natural world was a theoretical idea. It had no relationship to the external world, since it did not rely on scientific observation (although this wasn't quite true, as we shall see), and it was expressed through philosophical terms such as *natura* and *physica*.

Such a lack of intellectual engagement with the physical world is all the more puzzling in the light of the close *practical* relationship between early medieval communities and the land on which they lived. Early medieval people knew about crop cultivation and rotation, including complex farm system management and the care of delicate plants such as peach trees.[40] They knew how to care for and manage woodland and farmland, and the associated wild animals and livestock.[41] This extended to the adjustment of various rents and other payments as required in line with seasonal

**34** Stock, *The Implications of Literacy*, 473.

**35** Le Goff, "The Wilderness"; White, "The Forms of Wildness." The revisionary studies include Whitney, *Paradise Restored*; Whitney, *Medieval Science and Technology*; Hoffmann, "Homo et Natura, Homo in Natura"; Hoffmann, *An Environmental History*. On Hoffmann's impact on the field, see also Bruce, ed., *Ecologies and Economies*.

**36** Dale, *The Natural World*; Estes, *Anglo-Saxon Literary Landscapes*; Squatriti, *Water and Society*, *Landscape and Change*, and *Weeds and the Carolingians*; Siewers, *Strange Beauty*; McCracken, *In the Skin of a Beast*.

**37** See, for example, Blair, *Building Anglo-Saxon England*, and Bintley, *Settlements and Strongholds*.

**38** Jones, *The Medieval Natural World*, 3.

**39** Le Roy Ladurie, *Histoire du climat*; Glacken, *Traces on the Rhodian Shore*; Coates, *Nature*; Whited et al., eds., *Northern Europe*; Behringer, *Kulturgeschichte des Klimas*.

**40** Blan, "Charlemagne's Peaches."

**41** Salisbury, *The Beast Within*; Squatriti, *Landscape and Change*; Williamson, *Environment, Society and Landscape*.

yields.[42] Similarly, people were aware of the interconnectedness of water systems, and adjusted their use (or risked diverse penalties for over-use) of these systems in order to ensure sustainability and continuity in terms of landscape.[43] Climate change and natural disasters were a related and key concern in the early medieval period, although the terminology used to discuss these phenomena was quite different to that used today.[44]

However, the same—usually small and local—communities which managed and cared for early medieval landscapes rarely recorded their thoughts. Much of what we know about the practical aspects of early medieval life comes from archaeological and palaeoenvironmental studies that supplement textual sources. For this reason, until very recently, it has been assumed that there exists no evidence for the development of "living" thought about the natural world in the early Middle Ages, and that it was represented by the antique concept of *natura*. The early medieval copies of the *Physiologus*, within their manuscript context, provide precisely the kind of evidence that has been missing from this story.

In recent years, Carine van Rhijn and others have demonstrated the value of pastoral compendia—the kinds of manuscripts within which the early medieval *Physiologus* was copied—for studying the spiritual and intellectual life of the anonymous majority of the early medieval population.[45] Such compendia have the potential to transform our understanding of what pastoral care and education for both priests and the communities that they served looked like in the early Middle Ages. Unlike normative and prescriptive texts such as capitularies, penitentials, and canons, pastoral compendia were compiled by and for local people. They therefore reflect pastoral and educational practice at a localized level in all its diversity.

These compendia have been very easy to overlook, not only because they consist largely of short texts and excerpts that many older catalogues frequently simply described in batches, but also because these texts are very numerous and appear to be rather basic: regardless of their genre (liturgical, educational, penitential etc.), their contents are generally short and comprehensive, or address a very specific point. Explanations of the Mass or the Lord's Prayer are common, for example, as are computistical tables and short sermons or sermon models. The impression that these manuscripts were *ad hoc* compilations has only begun to be dispelled in the past few decades, particularly with the advent of the idea of the archaeology of the book. Deceptively simple in content, these compendia are frequently complex in terms of their codicological structures. By mapping the copying of their texts onto their suc-

**42** Kreiner, *Legions of Pigs*.

**43** Squatriti, *Water and Society* and *Landscape and Change*; Guillerme, *The Age of Water*; Hoffmann, "The Protohistory of Pike"; Oosthuizen, "Anglo-Saxon Fields," 302–05, *Tradition and Transformation*, and "Recognizing and Moving On"; Küster, *Geschichte der Landschaft*.

**44** Palmer, "Climates of Crisis." See also Devroey, *La nature et le roi*.

**45** Burridge, *Carolingian Medical Knowledge and Practice*; van Rhijn, *Leading the Way to Heaven*; Patzold and Van Rhijn, eds., *Men in the Middle*; Patzold, *Presbyter*; Waagmeester, "Pastoral Works." See also Keefe, *Water and the Word*, and *A Catalogue of Works*.

cessive layers, we can see the thought and planning that went into their creation.[46] They were among the most innovative kinds of written sources produced in the early Middle Ages.

Until now, the most comprehensive studies of these novel sources, by van Rhijn, Patzold, and Keefe, have focused on pastoral and baptismal manuscripts. The set of codices in which the *Physiologus* is found is a much narrower sample, but it does take us beyond these categories, and raises the question of classification. Thus far, I have used the word "compendium," but a long list of terms has been used to describe such miscellaneous books: among them *collectaneum*, collection, commonplace-book, compilation, dossier, handbook, miscellany, reader, *recueil*, *Sammelhandschrift*, scrapbook, sourcebook, and vademecum. To avoid the problem of genre, Rosamond McKitterick has coined the term "glossary chrestomathy" to describe those books that contain collections of glossaries.[47] Similarly, Susan Keefe, in describing the different kinds of credal texts, has labelled them "EF" for "explanations of the faith," "PF" for "professions of the faith" and "DF" for "defense of the faith."[48] The diverse nature of the compilations in question requires diverse language and diverse reactions: there can be no catch-all solution. I have stated that the *Physiologus* was copied in pastoral compendia; strictly speaking, van Rhijn's definition excludes bishops' handbooks from this category. Yet the *Physiologus* is found in at least one compilation that was demonstrably used within the circle of an early medieval bishop (Bern, Burgerbibliothek, MS lat. 611 + Paris, Bibliothèque nationale de France, MS lat. 10756). For the purposes of this study, therefore, I shall not attempt to categorize early medieval compilations. Instead, I treat them first and foremost as local books, produced by and for regional communities, and I attempt to take into account their unique features and contexts, in so far as they are apparent. I will use generic terms, including compilation and miscellany, to describe these manuscripts.

Early medieval miscellanies tell a completely new story about the early medieval natural world. The copies of the *Physiologus*, within their individual codicological and textual contexts, show how early medieval scholars and compilers used new grammatical and allegorical strategies to create connections: between the different understandings of the physical world local to individual communities, and the intangible, aspirational world of Heaven as presented in the Bible and authoritative works. The constant recopying and correction of texts in miscellanies, and the judicious textual selection and juxtaposition, created a kind of contemporary dialogue around these themes. The precise mechanisms for this are discussed in Chapters 3 and 4. And this was not just an eighth- and ninth-century phenomenon. There is evidence that both the *Physiologus*, and the treatment of the natural world in miscellanies, were part of a

**46** I have written about the historiography and codicological specificities of such compendia extensively elsewhere. See Dorofeeva, "Reading Early Medieval Miscellanies," "Strategies of Knowledge Organisation," "Visualizing Codicologically and Textually Complex Manuscripts," and "What is a Vademecum?".

**47** McKitterick, "Glossaries and Other Innovations."

**48** Keefe, *A Catalogue of Works*, 10–11.

wider early medieval desire to write an all-encompassing spiritual explanation of the physical world.

## A New Cosmography

The early Middle Ages had not inherited many texts from antiquity that explained the unity, order, and reason of natural things, from everyday plants and animals to the movement of celestial bodies, in Christian terms. Only a very few of these even touched on these themes. They included Boethius' *De consolatione philosophiae* (*The Consolation of Philosophy*), Martianus Capella's *De nuptiis Philologiae et Mercurii* (*On the Marriage of Philology and Mercury*), and Prudentius' *Psychomachia* (*Battle of the Soul*): works that became part of the early medieval literature canon. The *Physiologus* was among these few inherited texts on these topics. But additional works seem to have been needed. From around the seventh century, a range of cosmographic texts were written. Two of the earliest of these were composed in Ireland: *De mirabilibus sacrae scripturae*, attributed to a writer known as the Irish Augustine (Augustinus Hibernicus), and the pseudo-Isidorean *De ordine creaturarum*, which took *De mirabilibus* as its source. *De mirabilibus sacrae scripturae* was composed around 655 and was unusual in that it saw biblical miracles as "natural events of such magnitude that they were recorded in the scriptures."[49] For example, when explaining the ways by which salty water can be made sweet and vice versa, the Irish Augustine states:

> For waters have within themselves, by their very nature, this ability to change into one another and they show it much faster at the command of the creator than through the careful efforts of men or even through ministration by things...Now the creator and ruler of all creation can reveal in all things that hidden nature which would normally be manifested through the agency of some other thing.[50]

For the Irish Augustine, Creation was a landscape of hidden natural functions, which we understand slowly over time, and which only the Creator can reveal instantaneously through miracles. He used direct observation of natural phenomena to show this in his examples. For instance, he is credited with being the first known writer to explain the presence of large mammals on the island of Ireland by the existence of land bridges, which gradually disappeared—a hypothesis which we now know to have been correct, and which was not put forward again until the nineteenth century. The systematic application of direct observation is a unique feature of this text, but direct observation of natural phenomena in general was not uncommon in the early Middle

**49** Moriarty, "The Early Naturalists," 72. Michael Gorman argued that this text needed further study as it may have been composed well before the seventh century, on the basis that it ignores the Vulgate, takes nothing from Isidore, and cites Augustine's *De Genesi ad litteram*. Gorman, "A Critique of Bischoff's Theory," 192n29. For a bibliography of the wide-ranging response to Gorman's theory, see Flechner and Meeder, eds., *The Irish in Early Medieval Europe*. According to Marina Smyth, however, it does date to the later seventh century; Smyth, "The Word of God," 114, 125.

**50** Smyth, "The Seventh-Century Hiberno-Latin Treatise," 144.

Ages, as numerous studies of science in this period have demonstrated.[51] It is only that direct observation, however accurate the knowledge it imparted, was not considered in literate monastic circles to be especially useful for really understanding the natural world.[52] Early medieval geographers, for example, believed that the tangible and visible world had "no self-sufficient and true reality."[53] For many early medieval thinkers both in Ireland and in mainland Europe, such reality existed only in God.

It is for this reason that texts such as *De ordine creaturarum*, composed in Ireland around the third quarter of the seventh century, methodically make the link between God and the physical world by describing the universe in Christian terms. In *De ordine creaturarum*, the universe—in accordance with the Greco-Roman worldview—consists of "the supercelestial waters, the firmament, the sun and the moon, the higher space of air (and the celestial Paradise where the souls of the truly good await the final resurrection), the lower space of air immediately above the earth (the domain of the fallen angels), the layer of water, the earth (where humans once dwelt in the garden of Eden), and finally Hell (where the souls of the truly evil are punished immediately after death)." The text also clarifies the key points of doctrine, including the Trinity, redemption and sin.[54] *De ordine creaturarum* had very limited circulation, particularly outside Ireland, though it was one of the sources of Bede's *De natura rerum*.[55] But the fact that such texts were produced, and that they attempted to explain the structure of the universe in an ordered way, shows that there was a growing desire in the early medieval West to arrange the available knowledge about the natural world into a distinctive scheme. This desire elevated the understanding that nature was God's Creation—an idea that had long ago been expressed by Church Fathers such as Origen—to a model or paradigm that, taken up many times by different thinkers, became part of a uniquely medieval cosmography.

Two other seventh-century texts, Isidore of Seville's *De natura rerum*, and Bede's reworking of it, were also written to explain the structure and operation of the physical world for Christians. Isidore's treatise was composed at the request of King Sisebut of Spain, as a response to seemingly superstitious ideas arising among both the clergy and the wider population, prompted by the unusually frequent solar and lunar eclipses of 611–612 CE.[56] Isidore certainly knew the first-century BCE work *De rerum natura* by Lucretius, which he cited and whose title he adopted for his own text, but his more major sources were Virgil and Lucan.[57] He divided his text into three principal parts: a

---

51 Among them: Eastwood, *Ordering the Heavens*, and *The Revival of Planetary Astronomy*; Flechner and Meeder, eds., *The Irish in Early Medieval Europe*; Kelly and Doherty, eds., *Music and the Stars*; Lozovsky, *The Earth is Our Book*; Ramírez-Weaver, "Carolingian Innovation and Observation," and *A Saving Science*.

52 Crane, *Animal Encounters*, 82.

53 Lozovsky, "Carolingian Geographical Tradition," 35 and 42.

54 Smyth, "The Seventh-Century Hiberno-Latin Treatise," 138.

55 Smyth, "The Seventh-Century Hiberno-Latin Treatise," 156.

56 Isidore of Seville, *On the Nature of Things*, trans. Kendall and Wallis, 16–17.

57 On Lucretius, see Butterfield, *The Early Textual History*.

hemerology, or explanation of time, a cosmography, or explanation of the structure of the universe, and a discussion of meteorology, or the weather.[58] The cosmographical and meteorological parts of the text derived from antique texts (including Aristotle's *Meteorologica* and book two of Pliny's *Natural History*) and presented essentially the same model as in *De ordine creaturarum*: a universe starting at the "top" with the heavens, followed by the atmosphere and the earth. Each topic was given allegorical meanings: for example, the stars were understood to be holy men (*De natura rerum* 24.2). Bede did not significantly alter this structure in his adaptation of Isidore's work, supplementing it with Pliny's *Natural History* and with *De ordine creaturarum*. His text covered "the four basic elements—earth, air, fire, and water, the heavenly bodies and their orbits, meteorological phenomena like thunder and lightning, rainbows, hail and snow, apparent disruptions of the natural order like eclipses, earthquakes and volcanoes, and plagues, and the fact that the earth is a globe, and its zones and climates."[59] Bede did, however, separate Isidore's hemerology, turning into a separate book, *On Times*: a deeply original and important mathematical and computistical work. He also excised Isidore's allegorical interpretations, apparently because he intended both his works to be textbooks for helping his students understand physical phenomena and so correctly work out the ecclesiastical calendar.[60] (The importance of this task is also embodied by the nocturnal, or *horologium nocturnum*, a device for measuring time at night using the stars, allegedly invented in the eighth or ninth century by Pacificus of Verona to maintain the canonical liturgy at night.)[61]

From around the seventh century, then, there emerged new texts that amalgamated the Christian worldview with explanations of the structure and function of the physical world.[62] Iconographic and diagrammatic representations of this Christian universe also abounded. Among them were medieval *mappae mundi*, such as the ninth-century map of the Holy Land in Paris, Bibliothèque nationale de France, MS lat. 11561 (fol. 43v), and T-O or Y-O-shaped maps that originated with Isidore of Seville in his *Etymologies* and *De natura rerum* (for example, in the ninth-century manuscripts St. Gallen, Stiftsbibliothek, MS 236; and St. Gallen, Stiftsbibliothek, Cod. Sang. 237, pp. 1 and 219).[63] More famous later medieval examples include the Hertford and Ebstorf *mappae mundi*. Such maps were not intended as accurate representations of the Earth, but rather as schematic visions of a Christian world, often with Jerusalem or Eden in the centre or at the top. Early medieval wind diagrams had an analogous function.[64]

**58** Isidore of Seville, *On the Nature of Things*, trans. Kendall and Wallis, 14.

**59** Bede, *On the Nature of Things and On Times*, 3.

**60** Bede, *On the Nature of Things and On Times*, 12. On Bede's understanding of nature, see Ahern, *Bede and the Cosmos* and MacCarron, *Bede and Time*.

**61** St. Gallen, Stiftsbibliothek, Cod. Sang. 18, 43r (https://www.e-codices.ch/en/list/one/csg/0237)

**62** A wide range of additional examples of such medieval texts is provided in Goetz, *Gott und die Welt*, chap. 3.

**63** For an introduction to medieval maps, see Scafi, *Mapping Paradise*.

**64** Obrist, "Wind Diagrams."

More complex T-O maps did exist, however, showing various relationships between the Christian world and the physical one: in London, British Library, MS Harley 3017, fol. 135r (second half of the ninth century and first quarter of the tenth century), the T-O map is surrounded by wheels showing the moon phases, and the association between the lunar cycle and the tides. Similarly, the Cloth of the Ewaldi found at Cologne, which may have been used as an altar-cloth before the tenth century, depicts the zodiac, the day and night, the year, personifications of ocean and earth, and the words "The entire nation, which looks upon the product of art." Here, "the product of art," in the words of the art historian Benjamin Anderson, has a double meaning: it is "the ordered universe, shaped by the divine art of its creator," as well as the Cloth of the Ewaldi itself.[65] The eighth-century Horologium of Willibrord goes beyond computistical use by—among other things—charting the different positions of the sun at different times of the year, as if following the heavenly sphere, and by furnishing the four compass points with etymological glosses linking them to various aspects of humanity.[66] A more symbolic Christian cosmography is represented in the meditative *carmina figurata* created by Hrabanus Maurus in the ninth century for his set of poems entitled *Veneration of the Holy Cross* (*De laudibus sanctae crucis*). One poem, for example, depicts wheels or rings enclosing verses on the four seasons, elements, parts of the world, and quadrants of the natural day.[67] There are many other examples of both texts and images that describe or depict such a Christian view of the world; they cannot be exhaustively listed here. The full evidence for this clearly intense interest in the manifestation of God through the natural world remains to be gathered and examined.

The early medieval emphasis on a meaningful world may, in part, derive from the rich tapestry of beliefs and ideas about nature that circulated in the post-Roman Latin West. These were not necessarily rooted in "paganism," a notion that in any case fails to convey adequately the complexity of the contemporary religious landscape or the variety inherent within Christianity itself. Rather, early medieval monastic teachers faced the challenge of integrating a plurality of views about the physical world, among them a plurality of Christian views. Well-known examples of this are the ninth-century Merseburg charms and the Trier blessings from the late tenth or early eleventh century, which invoke divine power for the healing of horses and the obedience of bees.[68] The principal challenge faced by the early medieval Church was not to root out ideas

---

**65** "Populus qui conspicit omnis arte laboratum"; Anderson, *Cosmos and Community*, 76–77. The Cloth of the Ewaldi is kept as part of the relics of the two Saints Ewald at the Church of St. Kunibert in Cologne.

**66** Paris, Bibliothèque nationale de France, MS lat. 10837, fol. 42r (https://archivesetmanuscrits.bnf.fr/ark:/12148/cc72512p).

**67** Vienna, Österreichische Nationalbibliothek, MS 652, fol. 12v. Sears, "Word and Image."

**68** "Merseburger Zaubersprüche," Merseburg, Domstiftsbibliothek, MS 136, fol. 85r (https://archive.thulb.uni-jena.de/korax/receive/Korax_cbu_00000880); "Trierer Segenssprüche," Trier, Stadtbibliothek, MS 40/1018 8°, fols. 19v and 36v–37v. On the context of these charms, see Schulz, *Beschwörungen im Mittelalter*; Embach, "Trierer Zauber- und Segenssprüche"; and Haeseli, *Magische Performativität*.

and practices that it labelled "pagan" or "heretical"—though it did this too—but to systematize the faith and unify the faithful. Even populations regarded by ecclesiastical authorities as converted were liable to develop local forms of religion, which were sometimes not in harmony with Church teaching, and required correction. That this was the case is evident, among other sources, in early medieval works such as the *Duplex legationis edictum* and the *Homilia de sacrilegiis*, which forbade divination using the Psalter, Gospel and other sacred texts, and in lists such as the *Small Index of Superstitions and Paganism*.[69]

Many of the important early medieval cosmographical works represent efforts at systematic explanation of the world in response to such heterodox ideas, from Isidore's *De natura rerum* in the seventh century to Agobard of Lyon's long treatise on hail and thunder in the ninth century—both written to combat popular attributions of natural phenomena to agents other than God.[70] The natural world therefore came to play an important role in the teaching of orthodox belief in the early Middle Ages.

It is within this context of continuous Christian teaching of the general population, for the sake of its salvation but also within an institutional framework of ideological control, that the *Physiologus* took its place. Its vision of the natural world was steeped in biblical allegorical imagery, and perfectly suited the correction of faith and knowledge that was a defining characteristic of the Carolingian cultural and intellectual renewal. One of its essential functions, therefore, was to spread and reinforce a Church-sanctioned view of the physical world.

## Allegory, Etymology, and the Emblematic Worldview

The new works of Christian cosmography produced in the early Middle Ages emerged out of different cultural and intellectual contexts, at different times, pursued different aims, and sometimes operated at the margins of canonical texts or mainstream social groups. As a result, they did not always transmit the same message. To some extent this is true of each of the copies of the early medieval *Physiologus*, as we shall see in later chapters. Neither were these works new in the sense that they were wholly original: all relied to some degree on inherited knowledge, particularly the Greco-Roman conceptualization of the universe as layers of spheres around the Earth, and the fourfold Aristotelian scheme of elements.[71] Late antique geographical texts, such as the *Cosmographia* of Julius Honorius, were also popular in the early Middle Ages and occupied a similar niche in the intellectual landscape of the time.

But all these works were also part of a new focus on biblical learning, which had become concentrated in key monastic centres across the early medieval West from around the fifth century. These centres began to exercise their own wide-ranging intel-

69 *Indiculus superstitionum et paganiarum*, extant in Vatican City, Biblioteca Apostolica Vaticana, MS Pal. lat. 577, 7r (before 800). McNeill and Gamer, ed. and trans., *Medieval Handbooks of Penance*, 419–21. See also Filotas, *Pagan Survivals*, 246.

70 Agobard of Lyon, *Liber contra insulsam vulgi opinionem de grandine et tonitruis*, PL 104.147.

71 Obrist, "Wind Diagrams," 35.

lectual influence from around the seventh century. Among the more famous were the school of Theodore and Hadrian at Canterbury, Bobbio in Italy, Luxeuil in France, and Iona off the coast of Scotland, but there were many others. A century and a half later, the Carolingian project took place: large-scale reforms (or attempts at reform) of coinage, education, law, liturgy, and monastic life; a re-shaping of the rhetoric of kingship along Christian lines; an ambitious building program; the association of a new and more legible script with the written output of the empire; and the promotion of scholarship. Early medieval thought about the natural world was fundamentally affected by this activity, which revolved around all aspects of the written word: language and grammar, script, Christianity as a religion of the book, documentary exercise of power, and monastic reading or *lectio divina*.[72] The seeds of this attention to the written word were sown in the classroom. In a dialogue on the eight parts of speech, for example, two pupils are described as entering the "thickets of grammatical density" ("spineta grammaticae densitatis"); and when asked by their school-master where they should begin, they reply: "Where else but at the letter?" ("Unde nisi a littera?").[73] In a tenth-century manuscript, grammar, dialectic, and rhetoric are depicted as schoolroom teachers, highlighting the close link between language and monastic education.[74] As Martin Irvine has shown, this special focus on language arose out of the view by antique and late antique writers that grammar—in the sense of *grammatica* or "an art based on universal and systematic principles capable of being reduced to formal rules"—was the entry-point to all the liberal arts, and from the use of grammatical methods of interpretation by Church Fathers. This meant that readings of texts that focused on their *grammatica*—as exemplified in the terms exegesis or "leading out of the text," and *enarratio* or "from the narrative"—became a key feature of Christian interpretation. As a result, grammar also became a key feature of understanding and interpreting the natural world. The two fundamental components of this interpretative technique were allegory—in the sense of *allegoria*, a kind of metaphor assumed in ancient grammar and rhetoric to be an invariable component of understanding texts—and etymology.[75]

As well as being a component of grammar, and therefore of the trivium, in the ancient world, allegory was adopted by early medieval exegetes. It quickly became one of three levels of interpretation, perhaps under the influence of Neoplatonic ideas such as Plotinus' three-fold division of reality into three hypostases (substances or essences): Soul, Intellect, and One. The third-century Church Father Origen used three levels of interpretation for scriptural reading: the flesh, the soul, and the spirit (that is, the literal, moral, and spiritual). It is debated whether or not Origen (together with his contemporary exegete Clement of Alexandria) was influenced by Platonism or

---

**72** Robertson, *Lectio Divina*.

**73** Alcuin, *Dialogus Franconis et Saxonis de octo partibus orationis*.

**74** Paris, Bibliothèque nationale de France, MS lat. 7900A.

**75** Irvine, *The Making of Textual Culture*, 63–66, 244–45. See also Law, *The Insular Latin Grammarians*. On Christian grammar in the classroom in antiquity, see Nelson, "The Classroom of Didymus the Blind," and Stefaniw, *Christian Reading*.

Neoplatonism—and his use of this exegetical method was unsystematic, fluctuating between two, three, and four levels—but the idea that there are several possible levels of interpretation is undeniably common to both traditions, and it permeated the hermeneutics of thinkers such as Gregory the Great, whose work was widely read in the early Middle Ages.[76] The many illustrations of the Stuttgart Psalter, created ca. 820 at St. Germain-des-Prés in Paris, also provide historical, spiritual, and moral readings.[77] In Ireland, literal exegesis was more common, but allegorical and spiritual meanings are also attested. In the Litany of the Trinity, for example, Christ is man, lion, calf, and eagle, corresponding to the four Gospels and their interpretations of Christ as Lion of Judah, servant, sacrifice, and God. The Christian levels of interpretation were eventually formalized and set at four—the literal, allegorical, spiritual (anagogical), and moral or figurative (tropological). These four levels were also influenced by antique reading practices: for the first-century BCE Roman scholar Marcus Terentius Varro, for example, grammar could be divided into *lectio* (reading), *enarratio* (exposition), *emendatio* (correction), and *iudicium* (critical assessment).[78] The fourfold method of exegesis continued well into the central and late Middle Ages. Its order was memorized by monks, as we know from the mnemonic rhymed distich by the Dominican monk Augustine of Dacia (d. 1282):

> The literal sense teaches us what happened, the allegorical what to believe
> The moral how to act, the anagogical what to hope for.[79]

Allegorical exegesis or allegoresis was therefore part of a widespread hermeneutic method from the early Middle Ages onwards in the Latin West. As we shall see in later chapters, allegorical exegesis shaped the meaning of both the *Physiologus* and the other texts with which its copies were collocated and had wide-ranging implications for the understanding of nature.

In addition to allegory, etymology was a key component of interpretation based on *grammatica*. Although etymology was not unknown to late antique Christian grammarians and scholars such as Cassiodorus, it was Isidore of Seville who popularized it in the *Etymologiae*. The impact of this encyclopedia on early medieval thought was so extensive that it has not yet been fully assessed.[80] It was re-worked by the impor-

---

**76** Origen's trichotomy ultimately derives from St. Paul; Chadwick, "Origen," 183. Origen's levels of interpretation, and their later development, are examined in Lubac, *Histoire et esprit*, 139–49. Robertson, *Lectio Divina*, 16–19, summarizes and assesses Lubac's study. Gregory the Great discusses the different levels of scriptural interpretation in *Moralia in Iob, Ad Leandrum* 4, 173–78 (CCSL 143.6). See also Collins, *The Carolingian Debate*.

**77** Stuttgart, Württembergische Landesbibliothek, MS Bibl. fol. 23 (https://digital.wlb-stuttgart.de/index.php?id=6&tx_dlf%5Bid%5D=8680&tx_dlf%5Bpage%5D=1).

**78** Diomedes, *Ars grammatica*: "Grammaticae officia, ut adserit Varro, constant in partibus quattuor, lectione enarratione emendatione iudicio"; Keil, "Diomedes. Ars grammatica," 426.

**79** "Littera gesta docet, quid credas allegoria / Moralis quid agas, quo tendas anagogia." Parmentier, *L'Écriture vive*, 40–42. See also Michaud, "Des quatre sens." This distich is often attributed to Nicholas of Lyra (d. 1349).

**80** There exists no comprehensive study of Isidore's encyclopedia and its impact on the early

tant early medieval scholar and exegete Hrabanus Maurus, who added allegorical interpretations to Isidore's etymological ones. According to Natalia Lozovsky, "Hrabanus, quite in the Isidorean tradition, treats the created world as a text, but whereas Isidore focuses on its grammar, Hrabanus tries to uncover its symbolical meaning."[81] Yet Isidore also looked for the symbolic meaning inherent in the things he described. He merely used a different tool for doing this, etymology, which he raised to a level beyond that of simple grammar. In his work, as Jacques Fontaine noted, "l'étymologie… est devenue la démarche essentielle de toute connaissance."[82] Isidore stated this in a much-quoted passage in the *Etymologiae*: "Letters are the tokens of things, the signs of words, and they have so much force that the utterances of those who are absent speak to us without a voice, for they present words through the eyes, not through the ears."[83] This function of letters and words makes etymology—in the Isidorean sense, as a means of discovering a higher truth—a tool of allegoresis. But the function of written words and signs as visual symbols of a hidden meaning is so embedded in the presentation and interpretation of texts in the early Middle Ages that etymology is made equal in importance to allegory. Language itself becomes symbolic: things are known through signs, that is words, and are lost without them.[84] This, in turn, makes language the means by which the created world, and therefore God, are understood. As Stephen Harris argued, language "was considered an orderly natural phenomenon, like the disposition of the stars or the manifold variety of animals."[85] The written word is a particularly special component of language, since its meaning, like the meaning of the visible world, is accessed using the eyes. This idea permeated early medieval texts and guided not only their composition, if they were new, but also their selection and collocation, if like the *Physiologus* miscellanies they were copies of works composed in the past.

Allegory and etymology transcended *grammatica* as a simple schoolroom component of the trivium. They underpinned early medieval thought about nature, in ways

---

medieval cultural and intellectual worlds, although Evina Steinová has recently completed a project on Isidore's *Etymologiae* in the Carolingian period: "Innovating Knowledge: Isidore's *Etymologiae* in the Carolingian Period," Netherlands Organisation for Scientific Research Veni grant 2017–2021. This includes a manuscript database (https://innovatingknowledge.nl/?page_id=33). The seminal work on Isidore of Seville remains the three-volume study by Fontaine, *Isidore de Séville et la culture classique*. See also Fontaine, "Aux sources de la lexicographie médiévale," *Tradition et actualité*, and *Isidore de Séville: genèse et orginalité*; Henderson, *The Medieval World*, and "The Creation of Isidore's *Etymologies*." A review of the twentieth-century literature on Isidore up to 1975 was published by Hillgarth, "The Position of Isidorian Studies."

**81** Lozovsky, *The Earth is Our Book*, 111.

**82** "Etymology has become the essential starting point for all knowledge"; Fontaine, *Isidore de Séville et la culture classique* 1, 41.

**83** "Litterae autem sunt indices rerum, signa verborum, quibus tanta vis est, ut nobis dicta absentium sine voce loquantur. Verba enim per oculos non per aures introducunt"; *Etymologiae* 1.3.1.

**84** Irvine, *The Making of Textual Culture*, 223.

**85** Harris, "Anglo-Saxon Ciphers," 75.

that will be explored in the remainder of this book. Their effect can be described as a novel emphasis on symbolism through language, and a multiplicity of interpretations. There was no one truth, or rather, it was already known to be God; to attain that truth, one sought the array of meanings inherent in every created thing. This attitude, compounded from a set of partially integrated inherited traditions and the need to reconcile them, was something quite new. In its flexibility and variety, it paralleled what William Ashworth, writing about the Renaissance, termed the emblematic worldview: "The belief that every kind of thing in the cosmos has myriad hidden meanings and that knowledge consists of an attempt to comprehend as many of these as possible."[86]

## Conclusion

The natural world in the early Middle Ages can seem muddled because it is just one aspect of a general cultural attitude which had to embrace and assimilate a diversity of complex ideas and beliefs. Culturally and intellectually, early medieval Europe was undeniably an enormously intricate and surprising landscape. The *Physiologus* transmitted a view of the world influenced by Greco-Roman ideas, and it does not seem to have circulated in Islamic or Jewish textual contexts, which exerted their own cultural influence in the early medieval period, particularly in the Iberian Peninsula. These different beliefs and traditions highlight that understanding of the natural world in the early medieval West was far from uniform. The *Physiologus* can give us only a narrow view of this diverse setting. It seems clear, however, that the early medieval natural world was not one of dichotomies such as "us and them," "good and evil," or even "adversarial and harmonious," though we have been occasionally prone to interpret it in this way. Rather, the many possible meanings inherent in Creation were a pathway for approaching God. For those trained in this way of thinking in the early Middle Ages—both lay and Church people—this was done by means of the written word and the visible world, which were connected through Scripture.[87] With this in mind, the following chapters examine how writing and nature were integrated in the *Physiologus*: the works with which it was copied, the arrangements of its chapters, the presentation of its text, and its materiality.

**86** Ashworth, "Natural History," 312.

**87** Lozovsky, "Carolingian Geographical Tradition," 36.

**g.** "liber bestiarum" recorded in a list of books donated to the Benedictine house at Peterborough in 970 by Æthelwold, bishop of Winchester, preserved in London, Society of Antiquaries, MS 60, fol. 39v (the Black Book of Peterborough).[8]

**h.** "liber bestiarum et uolucrum" in a list of books belonging to the St. Emmeram (Regensburg) monk Waltherius in Munich, Bayerische Staatsbibliothek, MS Clm 14222, fol. 17r, copied in the tenth century (though the manuscript itself dates to the second quarter of the ninth century); and "de natura bestiarum et uolucrum" in the catalogue of 513 books belonging to St. Emmeram, compiled on the order of Abbot Ramuoldus or Ramwold (in office 975–1001) and completed by 993, though with subsequent additions. This catalogue is preserved in the tenth-century lectionary Pommersfelden, Gräflich Schönborn'sche Schlossbibliothek, MS 340 (2821), fols. 73v–75r.[9] The discussion below assumes that these two entries represented the same manuscript copy of a text on beasts and birds, in the personal possession of Waltherius in the tenth century and bequeathed by him to his community before 993.[10]

**i.** "Fisiologus" in a late tenth or early eleventh-century catalogue from eastern France or Belgium, preserved in Bern, Burgerbibliothek, MS 4, fol. 55v (line 10).[11]

---

**8** Digitized at https://collections.sal.org.uk/mss.0060.

**9** Becker, 128, no. 42 (the Pommersfelden manuscript), item 453; and 130, no. 44, item 10 (the Munich manuscript); MBK IV.1:142–49. The Munich manuscript is digitized at www.digitale-sammlungen.de/view/bsb00036222?page=1. The most complete description of the Pommersfelden manuscript is in Swarzenski, *Die Regensburger Buchmalerei*, 41–45. The private comital library of Schloss Weißenstein, in which the manuscript is kept, is also known as Graf von Schönborn Schlossbibliothek.

**10** This interpretation means that the date of Waltherius' booklist in Munich, Bayerische Staatsbibliothek, MS Clm 14222 is pushed back from the early eleventh century, as estimated by Swarzenski, to before 993. See Swarzenski, *Die Regensburger Buchmalerei*, 25.

**11** www.e-codices.unifr.ch/en/list/one/bbb/0004. Becker no. 29, item 47; MBK Ergänzungsband I.2, 743; Genest, Chalandon, and Genevois, *Bibliothèques*, no. 1934. The manuscript itself is a pandect Bible from Tours. The booklist was thought to be from Fleury by Cuissard, *Inventaire des manuscrits*, 209–11, Manitius, *Handschriften antiker Autoren*, 258, and others following them, but this is incorrect according to Mostert, *The Library of Fleury*, 48. Florian Mittenhuber suggested on e-codices that the booklist is from Lotharingia, based on the saints' lives that it lists (from Soissons, Saint-Quentin, Liège, and Maastricht). A provenance in Eastern France or Belgium was suggested by Munk Olsen, *L'étude des auteurs classiques latins*, 3.1 (1982), 283. Provenance in Alsace or Strasbourg has been suggested but is unproven: Krämer, *Handschriftenerbe des deutschen Mittelalters* 1.2, 743. Bischoff, *Katalog der festländischen Handschriften*, 1:103, gave a western German provenance, perhaps from the diocese of Cologne, though he also indicated Strasbourg (its own diocese and part of the archdiocese of Mainz) as a probable location. A recent conference report posited, without presenting the full argument, that the list was copied in the region of the cult of St. Romaric (on the basis of the inventory in the related codex Bern, Burgerbibliothek, MS 3), between Épinal, Verdun, Metz, and Strasbourg, and suggested Bonmoutier (Saint-Sauveur-en-Vosges) or Saint-Sauveur d'Andlau as probable places of origin; see Turcan-Verkerk, "Langue et littérature," 149. As regards the date of the list, only Cuissard, *Catalogue général* 12, iii, and Pellegrin, "Membra disiecta Floriacensia," dated it to the ninth century; all others follow Homburger in dating it to the

None of these booklists entries appears to refer to any of the still-extant *Physiologus* manuscripts. Early medieval books were frequently exchanged and given in gift, however, so we cannot discount the possibility that the above entries refer to codices originating at, or belonging to, other centres. Nevertheless, the high rate of loss of medieval manuscripts over time, and the wide geographical spread of the *Physiologus*, make it probable that most of the booklist entries denote unknown copies of the text that are now lost.

Of the nine entries, *b*, *c*, and *i* are not in doubt since they name the *Physiologus*. Entry *a* is almost certainly also a *Physiologus*, since John Chrysostom never wrote a text on the nature of animals and there appears to have been an early medieval tradition of crediting him with the authorship of the *Physiologus* (the *explicits* to the *Physiologus* in Munich, Bayerische Staatsbibliothek, Clm 14388 and Clm 19417 attribute their copies of the text to Chrysostom; see descriptions in Appendix I). Entries *c*, *f*, *g*, and *h* are the least certain, since they ambiguously refer to books about beasts or the nature of beasts and birds. Among such works were Isidore's *Etymologiae*, books 11–14, and Pliny's *Historia naturalis*, both of which contain information on animals, birds, and stones, were widely used and were so long that they were often copied in part rather than in full. However, the works of both Pliny and Isidore were well known and usually attributed to them by copyists: both the Murbach (*a*) and Alsace or Strasbourg (*i*) inventories, for example, list works attributed to Isidore under the headings "Ysidori/ Ysodori libri" before the presumed *Physiologus*. Listing the name or subject matter of a text without its author usually meant either that it was so familiar that the author's name was redundant, or that its author was unknown. The former seems unlikely in the case of the *Physiologus*. We can see from a catalogue of thirty-four books, made at Würzburg ca. 800, that this kind of abbreviation was reserved for widely read authors such as Gregory the Great, Augustine, and Bede.[12] Their respective works are listed simply as "dialogi," "enciridion" and "Historia anglorum."[13] Other early medieval catalogues also abbreviate the titles of famous works in this way. It is more probable, therefore, that entries *c*, *f*, *g*, and *h* are references to the *Physiologus*.[14] No other prose text on animals circulated in the early Middle Ages without an attributed author. The sole exception is the *Liber monstrorum*, but its subject matter—monsters or marvellous creatures—means that it was unlikely to have been listed as a "liber bestiarum."

The only truly doubtful booklist entries, then, are *e*, the "Liber Esopi de natura animalium," and *f*, the "libros bestiarum Ysiodori." Aesop's fables, or a version of these fables by Aelian, Babrius, or Phaedrus, were known in early medieval monastic libraries (the tenth-century library catalogue of St. Emmeram described in entry *h*, for

---

late tenth or early eleventh century; see Homburger, *Die illustrierten Handschriften*, 72–79. I have chosen to follow Mittenhuber and Munk Olsen as regards provenance and dating on the evidence of the saints' lives and the script, which I believe to be early eleventh century.

12 Oxford, Bodleian Library, MS Laud misc. 126, fol. 236r. This is a different booklist to the one in *a*.

13 McKitterick, *The Carolingians and the Written Word*, 171.

14 Entry *g* was also identified as a *Physiologus* in Lapidge, *The Anglo-Saxon Library*, 136.

example, lists an "Aesopus") and may have existed in a Fulda copy attested by entry *e*.[15] Book 12 of the *Etymologiae*, on beasts and birds, could have been behind entry *f*. Both entries could also, however, refer to the *Physiologus*, which was similar enough to both Aesop and Isidore's texts to be ascribed to these authors.

Even if we discount *e* and *f*, the remaining entries add to the picture of the circulation of the *Physiologus* in the early Middle Ages already provided by the extant manuscripts. By the mid-eighth century, the *Physiologus* was present in central France. By the ninth century it was being copied in the principal Frankish territories in France and Germany. But the history of the *Physiologus* in Europe is not well represented by the geographical spread of the manuscripts. The ninth-century witness from Catalonia, for example (see Chapter 5), cannot be said with certainty to have been made there, and it is the only known Latin *Physiologus* from the region. The Spanish bestiary tradition generally remained relatively weak throughout the Middle Ages. The earliest known version was translated into Aragonese, Castilian, and Catalan from the French-language *Livre dou Tresor* by Brunetto Latini (d. 1294).[16] A second translation, based on fourteenth-century Tuscan bestiaries, did not appear until the fifteenth and sixteenth centuries.[17] The *Physiologus* itself reappeared in Spain during the Renaissance. Gonzalo Ponce de León's translation of the Greek *Physiologus* into Latin was published in Rome in 1587 and subsequently translated into Castilian by Francisco Tejada Vizuete.[18]

Conversely, though the ninth-century manuscript from Cassino seems to suggest that it was an outlier, it is probable that the Latin *Physiologus* was disseminated from Italy. The earliest evidence for the text in the Latin West is from the Apennine Peninsula. Ambrose had access to either a Greek or a Latin version in the fourth century, while the Latin *Decretum Gelasianum*, a sixth-century Italian forgery also found in the *Physiologus* manuscript St. Gallen, Stiftsbibliothek, Cod. Sang. 230, explicitly prohibited the *Physiologus*. Later Italian copies of the *Physiologus* or a related version such as the *Physiologus Theobaldi* attest to a strong and continuous manuscript tradition, in both Latin and Greek.[19]

The history of the *Physiologus* in its Insular context is also more complex than the number of surviving copies or booklist mentions suggests. The text first appears to have been used by Aldhelm ca. 695 to compose riddles.[20] Aldhelm's use of the *Physio-*

---

**15** Becker no. 42, item 474.

**16** Baldwin, *The Medieval Castilian Bestiary*.

**17** Pascual, "La tradición animalística"; Salvat, "Notes sur les bestiaires catalans."

**18** Edited in Sebastián, *El Fisiólogo*. The Y recension of the *Physiologus* was most recently translated into Spanish by Guglielmi, ed., *El Fisiólogo*.

**19** For example, the Latin *Physiologus* in Florence, Biblioteca Medicea Laurenziana, MS San Marco 650 (s. xi), the Greek *Physiologus* in Milan, Biblioteca Ambrosiana, MS E 16 sup. (s. xi, https://ambrosiana.comperio.it/opac/detail/view/ambro:catalog:70286), and the Latin *Physiologus Theobaldi* in Fano, Archivio del Capitolo della Cattedrale, MS 5 (s. xiii).

**20** Manitius, *Geschichte der lateinischen Literatur*, 1:137n5; Milovanović-Barham, "Aldhelm's Enigmata," 51; Lendinara, *Anglo-Saxon Glosses*; Salvador-Bello, "Evidence of the Use of the *Physiologus*."

*logus* for riddles, and his proven use of the *Aenigmata Bernensia* in the Bern/Paris collection (parts 3 and 5, discussed below), indicate that he had read this manuscript or a related, now lost, copy. Aldhelm made several trips to the continent, including a pilgrimage to Rome with Cædwalla of Wessex ca. 688–89.[21] They travelled to Rome via France, and this was also the way taken by Aldhelm on his return journey. Joanna Story has shown that two ninth-century Rheims manuscripts preserve the unique text of the original Roman *tituli* or verse epigrams that served as a source for Aldhelm's verse compositions.[22] One of these manuscripts also contains a copy of Aldhelm's riddles.[23] This evidence seems to suggest that the *tituli* were copied by Aldhelm in Rome, that they were then transmitted by him to a centre in or near Rheims, and that this then enabled a manuscript anthology of Roman *tituli* to be compiled in the ninth century. It is not inconceivable, therefore, that Aldhelm spent enough time in France in the late seventh century to consult the exemplar for the Bern/Paris collection, or perhaps the relevant parts of the collection itself, which was created ca. 727 (see more detailed discussion of this manuscript in the second half of this chapter).

After Aldhelm, the evidence for the *Physiologus* in early medieval England remains sparse but clear. Hwaetberht (writing as Eusebius), abbot of Monkwearmouth-Jarrow in the first half of the eighth century, used either Aldhelm's compositions or the *Physiologus* itself for his own riddles.[24] The *Liber monstrorum*, an anonymous eighth-century text probably composed in England, also used the *Physiologus*.[25] However, the *Physiologus* material used by these texts may derive from the *Etymologiae*, which was a major source of the *Liber monstrorum*.[26] More text historical work is needed to establish their relationship.[27]

The fragmentary Old English metrical *Physiologus* in the Exeter Book (Exeter, Cathedral Library, MS 3501, fols. 95v–98r) also indicates that the *Physiologus* was known to Insular writers by the tenth century and was read with sufficient interest to be at least partially translated and re-worked in verse form. The Exeter Book was given to Exeter Cathedral by Leofric, its first bishop (d. 1072) and was produced in the seventy or eighty years before Leofric's elevation to his office, possibly ca. 970–90.[28]

---

**21** Lapidge, "The Career of Aldhelm," 52–64.

**22** Story, "Aldhelm and Old St Peter's."

**23** Milovanović-Barham, "Aldhelm's Enigmata," 51.

**24** Greenfield, Calder, and Lapidge, *A New Critical History*, 12. The identification of Eusebius with Hwaetberht has not been proven, but is probable; O'Brien, "Hwaetberht," 315.

**25** Lapidge, "Surviving Booklists," 55.

**26** Orchard, *Pride and Prodigies*, 86–115; see also Orchard, *A Critical Companion to Beowulf*, 134–35.

**27** On the Insular *Physiologus* tradition, see Orlandi, "La tradizione del *Physiologus*" and Frank, "Die Physiologus-Literatur." On the *Liber monstrorum* manuscripts, see Bologna, "La tradizione manoscritta."

**28** Krapp, ed., *The Exeter Book*, 10; Schubel, *Englische Literaturgeschichte*, 13. See also Conner, "On Dating Cynewulf." Leofric's donation is recorded in Exeter, Cathedral Library, MS 3501, fols. 1r–2v.

Its place of origin is uncertain.[29] The text consists of only three chapters (the whale, the panther, and the partridge) and a number of folios are missing.[30] This makes the source recension of the text difficult to determine. Various scholars have argued that it is related to the B version,[31] or the *Physiologus Theobaldi* (in fact a later work composed between 1022 and 1035),[32] and even that it has no connection to the main recensions (B, C, or Y).[33] Contamination across recensions may have ultimately influenced the Exeter Book text.[34]

The Exeter Book *Physiologus* is omitted from this book except as proof of the knowledge of the *Physiologus* in tenth-century England. As an original work of Old English poetry that focuses on the descriptive and aesthetic side of the tales rather than their moral lessons, the Exeter Book text is sufficiently different in form, function, and content from the Latin *Physiologus* in the early Middle Ages that it is clearly not part of the same context.[35] It is also a vernacular version, and as such deserves consideration on its own terms independently from the Latin recensions, and with reference to the Insular poetic and riddling traditions that informed it.

There is little to no evidence that the *Physiologus* was ever read in early medieval Wales or Scotland, though it may have made its way to Ireland. The eighth-century Irish Derrynaflan paten (a dish used to hold the Eucharist during Mass), found as part of a small hoard of liturgical vessels within the monastic enclosure in Derrynaflan, Co. Tipperary, depicts a stag and snake on one of the panels along its rim. It has been suggested that this imagery derived from the *Physiologus*, in which the stag spits water through cracks in the earth to drive out the snake before crushing it.[36] However, similar baptismal allegories involving these animals also appear in Cassiodorus' commentary on the Psalms (in relation to Psalm 42) and in Isidore's *Etymologiae*, among other texts borrowing from these.[37] The tangled-up nature of these sources means that

---

**29** The Exeter Book's origin was thought by Patrick Conner to be Exeter: Conner, "On Dating Cynewulf," 23–55, and *Anglo-Saxon Exeter*, 85–86. However, the consensus now is that the Exeter Book was not made there; see Scragg, "Exeter Book," and Gameson, "The Origin of the Exeter Book." The origin of the Exeter Book is also briefly discussed in Biggs, "The Eschatological Conclusion," and Drout, "'The Partridge' is a Phoenix."

**30** Drout suggested that the partridge was in fact the phoenix, but Pakis argued persuasively against this; Drout, "'The Partridge' is a Phoenix"; Pakis, "A Note in Defense."

**31** Lapidge, "Surviving Booklists," 55.

**32** Gastle, "The Beast Fable," 71. The *Physiologus Theobaldi* has been critically edited by Eden, *Theobaldi "Physiologus."*

**33** Frank, "Die Physiologus-Literatur," 36.

**34** Orlandi, "La tradizione del *Physiologus*," 1093.

**35** Letson, "The Old English *Physiologus*," 20.

**36** Ryan, "Some Aspects of Sequence and Style," 72, and *Early Irish Communion Vessels*, p. 39, fig. 14 and photo 16. See also Ryan, "The Menagerie." On evidence for *Physiologus* influence on Pictish symbolism, see Henderson, *Pictish Monsters*.

**37** As Psalm 41, in accordance with Cassiodorus' use of the Vulgate psalm numbering. CCSL 97; Cassiodorus, *Explanation of the Psalms* 1; *Etymologiae* 12.1.18–19. The story and its origins are described in Clark, *A Medieval Book of Beasts*, 134.

the ultimate origin of the imagery on the Derrynaflan paten cannot be ascribed with certainty to the *Physiologus*. Scandinavian influence is also possible: some Viking-age coins depict a stag and snake on their reverse (Figure 2.1).

Figure 2.1. Viking-age sceatta (reverse) depicting a stag apparently kissing a snake. Ribe, Sydvestjyske Museer Denmark, catalogue no. NM FP 14603.1. Available under CC-BY-SA licence.

Aldhelm and Hwaetberht's riddles, the Exeter Book and the Peterborough booklist demonstrate that the *Physiologus* was known to Insular writers from at least the second half of the seventh century. Some early medieval English coins may also have been influenced by the *Physiologus* in their animal depictions.[38] We can only speculate, given the lack of sources, about the spread of the text in the seventh century. It is unlikely that the *Physiologus* was transmitted to the continent from England, since the continental manuscript evidence is much stronger than the Insular, and since the text was used in Italy as early as the fourth and sixth centuries. The Latin *Physiologus* may have instead arrived in England from Rome with Theodore of Tarsus and Hadrian of Canterbury in the 660s or 670s, or with Aldhelm in the 690s.[39] Aldhelm was one of the pupils at the famous school founded by Theodore and Hadrian in Canterbury, which provides another conceivable locus for Aldhelm's introduction to the *Physiologus*.[40] While it is impossible to be certain during which of these moments of international exchange the *Physiologus* was transmitted to England, they are emblematic of decades of travel and pilgrimage from Insular settlements to Rome, other parts of continental Europe, and beyond, in the seventh century. This lively movement of people was ultimately responsible for the wide circulation of texts such as the *Physiologus*.

Regardless of the precise routes it took, the *Physiologus* clearly enjoyed reasonably widespread circulation in the Latin West, relatively soon after its translation into Latin in the fourth century. It made its way across the Channel before or during the

---

**38** Gannon, "Coins, Images and Tales." See also Gannon, *The Iconography of Early Anglo-Saxon Coinage.*

**39** On the context of the books of Augustine, Hadrian, and Theodore, see Gameson, "The Earliest Books of Christian Kent." Much less likely, though still possible, is the introduction of the *Physiologus* to England with Augustine in 597 or with another set of books sent to Augustine from Rome four years later.

**40** If Aldhelm used Byzantine riddles as one of his models, it is also possible that he accessed them through his teacher Theodore, either as part of oral tradition or as part of the book collection brought by Theodore to England. See Milovanović-Barham, "Aldhelm's Enigmata."

second half of the seventh century. Its textual tradition was well-established in Europe by the time the earliest extant copy of the Latin *Physiologus* came to be made. The early medieval Latin West therefore inherited a work that had at least one century of active monastic transmission behind it, and that would have been known to those copying and composing texts.

## The *Physiologus* in Early Medieval Libraries

Although all the *Physiologus* manuscripts mentioned by the extant booklists are now lost, the catalogues themselves tell us that the *Physiologus* was read in monastic and cathedral contexts in the early Middle Ages, both by clergy and by laypeople. Many more copies must have existed. In his 1961 lecture notes, Bernhard Bischoff estimated 500 ninth-century Carolingian libraries containing 200 to 300 volumes, which equals to at least 100,000 Carolingian manuscripts. Of these, only some 6 percent have survived. Various attempts have been made at using this and other figures to make more precise numerical estimates of manuscript survival.[41] They all indicate that the rate of loss was very high, and that we only have as many Carolingian sources as we do because of an explosion in text production. Merovingian sources are correspondingly much fewer in number. Charters from before 800 have a survival rate as low as 0.001 percent, by one estimate, due to the use of (easily degradable) papyrus in Merovingian chanceries.[42]

But even educated guesses at survival rates remain somewhat arbitrary. Too many factors remain unknown. Even if the estimates we have are correct, numbers say nothing about the distribution of books or about the cultural and intellectual importance of certain texts. By comparing book catalogues, we can conclude that the *Physiologus* was read as much as texts by Horace, Juvenal, Terence, and even Martianus Capella, who is mentioned only thirteen times in catalogues from the ninth and tenth centuries. Capella's *De nuptiis Philologiae et Mercurii* survives in much greater numbers in extant manuscripts; but more importantly, glossing evidence shows that it was read both widely and actively, especially in schools.[43] Contextual evidence is therefore much more valuable than numerical comparison for measuring the cultural significance of a text such as the *Physiologus*. The booklists are still valuable evidence in this respect, and are worth examining more closely (with the exception of lists *b* and *e*, of which we do not know enough to draw any real conclusions).

---

**41** Buringh, *Medieval Manuscript Production*, 237; Wood, "The Problem of Late Merovingian Culture," 202; McKitterick, *The Carolingians and the Written Word*, 167.

**42** Ganz and Goffart, "Charters Earlier than 800."

**43** Tahkokallio, "Manuscripts as Evidence," 39; Teeuwen, "Glossing in Close Co-Operation"; Teeuwen and O'Sullivan, *Carolingian Scholarship*.

## Iskar's List

Perhaps the longest library catalogue that mentions the *Physiologus* is the list made at the Abbey of Murbach (*b*). The manuscript in which it appeared is now lost and the catalogue exists only in a transcription made on paper by the Benedictine monk Sigismund Meisterlin in 1464.[44] The list was copied as a separate and subsequent addition to the main Abbey *registrum* of some 400 books, dated to ca. 840–42, edited by Wolfgang Milde.[45] According to Meisterlin's title, "Breuiarium librorum Isghteri Abbatis obmissis his qui in registro continentur pro parte," it was the catalogue of books belonging to Iskar, Abbot of Murbach around the middle of the ninth century.[46] The list was first edited by Hermann Bloch, and again more recently by Karl-Ernst Geith and Walter Berschin.[47]

Geith and Berschin believed that the words "obmissis...parte" meant that Meisterlin purposely omitted those titles which also appeared in the main catalogue (although some duplicates remain). He may also have omitted some titles not in the main catalogue. In addition, Meisterlin almost certainly transcribed in long lines what was originally a two-column list, which means that the books are not presented in their original order—as evident from the mixing of theological works with those pertaining to history and works related to the natural world, as discussed below.[48]

Like many other Carolingian abbots, Iskar donated his collection to the monastery on his death.[49] The contents of this collection are both interesting and unusual. Since they provide a potentially valuable context for a lost *Physiologus* copy owned by a named Carolingian abbot, they are reproduced in full here. Meisterlin's capitalization, abbreviations, and irregular treatment of *v* and *u* have been retained. Geith and Berschin identified the texts and associated some of the entries with extant manuscripts from Murbach. This information is provided beneath each entry:[50]

1. Epistole et canones diuersi volumen I
   Letters and lists from ecclesiastical councils. Gotha, Forschungsbibliothek, memb. I.85 or I.75.[51]

44 This transcription is in the manuscript Colmar, Archives Départementales du Haut-Rhin, Cartulaire Abbaye Murbach 1.

45 Milde did not edit the Iskar catalogue, although he reproduced it as an image alongside images of the main catalogue: Milde, *Der Bibliothekskatalog*.

46 "An abridged list of the books of Iskar the abbot, omitting those which are listed in the register in part."

47 Bloch, "Ein karolingischer Bibliotheks-Katalog"; Geith and Berschin, "Die Bibliothekskataloge."

48 Geith and Berschin, "Die Bibliothekskataloge," 86–87.

49 McKitterick, *The Carolingians and the Written Word*, 156.

50 Geith and Berschin, "Die Bibliothekskataloge," 68–84. With reference to Bloch's edition, 272–73, and to plate 12 in Milde, *Der Bibliothekskatalog*. I have followed Bloch's numbering of the entries.

51 https://dhb.thulb.uni-jena.de/rsc/viewer/ufb_derivate_00015136/Memb-I-00085_00001.tif.

2. Hebraicarum questionum et de XL mansionibus volumen I
   Jerome, *Liber Hebraicarum questionum in Genesim* and *De XLII mansionibus filiorum Israel in deserto (ep. 78)*. Colmar, Bibliothèque municipale, MS 33 (*olim* 41).[52]
3. Excerpta Iheronimi de Ethico philosopho
   Aethicus Ister, *Cosmographia*. Oxford, Bodleian Library, Junius 25.
4. Gesta pontificum et epistola Iehronimi [*sic*] de gradibus sacerdotum volumen I
   *Liber pontificalis* and St. Jerome, *ep.* 46 on the clerical grades.
5. Allexandri epistola ad Aristotilem et olimpiadem matrem suam
   Pseudo-Alexander the Great, probably the *Letter to Aristotle* and the *Letter to Olympia* about miracles in India, derived from Pseudo-Callisthenes.[53]
6. Orosius prouinciarum descripcio
   Orosius, *Historiae adversus paganos* bk. 1, chap. 2.
7. De eadem re Iheronimus
   The anonymous *Dimensuratio provinciarum*, falsely ascribed to Jerome.
8. Ysidorus de terra
   Isidore of Seville, *Etymologiae* bk. 14.[54]
9. Cosmographia Iulii cesaris
   Julius Honorius or Pseudo-Aethicus, *Cosmographia*, B recension.
10. Solinus de situ orbis volumen I
    Solinus, *Collectanea rerum memorabilium*.
11. Questiones albini in genesim
    Alcuin, *Interrogationes et responsiones in Genesin*.
12. Questiones Augustini et orosii in genesim
    Pseudo-Augustine and Pseudo-Orosius, *Dialogus quaestionum*.
13. Glose super regum
    Glosses or a commentary on the Book of Kings.
14. Bachiarius de reparacione lapsus
    Bachiarius, *Epistula ad Januariam seu De lapso*.
15. Exitium troianorum
    Dares Phrygius, *De excidio Troiae historia*.

---

**52** https://bvmm.irht.cnrs.fr/sommaire/sommaire.php?reproductionId=1946.

**53** Figueira, *The Hermeneutics of Suspicion*, 36.

**54** Thought by Bloch, "Ein karolingischer Bibliotheks-Katalog," 277, to refer to Isidore of Seville's *De natura rerum*, chap. 45–48. It is more likely, however, to refer to the *Etymologiae*, which was much more frequently copied.

16. Titus lucrecius de rerum natura volumen unum
    Lucretius, *De rerum natura.*

17. Explanatio Augustini in apostolum volumen I
    Augustine, *Expositio quarumdam propositionum ex epistola ad Romanos, Epistulae ad Romanos inchoata expositio, Epistula ad Galatas expositio.*

18. Rabanus in librum regum volumen unum
    Hrabanus Maurus, *Commentaria in libros IV Regum.*

19. De compoto Astrolabio de gramatica foci et arati et versus theodolfi volumen I
    Works on computus and the astrolabe, grammar of Phocas, *Phaenomena* of Aratus or a commentary on his work, poems of Theodulf of Orléans.

20. Rabanus in Iheremiam volumen I
    Hrabanus Maurus, *Commentaria in Ieremiam.*[55]

21. Geometrica et Iginius volumen I
    A work on geometry and Hyginus, *Astronomica.*

22. Partes donati maioris et minoris, declinacionis nominis et verbi volumen I
    Donatus, selections from the *Ars grammatica (Ars maior)* and *De partibus orationis ars minor* on nouns and verbs; or the works of Donatus and an anonymous work on nouns and verbs.

23. Rabanus de compoto
    Hrabanus Maurus, *Liber de computo.*

24. Beda de arte metrica
    Bede, *De arte metrica.*

25. Priscianus minor de scriptoribus diuinorum librorum[56]
    Isidore of Seville, *De ecclesiasticis officiis* bk. 1, chap. 12: *De scriptoribus sacrorum librorum.*

26. Beda de naturis rerum
    Bede, *De natura rerum.*

**55** This text is known to have been written ca. 840–42 and was therefore used by Bloch to date the catalogue. None of the other works written by Hrabanus after 840 are attested in this or the main Murbach catalogues.

**56** Bloch noted that this ought to read "Priscianus minor descriptio diversorum locorum," presumably because one of Priscian's works is a translation of the geographical poem *Periegenis* by Dionysius Periegetes. Geith and Berschin believed the phrase "Priscianus minor" refers to Priscian's *Institutionum grammaticarum* books 17–18 (on syntax), in addition to the *Periegenis* denoted by "de scriptoribus diuinorum librorum." It's more probable that the entry refers to the text by Isidore as listed here, since it (the full Isidore text) is also found under the rubric "scriptoribus divinorum librorum" in the ninth-century manuscript St. Gallen, Stiftsbibliothek, Cod. Sang. 878, p. 171 (www.e-codices.unifr.ch/de/csg/0878). Bloch, "Ein karolingischer Bibliotheks-Katalog," no. 25; Geith and Berschin, "Die Bibliothekskataloge," 77.

27. Ysidorus de accentibus et martirologium

    Isidore of Seville, *Etymologiae* bk. 1, chap. 18. "Martirologium" could refer to Isidore of Seville, *De ortu et obitu partum* (also mentioned in the main catalogue and extant in the Murbach manuscript Colmar, Bibliothèque municipale, MS 43 (*olim* 39), https://bvmm.irht.cnrs.fr/sommaire/sommaire.php?reproductionId=1935); or to a martyrology by Jerome, Bede, Florus of Lyons, or an anonymous author.

28. Epistola Ypocratis ad antiochum

    A letter of Pseudo-Hippocrates to Antioch on illnesses associated with the four seasons.

29. Epistola antimii medici ad titum imperatorem[57]

    Anthimus, *Epistula de observatione ciborum.*

30. Crisostomus de naturis animalium

    *Physiologus.*

31. Fabula auiani et esopi et phedri et allexandri et didimi

    Fables attributed to Avianus, Aesop and Phaedrus, and fictional letters of Alexander the Great to Dindimus, King of the Brahmans (*Epistulae* or *Collatio Alexandri et Dindimi*).

32. Ferrandus diaconus de formula vite

    Ferrandus of Carthage, *Ad Reginum comitem ep. VII.*

33. Gesta allexandri magni volumen vnum

    Julius Valerius Polemius, *Res gestae Alexandri Macedonis*?

34. Plinii Secundi volumina tria

    Pliny the Elder, *Historia naturalis.*

35. Lex ribuariorum et alamannorum

    Ripuarian and Alemannic law codes.

36. Cronica Severi libri ii

    Sulpicius Severus, *Chronica.*

---

**57** Bloch noted that this ought to read "Epistula Anthimi ad Theudericum regem Francorum." Both Titus and Theoderic are mentioned in the ninth-century Reichenau manuscript copy of this text. St. Gallen, Stiftsbibliothek, Cod. Sang. 878 (vademecum of Walahfrid Strabo (d. 849)), 352–65. The apparently unique reference to Titus (a confusion with the letter of Paul to Titus?) raises the possibility that this part of Cod. Sang. 878—codicological unit 5—is related to the copy mentioned in the Murbach catalogue, perhaps through a shared exemplar, or as antigraph and apograph. Since links between Reichenau and Murbach, both of which were founded by Pirmin, are well established, it may even represent the same copy. Two other manuscripts also mention Titus, but they are thirteenth and fourteenth century respectively: Paris, Bibliothèque nationale de France, MS lat. 14935, and Prague, Knihovna Národního muzea, MS XIV.A.12. In addition, several other texts in the catalogue correspond to texts in codicological units 1, 2 and 3 of MS 878: no. 23 on p. 178, no. 24 on p. 91, no. 25 on p. 171, no. 26 on p. 242, no. 28 on p. 327. These facts, together with the unique title of no. 25, strongly suggest a connection. On St. Gallen, Stiftsbibliothek, Cod. Sang. 878, see Bischoff, "Eine Sammelhandschrift."

37. Omelie origenis in leuiticum xvi
    Origen, *In Leviticum homiliae 1–16.*

38. Historie Iordanis libri ii
    Jordanes, *Getica* and *Romana*?

39. De instrumentis bellicis vegecii renati li [*sic*] iiii
    Vegetius, *Epitoma rei militaris.*

40. Liber achulfi de situ sanctorum locorum
    Adomnan, *De locis sanctis.*

41. De fide catholica Iustiniani imperatoris
    Justinian, *Edictum rectae fidei.*

42. Fulgencius Mirthologiarum
    Fulgentius, *Mythologiarum libri tres.*

43. Marcianus Felicis capelle
    Martianus Capella, *De nuptiis Philologiae et Mercurii*?

44. Claudius in Matheum
    Claudius of Turin, *Super Matthaeum.*

An important feature of this list is the kind of physical manuscript that each entry represents. Geith and Berschin believed that a number of these entries could be grouped together to form individual codices. They thought that texts 5–10, for example, were part of the same geographically focused manuscript, comparing it to volumes such as the tenth-century Leiden, Universiteitsbibliotheek, MS Voss. lat. Q.29 and the thirteenth-century Vatican City, Biblioteca Apostolica Vaticana, MS Pal. lat 1357, which contain combinations of these and other texts.[58] This would mean, however, that the Murbach cataloguer listed individual texts without regard for their physical supports, which is not the case. Some entries mention one or more *volumen*, referring to the number of bound volumes filled by the text in question. This is the case with Solinus' *Collectanea* (10). Those texts too short to fill an entire bound volume were listed as part of a miscellaneous bound volume (e.g. 19); or kept as loose or tacketed quires or booklets, a regular practice in early medieval manuscript culture.[59] The glosses on Kings (13), the letters of Pseudo-Hippocrates and Anthimus (28 and 29), and the *Physiologus* (30) may all have been stored in this way.

Early medieval library catalogues did regularly list texts, or simply their authors, without mentioning their physical format (thorough inventories such as Notker the Stammerer's famous catalogue in St. Gallen, Stiftsbibliothek, Cod. Sang. 728 are

---

**58** Geith and Berschin, "Die Bibliothekskataloge," 70.

**59** Gumbert, "Skins, Sheets and Quires." It should be noted that a miscellany, such as the one described in entry 19, could conceivably also have been a lightly bound or unbound set of quires or booklet rather than a hard-bound codex. On booklets, see Robinson, "The 'Booklet'"; Da Rold, "Making the Book"; and Gillespie, "Medieval Books."

exceptions).[60] But this does not preclude that they were kept as loose quires. Neither thematically similar content, nor the existence of thematic manuscript compilations, particularly those from a later period, can serve as proof that any set of texts in Iskar's list was bound together. This was certainly possible, and some of these quires or booklets do now survive in bound manuscript volumes—as Iskar's copy of the *Cosmographia* of Aethicus Ister (3) survives in the composite manuscript Oxford, Bodleian Library, MS Junius 25. But at the time Iskar's list was made, the majority of its entries probably represented unbound or tacketed quires, or booklets, unless otherwise indicated.

Meisterlin's omission of those texts on this list that also appear in the main catalogue, as well as the other difficulties described above, mean that it is impossible to evaluate the true extent and implications of Iskar's private library. But several things are clear. The catalogue contains no significant patristic works, and few liturgical or theological texts, except a number of biblical commentaries (2, 11, 13, 17, 18, 20, 44). Computus, grammar, law and matters of Church structure—subjects of general interest in the early Middle Ages—are represented by a few works. The catalogue's major focus, however, is geography and (to a lesser extent) history, subjects represented by an impressively comprehensive set of those pagan and Christian authors whose works were widely read in the early Middle Ages, among them Adomnan, Aethicus Ister, Dares Phrygius, Jordanes, Lucretius, Martianus Capella, Orosius, Pliny, Solinus, and Sulpicius Severus, as well as fictional or pseudepigraphic authors such as Alexander the Great.

The scholarly work that has been done on history and geography in the early Middle Ages shows that they were tools used for particular purposes: history for defining and forging identity, for achieving the objectives of political power and rulership, and for articulating shared memory; geography for understanding the world, not as a physical reality in the modern sense but primarily as a means of approaching the mysteries of the Creator and human salvation.[61] Consequently, since history and geography had no claim to objectivity, but were tools to an end, they resist definition. The very variety of uses to which historical and geographical material was put indicates, however, how interesting it was to Christian thinkers—perhaps because it was "human" science, which told the grand story of the human race. Ultimately, secular history and geography offered a way towards a better understanding of the histories and geographies of the Bible, and its eschatological, typological, and allegorical interpretation. Influenced by Augustine and Neoplatonism, the reading of scriptural truth in the visible world was a long-standing epistemological approach in Christian thought.[62] Iskar's interest in histories and topographies therefore fits within a long tradition; but the impressive range of the catalogue suggests that in the early Middle Ages, this interest reached new heights among some members of the intellectual elite. It indicates a possible context for the *Physiologus* as one of a range of texts sought out specifically for their commentary on Creation and the human place within it.

**60** Stansbury, "*Sammelhandschriften*."

**61** Lozovsky, "Carolingian Geographical Tradition," and *The Earth is Our Book*.

**62** Lozovsky, *The Earth is Our Book*, 142; Augustine, *De civitate Dei* XI.2.

### Waltherius' List

A text on beasts and birds, possibly a *Physiologus*, is also mentioned in one other personal booklist (*h*), belonging to the St. Emmeram monk Waltherius towards the end of the tenth century. It may have been written by Waltherius himself, since it not only lists his books but also an appended short list of liturgical vestments—alb, cincture, maniple, stole, and superhumerale (the chasuble isn't mentioned)—worn by priests to celebrate the Eucharist.[63] Nothing is known of Waltherius beyond this list. It is a reasonably standard catalogue of works useful for devotional study and worship, including grammatical, liturgical, and musical books, and saints' lives. A small range of poetic works, as well as a treatise on metrics, indicates that Waltherius had sophisticated Latin. The inclusion of the "libellus tagoberti" —possibly the *Gesta Dagoberti*, a life of the Merovingian king Dagobert I, which expounds on the duty and devotion of a lay ruler to the Church—may be a hint that Waltherius represented his monastery to powerful laypeople in some capacity, perhaps as an advisor, or even as a composer of sermons. Several other texts also suggest that Waltherius had a role in explaining the secular world from a Christian perspective. These texts include the *Physiologus*, a florilegium of the Psalms, and the *Commonitorium* or *Consultatio* against the Priscillianists and Origenists by Orosius.

If indeed the *Physiologus* is the text denoted by this catalogue entry, it is presented alongside a range of major works by widely read authors including Prudentius, Virgil, and Priscian. That the *Gesta Dagoberti* was composed in the mid-ninth century indicates that Waltherius had access to St. Emmeram's excellent library, and that he was attuned to recent moral political discourse. In this select group of texts, it is apparent that the *Physiologus* was part of the literary canon essential for a well-rounded education as a priest.

### Monastic Lists

The remaining monastic catalogues (*c*, *f*, *g*, and *i*) all present wholly different contexts for the *Physiologus*. The "liber bestiarum" (*g*) donated to the Benedictine house at Peterborough in 970 shows that the *Physiologus* was considered by Æthelwold to be one of a number of books suitable for literary study. As with other tenth and eleventh-century continental copies of the *Physiologus*, this Insular copy demonstrates a clear move towards the study of the liberal arts, and away from allegorical interpretation of nature (see Chapter 5). It had been donated alongside a glossary of Greek words ("liber de litteris grecorum"), which would have been useful for literary learning. (Æthelwold's didactic poem *Altercatio magistri et discipuli* similarly contains material from the *Physiologus*).

---

**63** It is difficult to determine from early medieval sources how the superhumerale relates to the rationale and the pallium, which are episcopal vestments, and whether their usage differed. The superhumerale mentioned here may refer to the simpler kind worn by priests, as opposed to the more elaborate version worn by bishops: see the early medieval commentary on the Pentateuch in Bischoff and Lapidge, *Biblical Commentaries*, 354.

Unfortunately, nothing more can be said about the list of donated books as a framework for the *Physiologus*, since nothing more is known about the circumstances of their selection. Equally little can be said about the "Liber I phisiologi" from the Würzburg cathedral library catalogue (*c*), written ca. 1000: This is a very long list with eleventh-century additions, compiled in no apparent order, which testifies only to the rich and varied library collection of the cathedral at the turn of the millennium. A further early eleventh-century catalogue (*i*) from a monastic house in eastern France or Belgium, which mentions a "Fisiologus" among the "auctores huius monasterii," is also evidence only of the presence of the *Physiologus* in early medieval libraries.[64] However, this one was much more modest than the one belonging to the cathedral in Würzburg.

## Madalwin's Donation

More interesting is a charter (*f*) from autumn 903 which records a transaction at a council convened by Burchard, the bishop of Passau, in which the *chorepiscopus* (missionary or suffragan bishop) Madalwin handed over his lands, vestments, and books in exchange for two life benefices. Madalwin perhaps sought to escape the Hungarian invasion of Bavaria, since he would have been in the region as part of his missionary activity among the Carinthian Slavs in lower Austria.[65] He may be the same Madalwin who appears as a notary in documents from the chancery of Carloman, King of Bavaria, between 876 and 879.[66] The charter lists fifty-six volumes, including a set of liturgical and theological books: a copy of the Gospels, commentaries, epistles, and saints' lives, a penitential, church council canons, a gradual, and a computus. This small collection is eminently suited for the needs of a missionary bishop. The bulk of the list, however, is headed "De arte grammatica," and these books are both more numerous and more diverse. They include standard authors read in the early Middle Ages as part of the liberal arts curriculum: Donatus, Bede, Sedulius Scottus, Boethius ("bene glosatum," "well-glossed"), Martianus Capella, and Prudentius. These texts represent *grammatica* broadly interpreted as an expression of early medieval literacy and literary culture. The strong emphasis on poetry, evident in Virgil's *Eclogues* and *Georgics* among other poetic works, shows that Madalwin, like Waltherius, had an interest in and high degree of skill in Latin. (While early medieval churchmen were generally expected to possess this skill, it was by no means universal.) The *Physiologus* was presented as part of this standard cultural reading-list. It appears in a set of texts brought together in a single volume: "Enigmata simphosii. et Althelmi et Ioseppi. et libros bestiarum ysiodori. in uno corpore."

64 "The authors of this monastery."

65 Hunyadi, "Signs of Conversion," 106.

66 Another Madalwin is named in a donation to the monastery at Prüm, giving fourteen manses and signing up as a monk in the 840s; chronologically this may be the same Madalwin. It is unclear how he may have come from Prüm to Austria or Pannonia, but that he did so is suggested by another text on the list, Wandalbert of Prüm's (now lost) work on the Mass. See MacLean, *History and Politics*, 37; and Manitius, *Geschichte der lateinischen Literatur*, 1:560.

The work by "Ioseppus" is probably the late antique *Hypomnesticon* of Josephus Christianus, so called in the early Middle Ages to distinguish him from the historian Flavius Josephus. In fact the *Hypomnesticon* is a collection of extracts from Flavius Josephus' works. It takes the form of chapters posed as questions on subjects from the Bible—Adam's tomb and name, the trumpet of Jericho, Joshua's twelve memorial stones, and Moses as the inventor of Hebrew letters—with a series of answers given to each. It is interesting both that it is listed as "enigmata" or riddles, and that the *Physiologus* was copied in a volume with such riddles. Several of the extant *Physiologus* manuscripts also contain similar wisdom dialogues and riddles, which increases the probability that this entry did represent the *Physiologus* and not book 12 of the *Etymologiae*.

Riddles played an important role in transmitting knowledge and fixing it in the memory, as part of a ludic tradition that had the power to engage audiences from school pupils to non-Christian laypeople. By asking about Adam, or about the trumpet of Jericho, such literature both provided set points from which to develop knowledge of the Bible, and created a social culture (perhaps even a popular culture) around this knowledge. Evidently the *Physiologus* was used for this purpose as well. Madalwin's booklist contains other works of this kind, indicating their importance: a text titled *De ratione anime* and attributed to Origen is more probably the anonymous question-and-answer *Disputatio de origine animae* between Augustine and Jerome, compiled from fourteen of their own works.[67]

## Eberhard and Gisela's Will

The final list, *d*, presents a special case of a *Physiologus* in the possession of a layperson. It appears in the will of Eberhard, Count of Friuli, and his wife Gisela, made in 863 or 864, which lists a *Liber bestiarum* among the books from their chapel that were bequeathed to their eldest son Unruoch.[68] Both the count and his wife belonged to the Carolingian higher aristocracy: Gisela was the daughter of Louis the Pious and sister of Charles the Bald, while Eberhard, like his father before him, was a high-ranking envoy (*missus*) of the king.[69] The will was a performative exercise of power, legally defining the couple's enormous estate and indicating its future.[70] Eberhard and Gisela's copy of the *Physiologus* was a chapel-book, since it was described as having been kept in the small book-collection of their personal chapel. It was therefore unlikely to represent their personal interests. Nonetheless, it was bequeathed to their heir and not to one of their eight other children, which included Adalard, Rudolph, and Gisela, all of whom entered the Church. Neither was it left to be donated in 865 with several other books

**67** Hennings, "Disputatio," 264 and 267.

**68** The will was edited by Schramm and Mütherich, *Denkmale der deutschen Könige und Kaiser*, from 93, and Becker, no. 12. See also La Rocca and Provero, "The Dead and Their Gifts," and Kershaw, "Eberhard of Friuli."

**69** This is discussed in detail in La Rocca and Provero, "The Dead and Their Gifts," 234–45.

**70** La Rocca and Provero, "The Dead and Their Gifts," 251.

from their chapel to the monastery they founded at Cysoing (near Lille). Yet the *Physiologus* stands out among other works inherited by Unruoch which were more obviously useful for the legal and military exercise of power, and the Christian devotion which he was expected to perform: Vegetius' *De re militari*; Alemannic, Bavarian, Frankish, Lombard, and Ripuarian law codes; a Psalter; and a Gospel book.

The appearance of the *Physiologus* in the inheritance of a lay magnate indicates that it was not read exclusively within a monastic or ecclesiastical context. In fact, the Carolingian *Physiologus* may have found its place within a culture of aristocratic hunting and the elite masculinity that it defined, as suggested in the recent work of Eric Goldberg.[71] The opening *Physiologus* story tells of the lion, which scents the hunter on the air and erases its tracks with its tail to prevent the hunter tracking it to its lair. As a compendium of often fantastical animals and their no less fantastical natures, the *Physiologus* was amusing, not only by virtue of its marvellous tales, but perhaps also because it was so clearly divorced from the practical realities of the hunt and the familiar animals that were hunted. The unreal beasts of the *Physiologus* may have also contributed to its suitability as a book for elite men, who hunted a different class of animal from that hunted by commoners. Its stories brought an element of imagination to the familiar and aligned the privileged masculine activity of hunting with the Christian faith. This alone made the *Physiologus* an appropriate work for Unruoch to own and read. His new copy may also have been richly decorated: the early medieval *Physiologus* had a strong pictorial tradition.

Early medieval book catalogues are not the satisfyingly comprehensive pieces of evidence we might at times wish them to be, but as a group the handful of lists that mention the *Physiologus* do show that it was a versatile text, highly valued both by monasteries and by aristocrats. It is found in the context of missionary activity, private devotion, public worship, literary and linguistic study, imaginative story-telling, and performative masculinity. The presence of the *Physiologus* in the private libraries of Iskar, Waltherius, Madalwin, and Unruoch confirms that teaching schoolchildren was far from exclusively its purpose in this period. The *Physiologus* was not only read, but read by a broad cross-section of contemporary literate people.

## The Earliest Extant Latin *Physiologus*

The earliest extant codex copy of the *Physiologus* in any language is Bern, Burgerbibliothek, MS 611 + Paris, Bibliothèque nationale de France, MS lat. 10756 (hereafter the Bern/Paris collection). Copied in the eighth century in France, this is a splendidly messy and complex set of booklets. Now split across two modern manuscripts, they were originally made in six different production stints, but are nevertheless very closely related. They contain a wealth of information about learning and administration in an ecclesiastical centre around the early eighth century and reveal a great deal about the collaboration of monastic scribes. This framework makes this

71 Goldberg, *In the Manner of the Franks*.

collection an important witness to the early context and use of the Latin *Physiologus*. It is worth exploring in detail.

The Bern/Paris collection was probably made in or near Bourges, as indicated by the formula of a mandate to register a donation there on fol. 64r.[72] A large and flourishing Roman settlement, Bourges remained an important centre throughout the early Middle Ages. According to Gregory of Tours, its first bishop was the third or fourth-century missionary Ursinus.[73] He may also have been the first archbishop; in any case, Bourges had long been a metropolitan see by the time the Bern/Paris collection was copied in the early eighth century. Surviving documentation from Bourges at that time is scant. We know the names of its archbishops, but little else about them: between ca. 662 and the mid-eighth century, they were Ado, Agosenus, Rochus, and Siginus.[74] These men, like their predecessors, were almost certainly members of the regional aristocracy, both able and required to maintain diplomatic relations with the local count, the dukes of Aquitaine and, most importantly, with the Frankish court(s). Since Bourges was situated at the northern edge of Aquitaine, near Neustria, its suffragan bishops had their dioceses in territories belonging not only to the dukes of Aquitaine, but also to the rulers of Neustria and the rulers of Austrasia (who had land in Aquitaine).[75] This situation required delicate management.

The Bern/Paris collection has been dated to 727. It is unusual to be able to assign so precise a year to so early a manuscript, but in this case a computistical text in the Paris codex provides an exact calendar. Parts of the collection may have been made several decades earlier, however (see Appendix I). The year 727 falls at the end of a period of stability for the city, before a series of attacks on Aquitaine by Charles Martel and his brief capture of Bourges in 731. Umayyad raids into Aquitaine during the 720s had left Bourges relatively unaffected. As demonstrated by the Bern/Paris collection, Bourges at this time was a very busy episcopal centre, which required many texts for both administrative and ecclesiastical purposes, and which was able to produce them at a rapid rate but without much concern for uniformity of style. Nevertheless, the Bern/Paris collection was not simply a bundle of booklets bound together for convenience; rather, its script and material structure indicate a great deal of planning and collaboration among its creators, who evidently aimed to make a compendium bound in codex form. Their choice of texts, and the manner in which these texts were added to the collection, shed light on how the first known Latin *Physiologus* was read. As later chapters will demonstrate, the Bern/Paris collection offers a contrast to the compendia produced in the ninth and tenth centuries, and shows what they might have looked like without the emerging interest in nature and Creation, which is not yet evident here.

---

**72** MGH Form., 166.

**73** Gregory of Tours, *Historia Francorum* bk. 1, chap. 30–31 (MGH SRM I.1: 23–24), and *Liber in gloria confessorum*, chap. 79 (MGH SRM I.2: 91–92).

**74** Péricard, *Ecclesia Bituricensis*.

**75** See, for example, the case of Bishop Desiderius of Cahors ca. 650: Esders, "The Merovingians and Byzantium," 357.

### Physical Structure and Script

The collection's complex material structure can be summarized as six production units made in the same environment and bound together within a century or less.[76] It is written in a variety of hands using pre-Caroline and Merovingian minuscules, as well as uncial.[77]

Each individual production unit was copied by a small number of scribes using a uniform page layout, but many blank pages and spaces were left, particularly at the end of each unit, which were filled with various excerpts by roughly contemporary hands. These additions can be found on the following folios, divided by codicological part:

Parts I/II: 19v–20r

Parts II/III: 40v–42r

Part III: 82r, 86r–92r (Bern manuscript) and 64r, 67v–69v (Paris manuscript)

Part IV: 114r–115r

The summary of the six different parts below gives an indication of their physical structure and palaeography, which are much more complex in some parts than others.

#### 1. Part I: Fols. 1–19 (quires 1–3)

The beginning of the first text is heavily damaged; a quire or more may therefore be lost. Ruled in twenty-six lines. Written in Merovingian minuscule by different hands.

#### 2. Part II: Fols. 20–41 (quires 4–6)

Ruled in sixteen lines. Written in Merovingian minuscule by different hands, with the exception of three lines of uncial on fol. 41v.

#### 3. Part III: Fols. 42–93 (quires 7–13; a single quire, originally found after fol. 72, is now bound in Paris, Bibliothèque nationale de France, MS lat. 10756).

Copied in a single hand writing uncial and Merovingian minuscule, except for additions in different hands on fols. 86r–92r. Ruled, except for the final quire, in eighteen lines. The extant quires of unit III contain medieval quire signatures, labelling each quire I–VIIII.

76 For a detailed examination of the manuscript's structure, see Dorofeeva, "Visualizing Codicologically and Textually Complex Manuscripts."

77 Here and throughout, "pre-Caroline" refers to a range of transitional scripts which are still "Merovingian" but have acquired "Caroline" features. Depending on manuscript and hand, these may include, for example, reduced ligatures, fewer allographs, and new letter shapes more commonly associated with the Caroline minuscule alphabet.

The order of these nine quires can be reconstructed as follows:

| | |
|---|---|
| I (quire 7): | fols. 42–49 |
| II (quire 8): | fols. 50–57 |
| III (quire 9): | fols. 58–66 (66 is an added leaf attached to the final leaf of the previous quire; the signature for this quire is on fol. 65v, indicating that leaf 66 was probably a conjugate of a lost leaf after fol. 72 in the following quire)[78] |
| IV (quire 10): | fols. 67–72 (final leaf now missing) |
| V: | Missing |
| VI: | Paris quire |
| VII (quire 11): | fols. 73–78 (first and final leaves and their text now missing) |
| VIII (quire 12): | fols. 79–86 |
| IX (quire 13): | fols. 87–93 |

David Ganz suggested that part III was limited to quires 7–12 (fols. 42–86), since these are the only quires to have signatures.[79] However, quire 13 does have a signature labelling it "VIIII" on the final page (93v), although it is very faded and, unlike the other signatures, placed on the left-hand side of the page rather than the right. A contemporary list of contents is also found in quire 13 (fols. 92v–93r). It only names texts in part III, which further suggests that this quire belonged to that part. Quire 13 was evidently an addition, however: its signature differs from the others, it was the only one not to be ruled, and only the list of contents at its end is written by the same scribe who copied the first twelve quires of part III.

Within this material context, the list of contents at the end of quire 13 is especially interesting. It is apparently complete, since the following page, 93v, is blank apart from a quire signature, some dry-point jottings that are probably not early medieval, and a wind diagram in its top half. It is possible to identify some of the texts in the list of contents with those extant in part III (see Table 2.2). An entire quire or more, which would have contained items III–X in the table of contents, has been lost after text ten, whose last page is also missing. The five formulae in the Paris quire—text eleven—then follow. The first of these formulae is numbered eleven, indicating that ten of them were lost, in addition to the lost texts III–X. Up until this point, the list of contents has matched the extant texts exactly, but they now diverge.

The scribe of the table of contents did not list the remainder of the texts in the Paris quire. They also left a gap after the formulae, which are followed by two items (XVI–XVII) that are not present in either the Bern manuscript or the Paris quire (Figure 2.2). No space was allowed for numbering these items, so that numbers eventually had to be inserted above the line or in the cramped space in the left margin.

---

**78** My thanks to Peter Kidd for pointing this out.

**79** Ganz, "In the Circle of the Bishop of Bourges," 269.

Table 2.2. List of contents for Bern, Burgerbibliothek, MS 611.

| Original Table of Contents | Text in Manuscript |
|---|---|
| I. Ars donati exposita ab aspero | 9. Asper, *Ars Asporii* |
| II. De notis uulgaribus | 10. Isidore of Seville, *Etymologiae* 1.22 |
| III. Quid est antifrasin, enigma, parabula, paradigma, prosa, bucolicum, epitalamia, trenos, epitafium, fabulas, sillogismi<br>IV. Confectio amforalis<br>V. Sermo de tribus magis<br>VI. De ponderibus et mensuris<br>VII. De drumeta uel citeris quaendam omnibus clarentis<br>VIII. Pauca nomina<br>VIIII. De trebus principalibus linguis quibus spiritus sanctus appellatur<br>X. Indicolos diuersos pauci | Missing quire/s |
| XI. Carta conmutationis<br>XII. Praecaria<br>XIII. Mandatum<br>XIIII. Securitas<br>XV. Ad archepresbyterum instituendum | 11. *Formulae Bituricenses* |
| XVI. Quid sanctus hieronimy de antidotis dixit<br>XVII. Differentias[80] | Missing texts |
| XVIII. De olla de lucerna de sale de mensa de calice de litteris | 19. *Aenigmata* |
| XVIIII. De arca noe | 21. List of measurements for Noah's Ark |
| XX. De stadiis | 22. List of various measurements and how they fit into the *stade* |
| XXI. Epistula gallieni de febribus | 24. Pseudo-Galen, *Epistula de febribus* |

After these two items, the table of contents once more matches the extant texts in quires eleven and twelve.[81] It seems that the scribe of the contents table was not certain which text(s) would be included after the formulae in the Paris quire, therefore leaving a gap in the table, and initially omitting the numbers of items XVI and XVII. In the end, these items were never included.[82]

**80** David Ganz has suggested this may be Isidore's *Differentiae*; Ganz, "In the Circle of the Bishop of Bourges," 273.

**81** See Appendix I. Texts 20 and 23 are not listed but they are later additions, as perhaps are 24 and 25.

**82** Fol. 89r, which contains this gap and the two missing items, appears to have an erasure in the lower third of the page, which was written over by the scribe of the list and is therefore eighth century or earlier. The upper third of the parchment is also scraped in places, suggesting that these are not corrections but the re-use of older sheets. Beneath the "m" of "mensa" in the third line from

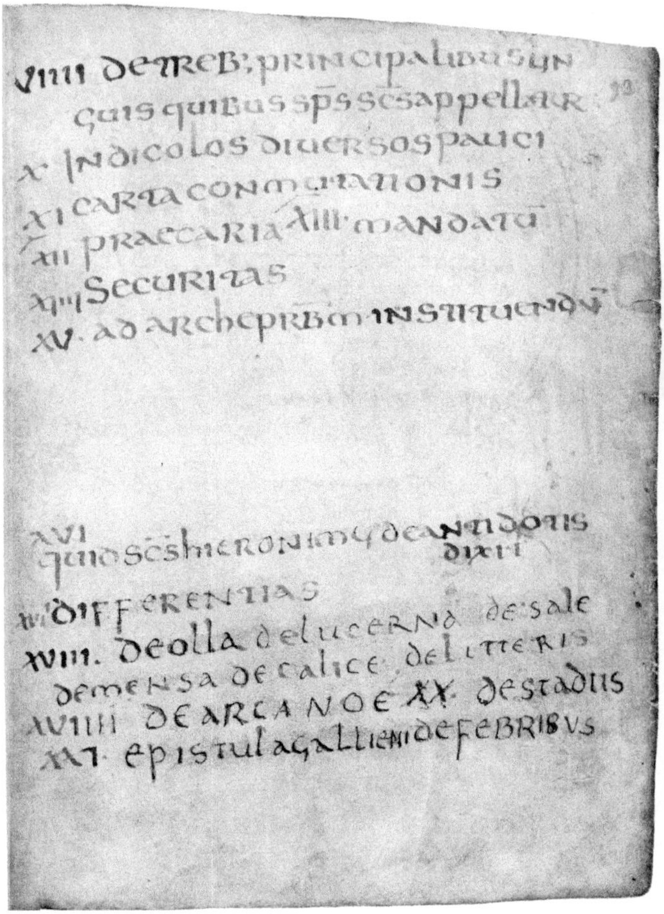

Figure 2.2. Planned programme of copying in Bern, Burgerbibliothek, MS 611, fol. 93r. Used with permission.

These complicated features indicate that the list of contents may not have been a list of contents at all, but rather a guide to the planned program of copying. This is supported by the fact that the list of contents itself is found at the end of the additional quire 13, and that it mentions none of the texts in that quire, which was evidently intended as spare space for the planned texts. But at some point the actual contents of the manuscript deviated from the planned program. The spare space in quire 13 was left blank, and eventually filled in with other texts by other hands.

This—like the many other textual and codicological additions to the Bern/Paris collection—is a symptom of a very busy and chaotic production environment. The collaboration of many scribes meant that they not only wrote different scripts, but also different versions of the same script. Tironian notes, uncial, half-uncial, minuscule,

---

the bottom there appears to be a faded obelus, with perhaps another further down the line. Since obeli were used to indicate obsolete or incorrect text, their presence also supports the possibility that this was re-used parchment. The tracing over the letter-strokes in the bottom half of the folio was almost certainly done by a later fifteenth- or sixteenth-century hand.

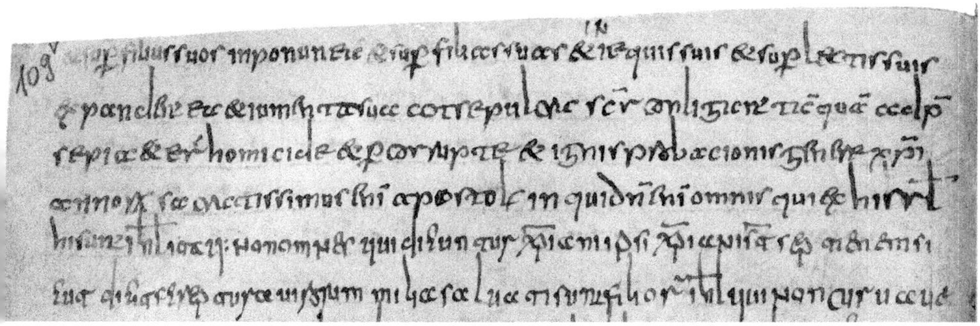

Figure 2.3. Change of hand in Bern, Burgerbibliothek, MS 611, 109v. Used with permission.

and cursive scripts are mixed without especial regard for where they are used and for what, and there is little attempt at consistency in the size and spacing of individual graphs across hands—no single "house style," though many of the hands are closely related. As a result, it would have been impossible to plan precisely how much text would fit into a single quire, a fact that may have contributed to the disparity between the plan and the contents. This may represent a distinctly Merovingian practice for community-produced books: Carolingian centres demonstrably planned each manuscript's contents to match the available space exactly.[83]

### 4. Part IV: Fols. 94–115 (quires 14–17)

Written in Merovingian minuscule and clearly ruled in twenty-three lines. The exception is the Pseudo-Methodius text, which alternates between pre-Caroline and Merovingian minuscules. This change is especially evident on fol. 109v (line five) and fol. 110v (line six); see Figure 2.3. It shows that two scribes trained in these scripts worked together on copying this text.

Quire 17 (fols. 114–15) may represent a new production unit, though this is debatable. It is written on a single bifolium, in a distinct right-leaning and well-spaced (both horizontally and vertically) Merovingian chancery minuscule not present elsewhere in the manuscript, with very long ascenders and descenders. Each page has 16 lines. Lowe noted that this bifolium "seems originally to have been ruled for a charter": that is, the wide-spaced ruling runs vertically down the pages, which were clearly intended to form a larger sheet before being cut up into the present (otherwise unruled) bifolium.[84]

However, the parchment for the Bern/Paris collection was gathered from a variety of sources, as indicated by the palimpsest folios, the irregular quire structures, the varied rulings, and the very scrappy quality of some of the leaves. Quire 17 fits very well within these parameters. Its unique hand is also not unusual, since other produc-

**83** As demonstrated in many studies of manuscripts. Some of the Carolingian practices are outlined in Gumbert, "Skins, Sheets and Quires."

**84** CLA VII.604c.

tion units also show signs of an administrative context. It is therefore presented here as belonging to part IV.

### 5. Part V: Fols. 116–145 (quires 18–22)

Part V contains the *Physiologus* and is entirely palimpsest of Italian origin. It is written in Merovingian minuscule throughout, with the same uncial titles as the list of contents in unit III, and by a very similar hand. Lowe described the minuscule script as "barely distinguishable from charter-hand."[85] The pages are unruled, but the text is consistently presented in 18 lines.

### 6. Part VI: Fols. 146–153 (quire 23)

This last part consists of a single palaeographically distinct quire, ruled in 25 lines. Lowe described its script as "a crude pre-Caroline minuscule of French type, with some features recalling Luxeuil and some Corbie."[86] It is more precisely characterized as a hybrid minuscule with Merovingian and strong pre-Caroline features, being clearly punctuated, upright, rounded and lacking many of the ligatures found in Merovingian minuscule.

## Summary

Despite the variety of hands and the complex codicological history of this collection, there is clear evidence that the different units were produced within the same context. Its original contents and hands are more or less contemporary, and connections exit between the different booklets. By the later eighth century, when the birth lunarium joining parts I and II (19v–20r) was copied, these parts must have been in their present order. Parts II, IV, and V are linked by the same hand, which copied sections on fols. 40v–41v, 98r–99r, and 138v–140v.[87]

As discussed above, part III is a special case. Its structure and length suggest that it may have been planned as an independent codex, or at least as the core of the collection. Its principal hand resembles others in the collection, however, particularly the hand, script hierarchy and ruling of part V. The additions to its final quire, like other additions to empty pages throughout the collection, were probably made once all the production parts had been brought together. Finally, a textual connection may also exist between parts III and V, as they contain works that eventually went into the Corbie recension of the *Collectio Vetus Gallica* (see below). For these reasons, all the parts that comprise the Bern/Paris collection are discussed in this book as the product of a single centre and period.

---

**85** CLA VII.604d.

**86** CLA VII.604e.

**87** Ganz, "In the Circle of the Bishop of Bourges," 269.

## Eighth-Century Knowledge Networks

It has been important to demonstrate that the Bern/Paris collection was executed in a series of related production stints, because this has implications for how the collection can be interpreted. A tendency towards compilation (that is, the copying of texts and excerpts from otherwise unrelated works together) began in late Antiquity and grew into the miscellany, one of the most—perhaps the most—common forms of codex in the early Middle Ages (see Chapter 3).[88] As I have argued elsewhere, such miscellanies were not slapdash gatherings of irrelevant texts, but useful and internally cohesive collections.[89] Their compilation also required a great deal of independent and creative thought: what we mean today when we speak of originality. This is the case even for books whose contents consist entirely of non-original (inherited antique) works. Early medieval compilers used a range of sophisticated authorial and editorial techniques, including but not limited to excerpting, alphabetization, abbreviation, commentary, and juxtaposition; and they applied them to the creation of manuals tailored to the specific requirements of an individual or community. Often, the form and combination of texts in early medieval miscellanies represented highly original solutions to the demand for certain kinds of content, within the (often quite narrow) constraints of available literature. As a result, the study of miscellanies as complete collections can shed light on a range of aspects related to early medieval intellectual life, including textual interpretation, links between ideas, new developments in politics or culture, and special local or regional concerns.

Until recently, however, miscellanies have been dismissed as basic or unimportant. In his influential 1970 study of medieval school curricula, Günter Glauche suggested that manuscripts containing works by respected writers (*Autoren-Sammelcodices*) and transmitting multiple schoolroom texts—including but not limited to commentaries, glosses, and grammatical works—should be understood as schoolbooks.[90] But this definition encompasses most early medieval miscellanies. While manuscripts containing such texts may have been used in schools, there is nothing to suggest that they were not also used elsewhere. As a class of book, miscellanies cannot have been confined only to a formal teaching environment. They were produced in far too significant numbers across early medieval Western Europe, and their contents were too heterogeneous. As discussed in Chapter 1, current research is changing our perception of these interesting books.[91]

It is difficult to judge to what extent the Bern/Paris collection represents early eighth-century compilation: whether it is a standard product of those monastic centres that made books prior to changes introduced by the Carolingians, or whether it

88 On the late Antique context of miscellanies, see Petrucci, "From Unitary Book to Miscellany."

89 Doroteeva, "Reading Early Medieval Miscellanies," "Strategies of Knowledge Organisation," and "Miscellanies."

90 Glauche, *Schullektüre im Mittelalter*, 23–35.

91 van Rhijn, *Leading the Way to Heaven*; Patzold and van Rhijn, eds., *Men in the Middle*; Chazelle and Name Edwards, ed., *The Study of the Bible*.

represented the flourishing of written culture in Bourges thanks to several centuries of prosperity and stability. Much of the context for this period is poorly documented or disputed. A number of the features of the Bern/Paris collection can, however, be teased out based on its contents and appearance.

The first and perhaps the most significant of these features is the extent to which the collection participated in contemporary intellectual culture. As David Ganz observed, some two-thirds of the texts in this manuscript were composed no more than a century before this collection was copied.[92] An example of this is the *Apocalypse* of Pseudo-Methodius (101r–113r). It was written in reaction to the Islamic Caliphate's second fitna or civil war (680–692 CE), meaning that it was composed in northern Mesopotamia, translated from Syriac into Greek, the Greek translated into Latin (possibly at St. Catherine's Monastery in Sinai) and the Latin brought to Bourges (possibly via Marseilles, which might have had a direct link with Sinai), all in the short period between ca. 692 and ca. 727.[93] Three other eighth-century copies of the Apocalypse are known, all of them from Gaul.[94] This indicates a remarkable degree of connection both to contemporary intellectual networks, and to distant monastic centres.

A similar observation can be made about the *Collectio Bernensis*, a unique version of twenty-two Greek and African church canons (fols. 138v–140r). The compiler of the *Collectio Bernesis* used the *Collectio Vetus Galllica*—the earliest Merovingian canon law collection, which became widely influential in the Carolingian period—as a kind of mood-board for the themes and structure of his collection.[95] He used another manuscript containing church council canons as the actual source for the *Collectio Bernensis*. While the connection between the *Collectio Bernensis* and the *Vetus Gallica* is thus mainly one of form, the link between them is clear. In addition, the version of the *Vetus Gallica* consulted for this manuscript did not yet contain the letter by Gregory the Great to Queen Brunhilda of Austrasia, which was first abridged and inserted in the Corbie recension of the *Vetus Gallica*, made after about 721. But the letter is present,

**92** Ganz, "In the Circle of the Bishop of Bourges," 273.

**93** Internal textual evidence indicates that the Latin translation depended on the Greek. See Bonura, "A Forgotten Translation."

**94** Paris, Bibliothèque nationale de France, MS lat. 13348, fols. 93v–110v; Vatican City, Biblioteca Apostolica Vaticana, MS Barb. lat. 671, fols. 171r–174v; and St. Gallen, Stiftsbibliothek, Cod. Sang. 225, pp. 384–439.

**95** "...diente dem Verfasser der Collectio Bernensis jedoch nur als Vorlage für die Gestaltung der äußeren Form seines Werkes sowie als Anregung für seine Themen und ihre Quellen; direkt entnahm er ihr keine Kanones, bediente sich vielmehr zur Textherstellung eines zweiten Codex canonum zweifellos der historischen Ordnung, dessen Wortlaut er ganz offensichtlich mehr vertraute" ("it served the compiler of the *Collectio Bernensis* only as a model for the design of the outer form of his work, as well as a suggestion for his themes and their sources; he took no canons directly from it, relying much more on a second *codex canonum* for the production of the text, a codex doubtless following a chronological order, whose wording he quite obviously trusted more"); Mordek, *Kirchenrecht und Reform*, 108. The *Collectio Vetus Gallica* is also sometimes known as the *Collectio Andegavensis*, after one of its copies in an Angers manuscript (Mordek, *Kirchenrecht und Reform*, 19).

in the same abridged form, in the Bern manuscript (fols. 98r–99r). It was, moreover, copied by the same scribe who copied the *Collectio Bernensis*, in a hand with Corbie or Luxeuil features. This suggests that Bern, Burgerbibliothek, MS 611 was one of the source manuscripts, and the source of Gregory's letter, of the Corbie recension of the *Vetus Gallica*.[96] There was therefore a close relationship between the centre that made the Bern/Paris collection, Corbie and possibly also Luxeuil.

We can guess at other connections as well. The *Ars Asporii* (42v–72v), a grammar composed in the sixth or seventh century, refers to Lyon and Autun, and its hand also recalls those of Luxeuil or Corbie.[97] A southern European connection is indicated by the Italian-origin palimpsest that is the support of part V, and by the *Carmen de ventis* (42r), composed in the early seventh century in Iberia or Italy and included in this collection only a few years later.[98] The *Sententiae* of Taio of Saragossa (d. 683) in this collection is the earliest evidence that Taio's work was read, which suggests a direct link with Spain, perhaps via refugees from the Arab conquest. It is also the probable source of the excerpts from Gregory the Great, which indicates that the compilers engaged closely with the texts and mined them for useful resources.[99] In doing so, the compilers were evidently restricted to the available free space in existing booklets (parts II, III, and IV): a full empty set of pages was not available.

These features offer a frustratingly brief glimpse into what must have been an incredibly active and far-flung network of knowledge exchange both within and beyond Gaul. This network cannot have been an invention of the early eighth century; instead, it must have existed for a long time by ca. 727, to enable the seventh-century texts in this collection to be disseminated. Such a network would have rivalled anything the Carolingians built up in the following two hundred years. Conceivably, the difference in the eighth century was only the lack of a centralized ideological program focused on textual culture. Bourges had the resources and people to create texts for education, liturgy, and administration, but this was not true for many monastic centres.

It is significant that the earliest known *Physiologus* in any language is included in such a collection. It represents an effort, in a major Merovingian ecclesiastical centre, to create a highly modern compilation of works that would not only meet various administrative, liturgical, and learning needs, but perhaps also keep Bourges up to date with what was being read and discussed within the wider Church to which it kept up connections. The *Physiologus* fits well within this context, since it was also an imported text, translated relatively recently into Latin, and only just starting to be widely read in Western Europe. The *Physiologus* must have seemed full of exciting potential, since it did something completely new: combine the natural world with Christian moral interpretations. In fact, there was never again anything quite like it,

**96** As noted by Mordek, *Kirchenrecht und Reform*, 109.

**97** Homburger, *Die illustrierten Handschriften*, 21–23.

**98** Alberto, "The Textual Tradition," 363.

**99** Ganz, "In the Circle of the Bishop of Bourges," 273.

for every subsequent Latin text that did the same was ultimately based on the *Physiologus.*

A second overarching feature of the Bern/Paris collection is its focus on administration, especially evident in its formulae. There remain only five of the original fifteen formulae, with an additional, slightly later formula squeezed in at the end.[100] The addition shows that this was a working collection, which eventually, within a century of its creation, required supplementation. The formulae are a reminder that Merovingian monasteries inherited much of the civic administration that had been undertaken by the Roman state. The palaeography of the collection confirms this: "Both script and shorthand are remarkably close to the chancery, especially in the final section, ff. 116–45, where the *b* with a bow that turns back on itself and a *g* with a tail that seems to be "hinged" at the sharply angled base are used: These letters are both rare in book script."[101]

In the context of the other texts in the Bern/Paris collection, this administrative element indicates that writing, and the reading of that writing in the eighth century, were not necessarily restricted to particular spheres. An example of this can be found in the *In aurium apertione* text (Appendix I, no. 31), the third of seven scrutinies for the induction of catechumens.[102] Copied out in chancery shorthand, it was nevertheless a text used for an internal church ceremony that had no relationship to public worship or to public administration. Education was also one of the concerns in the collection, as suggested by the question-and-answer form of the computus in part IV. David Ganz has pointed out other administrative manuscripts whose "annotations show that they were studied, and that copy was collated with exemplar."[103]

This was truly a collection for an entire community, then, containing a range of works appropriate for the intellectual activity of the entourage of the bishop of Bourges, which included monks but also a range of other clerical and lay people. They were closely involved in the running of the diocese, civic administration, public worship, education (probably of both children and adults), and monastic life.[104] Text composition was also an important concern, not just for administrative purposes but also for sermons, for example. Both the *Apocalypse* of Pseudo-Methodius (101r–113r) and the following excerpt from Jerome's epistle 22 to Eustochium (114r–115r) use extravagant, violent language to describe the dangers posed by pagans—the Arabs in the first text, classical Greek and Latin writers in the second—and the triumph of Christianity. Used as inspiration, these texts would have made for effective sermons. Similarly, the collection of epitaphs on fol. 86v, perhaps copied from real examples, provided an excellent set of Latin models.

---

**100** Brown, "The *gesta municipalia*," 108; Rio, *Legal Practice and the Written Word.*

**101** Ganz, "Bureaucratic Shorthand," 69.

**102** Brown, *The Book of Cerne*, 110.

**103** Brown, *The Book of Cerne*, 70.

**104** On the uses of epitaphs in manuscripts, see Handley, "Epitaphs, Models and Texts."

The collection is strikingly practical, omitting theological, patristic, polemical, and philosophical works. Instead, the concerns of its compilers ranged from the dietary effect of certain foods to ordination. This may have been a direct inheritance of antiquity, which had long had set forms for compilations of important Christian texts—catenae, florilegia, collectanea, and so on. The kind of practical collection represented by the Bern/Paris manuscripts is therefore still unusual at this time, but it is testament to the resourcefulness and pragmatism of early eighth-century Christian communities in the Latin West.

## Conclusion

The Bern/Paris collection is a glimpse into the realities of book production in the eighth century, before the multiple reforms and gradual systematization of monastic life introduced by the Carolingians. It is written on poor-quality, sometimes re-used parchment, by many people, without much care for uniformity in terms of the written space or in terms of script type, size or hierarchy. Its planning was somewhat haphazard: the quires contain both added and blank pages, some ruled and others not; the content design was not fulfilled; many texts were added in empty spaces rather than pre-planned; and the production units are semi-independent of each other. But it also shows evidence of the astonishing international interconnectedness of seventh- and eighth-century monastic knowledge networks, the speed at which they operated despite great distances and violent conflicts, and most of all the great interest of the Bern/Paris compilers in the world beyond their own circle. Perhaps the value of the *Physiologus* as an international work was not the least of the reasons for its inclusion in the earliest early medieval Latin manuscripts.

Codicologically uneven production largely disappeared from manuscript books in later centuries, as the Carolingian concern with the written word led to better funding for scriptoria and introduced more rigorous practices. But some elements remained the same throughout the early Middle Ages: miscellany manuscripts were made in great numbers, often collaboratively; the texts copied were diverse, and generally guided by practical concerns; and compilation remained a symptom of busy communal environments.

As the Bern/Paris collection demonstrates, the principal difference between Merovingian and Carolingian compilation lies in the intellectual practices that underpinned it. In the eighth century, miscellanies were purely practical books. With the advent of the Carolingian reforms, every element of book production became an ideological matter. This applied just as much to excerpting as it did to page layout, script, and decoration. Like all books, miscellany manuscripts acquired a significance that went beyond their contents. As a result, more thought was put into the presentation of the *Physiologus*, and its material on the natural world, within the framework of surrounding texts. The following chapters examine the ways in which this was done, and how it led to the evolution of early medieval ideas about the physical world.

Chapter 3

# MISCELLANIES AND COMMUNITIES

RECENT RESEARCH BY Rutger Kramer and Carine van Rhijn has demonstrated the astonishing extent to which the Carolingians were concerned with the greater good.[1] The ultimate goal was, of course, spiritual salvation, but the road to this salvation was long and difficult. Above all, it was a social and communal enterprise that necessitated the willing participation of those involved in the day-to-day running of the empire. As Kramer has shown, during Louis the Pious' reign in particular there developed a kind of public conscience among the educated elite that was characterized by watchful self-consciousness, surveillance of others, and constant dialogue. Kramer put forward the key idea that constant conversations, rather than top-down decrees, were at the heart of the idea of the Carolingian reforms. This, ultimately, meant that everyone who could write was implicated in the reform effort. Compilers and editors—whether or not they were also the scribes—came to have a great deal of influence. This was because the constant compilation and re-compilation of books and texts, which proceeded at an enormous pace, contributed to the conversation. It *was* the conversation.[2]

As part of this cultural development, Carolingian miscellanies represented the rise of a new type of book. Miscellanies had their antecedents in antiquity but they acquired a new significance in the eighth and ninth centuries—like so many other aspects of written culture in this period. This significance was enshrined in their social function. Carolingian miscellanies were, first and foremost, living and flexible books. This meant that their textual contents were in the majority of cases excerpted or abbreviated in some way, often extremely so, as with a simple list of eight words from Isidore's *Etymologiae* in Wolfenbüttel, Herzog August Bibliothek, MS Gud. lat. 148 (fol. 60r). This served two purposes: economy of space coupled with making as broad a use of the available library resources as possible; and the creation of a volume that was laser-focused in terms of the pastoral and educational needs of the geographically restricted community for which it was being created.

This chapter examines four eighth- and ninth-century miscellanies containing the *Physiologus* as case-studies of this new type of early medieval book. It focuses on the unique features and production circumstances of each miscellany, and examines each in terms of its implications for the dissemination of Christian learning. As books for use by local priests and monks within a restricted geographical range, miscellanies were able to develop into textual tool-kits for helping to consolidate faith among small or geographically restricted communities, and to promote the aims of Carolingian authority. To do so, the contents of these manuscripts became extraordinarily focused on the natural world as represented in the Bible, and on grammar and language.

1 Kramer, *Rethinking Authority*; van Rhijn, *Leading the Way to Heaven*. See also Jebe, "Regeln, Schrift, Correctio."

2 On this process, see also Reimitz, *History, Frankish Identity and the Framing of Western Ethnicity*.

## The Whole Book

The origin of miscellany manuscripts in antiquity was demonstrated by Armando Petrucci, who observed that antique compilations were "organic" and "coherent."[3] The monastic practice of assembling miscellanies seems to have been begun by Cassiodorus at Vivarium. Cassiodorus described collections of primarily patristic and scriptural, but also medical, historical, rhetorical, and philosophical texts in one material "body" or volume ("in uno corpore").[4] According to Petrucci, a less "coherent" kind of miscellany—from which he excluded anthologies of excerpts or authoritative citations, as well as liturgical and composite manuscripts—came to the fore from the sixth century, partly as a result of new book models introduced by Irish *peregrini* (travelling monks).[5] This new kind of miscellany was heterogeneous and therefore "unorganic":

> This phenomenon seems of great importance because the unorganic miscellaneous book ends up being one of the greatest and most significant novelties of book production in the eighth century. The most exasperating forms, which incorporated liturgical texts and patristic excerpts together with medical works and historical or grammatical compilations, covered almost the whole spectrum of contemporary monastic (more than ecclesiastical) culture.[6]

Petrucci made the important observation that early medieval miscellanies were a new and significant genre of manuscript, and that there was no way to categorize them all neatly. However, he saw the development of the miscellany as a natural, spontaneous, and unforced process. This was not the case: medieval compilation was highly controlled, a quality that it may well have inherited from antique practice. The "exasperating" diversity of content in medieval miscellanies was not due to reduced coherence when compared to antique compilations, but partly to new criteria for coherence, as well as a much greater diversity of use.

Perhaps the most significant shift in the production of books between antique and Merovingian monasteries, and those in the Carolingian period, was a conscious awareness of book-making as a political statement. The ecclesiastical and monastic centres that produced books and controlled script were, by the Carolingian period, using them to communicate specific identities, through the cultivation of house and regional styles. In many places, books and script became one of the means through which it was possible to exercise power. Texts and texts extracts, as found within miscellanies,

---

**3** Petrucci, "Dal libro unitario al libro miscellaneo," 185; Radding, trans., *Writers and Readers*, 16.

**4** Cassiodorus, *De institutione divinarum litterarum* (ed. Halporn and Vessey as *Institutions of Divine and Secular Learning*). See also Traube, *Vorlesungen und Abhandlungen*; Baldacchini, "*In uno corpore continentur.*"

**5** Petrucci, "From Unitary Book to Miscellany."

**6** "Il fenomeno appare di grande importanza, in quanto il libro miscellaneo disorganico finirà per constituire una delle maggiori e più significative novità della produzione libraria del secolo VIII e proprio nelle forme più esasperate, che accorpavano testi liturgici a escerti patristici a opere mediche a compilazioni storiche o grammaticali, coprendo quindi quasi l'intero spettro della cultura monastica (più che ecclesiastica) contemporanea." Petrucci, "Dal libro unitario al libro miscellaneo," 185; Radding, trans., *Writers and Readers*, 16.

played a similar role in this period. It is well-established that the whole book—encompassing the contents, the script, and the overall look (from the size of the margins to the cover decoration)—became part of a broader statement about the identity, faith, and authority of the Frankish, Carolingian-controlled Church, within the context of the whole books in which they were copied and as part of the intellectual networks these books characterized.[7]

This socio-political re-orientation of the function of books is reflected in their numbers. There was a huge increase in manuscript production under the Carolingian rulers and their successors, who deliberately sponsored a wide-reaching program of study and literary culture within religious institutions to promote their authority. The often-cited figure to illustrate this comes from modern catalogues. We have ca. 1,800 manuscript codices or fragments for the entirety of continental Western Europe from before 800 CE. Conversely, we have over 9,000 manuscript codices or fragments from the 100 years between 800 to 900 CE.[8] The figures during the early Middle Ages will have been significantly higher: as discussed in Chapter 2, much has been lost during the intervening centuries.[9]

It is not clear how many of these Carolingian manuscripts are miscellanies, since, as discussed in Chapter 1, the criteria for identifying a miscellany are fuzzy and must remain so. Nevertheless, it must have been a fair proportion of the overall number of books produced. Their contents are, as Petrucci observed, very variable. Some miscellanies were intended primarily as teaching tools; some had a more pastoral outlook; others were centred around key individuals, such as bishops, and their varied duties; still others were tools for an entire community, which needed a range from texts from the liturgical to the political. What brought their contents together, and made them a category of book in their own right, was two things arising out of the political re-orientation of manuscript production discussed above: participation in the conversation about salvation and reform, through the inclusion of relevant texts; and treatment of miscellanies as text collections "in uno corpore," through the seamless integration of both full texts and extracts—which would occasionally form entirely new sense-units—using diverse strategies and insofar as diverse planning and production circumstances would permit. This is demonstrated using concrete manuscript examples below. The selection and arrangement of the text contents in miscellanies was, it should be emphasized, not governed by subject matter (as it was in eleventh and

7 The seminal work on this is McKitterick, *The Carolingians and the Written Word*, but there are many others; see the Introduction.

8 This figure is based on CLA and Bischoff's *Katalog*, as well as Bischoff's *Paläographie des römischen Altertums*, 271. It has been cited with minor differences in Davis and McCormick, "The Early Middle Ages," 2; Costambeys, Innes, and MacLean, *The Carolingian World*, 16; Ganz, "Book Production," 786; Brown, "Introduction: The Carolingian Renaissance," 34; and McKitterick, *The Carolingians and the Written Word*, 163-4. The figure for the Carolingian period excludes Visigothic manuscripts.

9 Buringh, *Medieval Manuscript Production*.

twelfth-century encyclopedias), but rather by the politically determined and locally variable ideas of what it was essential for Christians to know.

Since the early medieval approach to the codex was, in general, to treat it as a cohesive product, in which every part contributed to the formation of a single whole, it must be examined as such. That the complexity of medieval books in general requires a specific approach has long been acknowledged by codicologists, who have developed the concept of an archaeology of the book. It views the stages of production and stages of use as layers, sometimes with complex overlaps. This stratigraphy means that we end up not with a static set of pages between definitive hard cover boundaries, but rather with a still-evolving object whose layers accumulate over time, continually changing the whole.[10] The codicological structure of medieval manuscripts has also been framed in terms of single-text and multiple-text mono-block and multi-block codices, which rightly foregrounds the book production process.[11] This approach to manuscripts is crucial for understanding miscellanies, which were sometimes very complex objects that could evolve in both structure and content over a decade or more, yet remain focused and coherent collections.[12] As a result, the manuscript analysis presented in this and subsequent chapters sometimes omits those units of multi-block codices which are not early medieval, although the descriptive catalogue in Appendix I includes the full contents.

## Grammar

A key component of many miscellanies was grammatical texts. Together with logic and rhetoric, grammar was one of the pillars of the trivium, the foundational concept of antique and early medieval education. An elementary constituent of grammatical learning in this period was the idea (derived from ancient grammarians like Donatus and Priscian) that each letter had a shape (*figura*), a name (*nomen*), and sound (*potestas*). Every major work of grammar or metric, including those by Isidore and Bede, began with an explanation of these attributes.[13] In other early medieval grammatical treatises this basic training was then supplemented with more allegorical explanations. Some texts, for example, give numerological interpretations for the number of

---

**10** The term "stratigraphy" was first used for medieval manuscripts by Gumbert, "Codicological Units." For an annotated bibliography of the relevant scholarship, see Andrist, Canart, and Maniaci, *La syntaxe du codex*, 11–41. See also Shailor, "A Cataloger's View." For a more detailed discussion of the codicology of complex miscellanies, see Dorofeeva, "Visualizing Codicologically and Textually Complex Manuscripts." On manuscripts in the modern age, see Treharne, *Perceptions of Medieval Manuscripts*. For critical bibliographical approaches, see Nichols and Wenzel, ed., *The Whole Book*, and the double special issue of *Criticism* on "New Approaches to Critical Bibliography and the Material Text," ed. Ozment and Maruca, especially Sargan, "What Could a Trans Book History Look Like?," 571–86.

**11** Maniaci, "The Medieval Codex as a Complex Container."

**12** Dorofeeva, "Visualizing Codicologically and Textually Complex Manuscripts"; Dorofeeva, "What is a Vademecum?"

**13** Bede, *De arte metrica*; Isidore of Seville, *Etymologiae* 1.4.16.

strokes needed to copy each letter: the three strokes of A denote the three persons of the Trinity, the two strokes of b the two Testaments, the single stroke of C the Church, and so on.[14] Especially during the early Middle Ages, these and other grammatical concepts were raised to a semiotic art, in which allegorical and spiritual interpretation based on grammatical and etymological evidence was widely taught and practiced.[15]

One of the most innovative aspects of this in miscellanies is that extracts on the natural world also began to be integrated with grammatical texts into independent sense-units. These collocations reflect an Isidorean "hyperliterate experience of the world through texts."[16] Early medieval compilers, who had themselves undergone grammatical training, found it both easy and natural to apply the interpretative methods of *grammatica* to find the symbolism of the natural world in the Bible. To understand the meaning of language was also to comprehend the essence of the world as created by God.[17] The close connection between the natural world and grammar is exemplified in two eighth- and ninth-century manuscripts from Montecassino and Lotharingia which depict the disciplines of the liberal arts and their different parts mainly as animals and plants.[18] This textual method, inherited by the Carolingians from ancient grammarians and the profound influence of the thought of Isidore of Seville, enabled scholars to bridge the biblical and real worlds for their audiences. It is demonstrated in the close manuscript readings in this and the following chapter.

Some of the clearest evidence for the use of the *Physiologus* in the early Middle Ages is associated with this kind of grammatical interpretation. It is found in glosses to Priscian's grammar in a Bern manuscript dated to the third quarter of the ninth century and originating near Paris.[19] Beside the chapter heading "On the divisions of gender" ("De divisionibus generum"), in the margin, is the following gloss: "The panther is a many-coloured but beautiful kind of animal. Physiologus [says] of it that it is the only enemy of the dragon. When it has eaten and has become full through various hunts."[20] This remark, paraphrased from the *Physiologus* description of the panther, accompanies the first section of book 5 on nouns, "On genders" ("De generibus"), in which panthers are discussed among other words that pose difficulties when identifying their grammatical gender. Turning the manuscript leaf, we also find the simple gloss "panthera" in the margin. It refers to the following sentence in the main text: "If he were on earth, Democritus would laugh, whether a mixed creature such as a pan-

14 Ganz, "The Preconditions for Caroline Minuscule," 42; Hagen, *Anecdota helvetica*.

15 Irvine, *The Making of Textual Culture*; Amsler, *Etymology and Grammatical Discourse*.

16 Amsler, *Etymology and Grammatical Discourse*, 134.

17 Nutton, "Early-Medieval Medicine," 335.

18 Bamberg, Staatsbibliothek, MS Msc. Patr. 61: https://bavarikon.de/object/bav:SBB-KHB-00000SBB00000157, and London, British Library, MS Harley 2637: www.bl.uk/manuscripts/FullDisplay.aspx?ref=Harley_MS_2637.

19 Bischoff, *Katalog der festländischen Handschriften*, 1:113. Priscian, *Institutiones*, bk. 5.

20 "Panthaera genus animalis varium habens colorem sed speciosus. Phisiologus de hoc quod inimicus est draconi solum. Cum ergo comederit et satiates fuerit diversis venationibus." Bern, Burgerbibliothek, MS 109, fol. 36v.

ther crossed with a camel, or a white elephant, attracted the eyes of the crowd."[21] This quotation from an epistle by Horace is a reference to human folly.

The Bern Priscian is a school text, but the annotations in this manuscript—most of them in Tironian notes—were done by one or more masters, one of whom may have been Heiric of Auxerre.[22] The above glosses show complex cross-referencing of the term "panther" between those places in Priscian's text when it is mentioned, and the *Physiologus*. The emphasis in the glosses goes beyond basic Latin grammar. Instead, the glossator is not only mining Priscian's text for its rich literary allusions, but also supplementing them with their own memory of particular topics—here, the panther. The associations between the panther and the moral and allegorical conclusions that arise from these notes are, nevertheless, rooted in basic grammatical knowledge. A clear link is established here between the natural world as represented by the *Physiologus*, *grammatica*, and compilation.

What is gradually becoming clearer from these sources, then, is that they embodied a specific kind of early medieval intellectual innovation. Building on the peculiarly Carolingian emphasis on linguistic, grammatical, and literary learning, compilers of miscellanies containing this material also began to incorporate the *Physiologus* and other texts referencing the natural world of the Bible. This meant that those with literary and grammatical training also learned to "read" the natural world. This is explored further in the examples below.

## A Communal Miscellany

The earliest full-length *Physiologus* miscellany, in the terms defined above, is St. Gallen, Stiftsbibliothek, Cod. Sang. 230. It was copied in the late eighth century at St. Gallen, in a single hand writing Alemannic minuscule. It is extraordinarily long, running to 561 pages. For these reasons it presents some of the typical problems of early medieval miscellanies: how to characterize its contents in a meaningful way, and how to understand its historical function based on the large and seemingly chaotic arrangement of texts. Nevertheless, it is evident that the miscellany served two purposes: firstly, and most importantly, to provide a toolkit for fourfold exegesis (see Chapter 4), particularly its second, allegorical, level; and secondly, to provide a practical source of essential short liturgical and pastoral texts—homilies, prayers, lists, and sentences.

There are several points of entry into the dense thicket of texts within this manuscript. On two occasions its contents are arranged by type: items 15 through 19 are sermons; items 20 through 23 are commentaries on the Lord's Prayer. (Here and throughout, item numbers refer to the descriptive catalogue in Appendix I.) The authorship of the commentaries is not acknowledged and is not identifiable except in the case

---

**21** "Si foret in terris rideret democritus seu diversum confuse genus panthera, camelo sive elephas albus vulgi converterit hora." Bern, Burgerbibliothek, MS 109, fol. 37v. Horace, *Ep.* 2.1.194–198. Trans. in Keane, *Figuring Genre*, 35. I am grateful to Meg Leja for her transcription of these notes, and for drawing my attention to them.

**22** Contreni, "What was Emperor Augustus Doing."

of item 21, which is by Cassian. These commentaries have evidently been grouped together because of their subject matter. The sermons, on the other hand, may have been grouped together in part because of Augustine's authorship, and in part because together they form a selection that concerns three successive Old Testament prophets: Abraham, Isaac, and Joseph. Item 19, extracts from Gregory the Great's Gospel homilies, may also belong to this group, to form a collection of sermon and homily material. Of the three possible principles of organization—authorship, biblical chronology, and text type—the latter seems to have been the most important, not only because of the addition of item 19 but also because item 14 apparently belongs to the group as well. This text is part of Jerome's commentary on Matthew, another chapter of which appears in item 40. The *incipits* of both items 14 and 40 show that the makers of the manuscript were aware of the text's author and title. It seems strange that the two items are not copied together in such a highly organized book, unless we again look at the *incipits*. Item 14 is listed as a sermon: "Sermon from the treatise of St Jerome the priest from the Gospel of Matthew."[23] Item 40, on the other hand, is given as a reading: "Reading from the treatise of St Jerome the priest on the Gospel of St Matthew."[24] The distinction is one of usage: the first is specifically intended as a public reading, while the second is simply a variant interpretation of that Gospel passage. *Sermo* and *lectio* are not being used synonymously, for the Carolingians normally used *sermo*, *omelia* (*humilia* in the case of St. Gallen, Stiftsbibliothek, Cod. Sang. 230) and *tractatus* to designate sermons, with the first term being predominant.[25] *Lectio* was, by contrast, a word specifically associated with the reading of a particular passage and its exposition.

It is, moreover, unlikely that item 14 was the result of a mistake based on a casual reading of the first line of the text: "The Gospel reading which has been recited by us today, dearest brothers."[26] This direct address is a suitable beginning for a sermon and could explain the presence of item 14 next to the group of sermons. However, each sermon—unlike the texts before and after this group—carries a marginal description of its subject. Items 15 and 16, for example, are both marked "de abraham" (pp. 361 and 364). Item 14 is annotated "adpropinquante iesu iherosolymis" ("On Christ's entry into Jerusalem") in the same hand (p. 348). The ink of these annotations is darker, and the hand has a very different ductus to that of the principal scribe; however, the script is contemporary Alemannic minuscule, which indicates that the annotations were made not long after the text had been copied. Evidently, the users of the manuscript thought of item 14 as part of the annotated group of sermons represented by items 15 through 19.

The question that arises is why, if the sermons and commentaries in these two sections were brought together by type, this was not also done for other texts. The deliberate composition of the two text groups suggests that the manuscript's contents

23 "Sermo de tractatu sancti hieronimi presbiteri ex euangelio mathei."

24 "Lectio de tractatu sancti hieronimi presbyteri de euangelii sancti Matthaei."

25 Mantello and Rigg, eds, *Medieval Latin*, 659.

26 "Euangelica lectio quae nobis hodie recitata est, fratres dilectissimi."

were selected and arranged in advance of the copying process. Why were the biblical commentaries in the manuscript, for example, not grouped in the same way? In St. Gallen, Stiftsbibliothek, Cod. Sang. 230, they—items 13, 14, 31, 40, and 45—are spread throughout the manuscript. However, item 14 aside, a look at the titles of the four additional commentaries suggests that they were simply not all thought to be the same kind of text: "Here begins the selection from the three Gospels" (13), "interpretation of the Song of Songs" (31), "a reading from the work on Jerome" (40), "likewise sentences of Isidore on Leviticus" (45).[27] The compiler has recognized the selective, expository nature of these texts, but does not see them as being part of the same genre. They are exegetical, but so are most of the texts in the manuscript. A further common feature, such as treatment of the same subject or passage, or the same specific form, seems to have been required.

It is possible, of course, that the two groups of texts in St. Gallen, Stiftsbibliothek, Cod. Sang. 230 have not been deliberately formed, but rather as a result of copying circumstances. The compiler may have made a selection from one or more exemplars that contained groups of texts pre-arranged in a similar manner, such as a sermon collection. This is all the more likely since such groupings are not usual among the manuscripts on the handlist. However, smaller groupings in St. Gallen, Stiftsbibliothek, Cod. Sang. 230 exist. Items 9 and 10 are questions and responses; 42, 43, and 44 concern psalms. Without knowing the exemplar or exemplars for this manuscript, it is impossible to judge the degree of organization; but the arrangement of texts into several groups, while not systematic, seems no accident of copying practices either.

Whatever the internal arrangement of the texts in St. Gallen, Stiftsbibliothek, Cod. Sang. 230, the whole manuscript is billed as a patristic collection in two particular lists. The first is its own list of contents on p. 1, which reads as follows:

> In the name of the Holy Trinity [here] begins the collection of divine books on diverse questions which were bequeathed to us by the Holy Fathers, most learned and most erudite men; first [those of] Saint Isidore, the bishop of Spain, after which follow the other, younger works of Isidore; after which the books of Eucherius, after which the books of Augustine the bishop, after which those of Jerome the priest and those of other Fathers, that is of Gregory.[28]

The last word, "gregorij," is an addition in the margin underneath the main text, whose end is signalled by a short multi-coloured ribbon at the end of the final line. Unlike the list of contents itself, which is written in large multi-coloured (green, purple, red, and yellow) capital letters, the word "gregorij" is written in much smaller rubricated capitals. It is centered in the middle of the line, suggesting that the artist wanted to preserve the proportions of the lines and the harmonious impression of the overall page,

---

**27** "Incipit træ euangelii excarpsum," "interpretatio de canticis canticorum," "lectio de tractatu sancti hieronimi," "item sententias hisidori super leuiticum."

**28** "In nomine sancte trinitatis incipit concollectio diuinorum librorum quae sancti patres doctissimi atque eruditissimi uiri tradiderunt de diuersis questionibus; in primis sancti hysidori hyspanensis episcopi post hunc secuntur libri alii hysidori iunioris; de inde eucherii de inde agustini episcopi; post hos hieronimi presbyteri; et ceterorum patrum; id est gregorij."

despite running out of space. This, in turn, implies that the word was a planned addition to the manuscript, just as the single Gregorian text itself (item 19). This internal list of contents is further evidence that the manuscript's texts were carefully selected.

The second list begins on p. 9 of St. Gallen, Stiftsbibliothek, Cod. Sang. 728, and is the earliest known (ninth-century) St. Gallen catalogue of books.[29] It is one of two lists preserved in St. Gallen, Stiftsbibliothek, Cod. Sang. 728. This catalogue lists St. Gallen, Stiftsbibliothek, Cod. Sang. 230 on p. 10 under the rubric "books of Isidore the bishop" ("de libris ysydori episcopi") and its contents are given in the last three lines as follows: "By the same, one book on God, and one book of *De officiis*; on the *Differentiae* of Eucherius, and on minor questions of Augustine, and on diverse excerpts, and on many other excerpts from small works of the Holy Fathers in one volume."[30] This list of contents attributes Isidore's *Differentiae* to Eucherius, who never wrote a work of that name, but otherwise it is a correct description of the texts in St. Gallen, Stiftsbibliothek, Cod. Sang. 230.

The emphasis in both the list of contents and the booklist in St. Gallen, Stiftsbibliothek, Cod. Sang. 728 is on the patristic authority of the principal authors represented in this miscellany: that is, Augustine, Eucherius, Isidore, and Jerome. The miscellany's existence is validated by the fact that its excerpts derive from the Holy Fathers. On the one hand, therefore, the contents of the manuscript have been selected and arranged, sometimes in groups, following a set of particular principles. On the other hand, they have also been selected on the basis of authoritative authorship. The *incipits* to many of the anonymous or pseudepigraphic texts—9 and 13, for example—attribute them to a patristic author. These factors together guaranteed both the utility and reliability of the miscellany.

However, it did not seem to have been well regarded by at least one person. The hand of Notker the Stammerer, who annotated the booklist in St. Gallen, Stiftsbibliothek, Cod. Sang. 728, marked this volume as extremely old, "uetustissimo," in the right-hand margin.[31] For comparison, the cataloguer of Glastonbury Abbey from 1247/1248 also marked books as "old but can still be read," "good but old," "old but useless."[32] Thus, for Notker, "antiquity was not necessarily a guarantee of quality."[33] But quality of what? Notker could not be questioning the orthodoxy of the texts in St. Gallen, Stiftsbibliothek, Cod. Sang. 230: the theology in texts attributed to Augustine, Eucherius, Isidore, and Jerome, as the contents of the book largely are, could not be "out of date."

---

**29** Stansbury, "*Sammelhandschriften*."

**30** "Eiusdem de deo liber i. et de officiis liber i. differentiarum eucherii. et de questiunculis sancti augustini. et de floratibus diuersis. et alia multa de sanctorum patrum opusculis excerpta in uolumine i." "De deo" refers to the chapter on God in the *Etymologiae*, which is part of item 5 in appendix I, and "de officiis" to item 49.

**31** On Notker's comments in this catalogue, see Nasmith, *Catalogus librorum manuscriptorum*; on his authorship see especially Rankin, "Ego itaque Notker scripsi."

**32** Lapidge, *The Anglo-Saxon Library*, 73. Cambridge, Trinity College, Wren Library, MS R.5.33, fols. 102r–103v: https://mss-cat.trin.cam.ac.uk/Manuscript/R.5.33.

**33** McKitterick, *The Carolingians and the Written Word*, 154.

Other notes in the book catalogue indicate that Notker marked those codices whose contents he believed to be false or falsified ("falsatis"), untruthful ("mendacium") or corrupt ("corruptus"). This was not the case with St. Gallen, Stiftsbibliothek, Cod. Sang. 230. The note does not refer to the physical condition of the manuscript, which is exceptionally well preserved even today. It seems unlikely that Notker was commenting on the Alemannic script (or on the script alone); although it was out of date by the end of the ninth century, it was perfectly legible, and well-copied. Probably Notker, whose intent to evaluate the utility of the books in the catalogue is clear, was referring to the relevance of the manuscript to the monastery. On reviewing the booklist and the library itself, he found that the selection of excerpts in the manuscript was entirely redundant.

We can only speculate on why this might have been the case. Too little is known about the status of miscellanies and the principles of their production generally, let alone at St. Gallen, to make an informed guess. Perhaps Notker's idea of what a collection of patristic excerpts should contain to fulfill the needs of his monastery simply differed from the idea of this miscellany's makers, especially if almost a hundred years had elapsed since the production of St. Gallen, Stiftsbibliothek, Cod. Sang. 230 at the end of the eighth century and Notker's correction of the booklist towards the end of the ninth century (he was born ca. 840). If so, this is an indication that requirements of and tastes in miscellanies changed relatively often. The partial dependence of St. Gallen, Stiftsbibliothek, Cod. Sang. 125 on St. Gallen, Stiftsbibliothek, Cod. Sang. 230 indicates that at the end of the eighth century at least, some of the contents of the latter were thought worthy of copying.

St. Gallen, Stiftsbibliothek, Cod. Sang. 230 has been characterized as confusing.[34] This is the case only from a modern point of view. Those texts within it that could be grouped together, from an eighth-century perspective, were grouped together. The others were copied in no particular order that we can now see. However, in the case of St. Gallen, Stiftsbibliothek, Cod. Sang. 230, this did not matter at all. This was a book whose entire contents were expected to be used; it was not a dip-in encyclopedia. A detailed index was no use without page numbers (and there were none in the eighth century). There was no need to arrange the texts in a logical order, especially when no clear order was evident. Today we would arrange the texts in the manuscript in alphabetical order by title or author; but many of them have no title or author. The only obvious groups are the sermons and the expositions of the Lord's Prayer, which do appear together. As regards finding one's way through the book, the texts are made easily legible and memorable using a clear system of colours and rubrication of new sections, questions, and responses. An example of the thought that went into this is in the list of *Physiologus* chapters on p. 510, which inserts the rubric "On birds" ("De auibus") above the next five chapters in the list, which are all about birds (the charadrius or caladrius, the pelican, the night raven, the eagle, and the dove): something that does not appear

34 "Unübersichtlich." Dorfbauer, ed., *Pseudo-Augustinus*, 232.

in the lists of *Physiologus* chapters in the other manuscripts.[35] Since it is numbered like the other chapters, this rubric has occasionally been mistaken as the name of a chapter which does not exist in the main text, making the *Physiologus* incomplete. It is, of course, simply a heading inserted by the scribe of this manuscript for greater clarity..

This clarity and coherence continue throughout the manuscript. The Isidorean extracts in item five are a good example of the range and logical arrangement of topics. The first part explains philosophical and theological subjects: God's immutability (p. 2), immenseness and omnipotence (p. 3), and invisibility (p. 4), the saints (p. 42), the signs of the coming of the Antichrist (p. 43), the one Trinity (p. 49), and so on. The second part covers astronomy and God, the saints, and Apostles. This first selection of extracts from Isidore is a very thorough compilation of information on all aspects of the order of being of God, humanity, and celestial bodies, as well as an explanation or summary of biblical, theological, and liturgical matters.

Many of the texts that follow are similar in content and arrangement. Two in particular are worth mentioning. The first is item 24b, a list of clerical grades also known as the Ordinal of Christ. It is a short list of ecclesiastical grades (among which are reader, exorcist, sub-deacon and priest) and how these relate to Christ's life. The Ordinals of Christ helped to explain the function and divine foundation of the ecclesiastical grades.[36] The same ordinal, in the same Hibernian version, also appears in Bern, Burgerbibliothek, MS lat. 225 + Bern, Burgerbibliothek, MS lat. 233 + Orléans, Médiathèque, MS 313 (the Bern/Orléans collection; see item 15 in this collection in Appendix I). This text offers a reference point for those learning to work within the hierarchy and theological framework prescribed by the Church.

The final text of interest for the definition of St. Gallen, Stiftsbibliothek, Cod. Sang. 230 as a collection is the copy of the apocryphal prefatory letters of Jerome and Damasus from the *Liber pontificalis*: the book of papal biographies. These also appear in the Bern/Orléans collection. Rosamond McKitterick suggested that one of the purposes served by the letters in the *Liber pontificalis* is to establish the importance of Rome as the centre of Christian power, papal authority "and the temporal role of the pope."[37] This may be the role played by the letters in the St. Gallen manuscript as well. Unlike in the Bern/Orléans collection, they are not simply part of the *Liber pontificalis* text, but have either been deliberately copied separately from the *Liber*, or copied from some other source in which they also appear on their own. In any case, the letters were deemed important enough to include them in St. Gallen, Stiftsbibliothek, Cod. Sang. 230. The presence of these two works—the Ordinal of Christ and the letters of Damasus and Jerome—in both this manuscript as well as the Bern/Orléans collection indicates similar contexts for both. They contain dogmatic, exegetical, moral, and other essential works useful for the education of priests. All these texts could also be used for teaching orthodox belief and to reinforce the authority of the Church.

**35** "De caradrio, de pellicano, de nictycorax, de aquila, de quodam arbore uel columba."

**36** Reynolds, *The Ordinals of Christ*, 1–2.

**37** McKitterick, "Roman Texts and Roman History," 27. See also McKitterick, *Rome and the Invention of the Papacy*.

The St. Gallen manuscript is, then, a true community miscellany. It represents one of the most impressively comprehensive early medieval compilations still extant, created to fulfil many of the fundamental textual needs at one of the largest known early medieval monastic houses. As evident from the table of contents, the arrangement of various texts groups as discussed above, and the shared features of layout, it was treated as a coherent volume. Interestingly, however, St. Gallen, Stiftsbibliothek, Cod. Sang. 230 does not yet display a clear integration of the natural world in its subject matter, although many of the texts already have a strong focus on teaching allegorical interpretation. This is true not only for the *Physiologus* and for biblical extracts, but also the ecclesiastical grades, law, the Divine Office, and the singing of Psalms. The early date of production of this manuscript may well account for a relative lack of content related to the natural world, when compared to later miscellanies.

## A Lexical Miscellany

In contrast to the St. Gallen miscellany, the ninth-century compilation Munich, Bayerische Staatsbibliothek, Clm 14388 is much scrappier and is indeed more thematically focused. While the St. Gallen manuscript offered a broad range of texts and extracts in answer to the Carolingian reforms—suitable for the use of a large community such as St. Gallen—the Munich manuscript is a narrower lexical reference work, which nevertheless demonstrates significant scope for a broad range of organization strategies. These make it a highly innovative and tightly organized volume.

The current manuscript is a composite in three parts, bound together in the second half of the fifteenth century. Part II (fols. 8–112), containing Jerome's commentary on the letters of Paul to the Ephesians and Titus, was copied from Munich, Bayerische Staatsbibliothek, Clm 13038 (fols. 262v–387v), which was made at St. Emmeram, Regensburg. The copy in Clm 14388was thought by Bischoff to have been copied not at St. Emmeram but in an unknown scriptorium in south-west Bavaria in the middle of the ninth century. The first quire of the manuscript (fols. 1–7), which constitutes part I, dates from the twelfth century and was added as a supplement to the otherwise incomplete part II. These two first parts of the manuscript will be omitted from the below discussion as they are not related codicologically to the third part.

Part III (fols. 113–238) dates from the middle of the ninth century and was tentatively located by Bischoff to north-west Germany. It was written throughout by one hand using a rapid, frequently rather cramped Caroline minuscule with regular abbreviations and ligatures. The manuscript is mostly very plain, except decorated pen-work initials, executed in the same brown ink as the main text and adorned with geometric patterns, natural motifs, and dots. The parchment has obvious hair and flesh sides, with frequent split holes. Occasionally the scribe has left spaces in the text to avoid sections of parchment that were particularly hairy and therefore difficult to write on (for example, on fol. 151v). These details together indicate that this was a utilitarian miscellany, created for the private use of the scribe or a small community around them.

The book contains five glossaries and one list of synonyms, together covering just under half (54 of 125) the folios in the original codicological unit. The first glossary, *Accipe*, has not been identified in any other manuscripts so far. According to its early twentieth-century editor Karl Thielo, however, it is probably an earlier and shorter version of the more widely used *Auctores antiquissimi (AA)* glossary from a cousin manuscript branch.[38] It received thirty-seven new glosses in Munich, Bayerische Staatsbibliothek, Clm 14388, which demonstrate specific early medieval strategies for organizing information. These glosses were written using the Greek alphabet. They are spread through *Accipe* and continue into the Hebrew/Latin onomastic word-list that follows. Notably, in *Accipe*, they appear to be additions to the beginning and end of each set of glosses on a particular letter. This placement coincides with poor alphabetization, in contrast to the middle of each chapter, where the glosses are highly alphabetized.[39] This suggests that a pre-existing, alphabetically arranged glossary was expanded from the Greek source.

This is borne out by the continuation of the Greek lemmata into the following glossary, an onomastic word-list based on the second book of the *Instructiones* by Eucherius of Lyon. The Greek words have been added in exactly the same way as in *Accipe*—that is, at the beginning or end of each group of words beginning with the same letter. For example, the section for the letter *r* ends with "rabbi magister," a gloss also found in Eucherius' *Instructiones* II.21. A further three glosses beginning with the letter *r* are then inserted, one of them Greek, from an unknown source.

This pattern is not quite so neat in other parts of the word-list. From the beginning of the first column on fol. 225v, for example, those glosses for the letter *s* that derive from the *Instructiones* are no longer arranged in the same order as the original text and are mixed in with glosses from another source. The *Instructiones* contain almost no nouns beginning with the letter *t*, moreover, so that section of the word-list is composed entirely of glosses from another source.

These features suggest that *Accipe* and the following onomastic list were paired together during an early phase in their transmission. Their shared pattern of expansion indicates that the additions to these two otherwise unrelated lists were made by the same person, who may have been working from an exemplar with notes inserted in available blank spaces, which were incorporated into the new copy. Judging from the copying errors evident throughout, these additions were made at a stage before the production of Munich, Bayerische Staatsbibliothek, Clm 14388, in one of its direct or indirect exemplars.

As well as weaving unrelated glossaries together through the addition of Greek words from a third source, this miscellany brought together unrelated texts to form a vocabulary on metrology (weights and measures). The first of these texts, a list of fifteen terms, is based on Eucherius' *Instructiones* and is arranged alphabetically on fol. 227v. This short list, from "talontum" to "siclum," gives the principal function and

38 Thielo, *De glossario*.

39 Dorofeeva, "Strategies of Knowledge Organisation."

features of weights and measures. The other texts derive from the *Etymologiae*. One of them is *De ponderibus*, a short chapter presented here under the heading "[here] begin the Hebrew names of weights."[40] It is followed on fol. 228r by five lines headed "On measures" ("De mensuris").[41] These two passages provide numerical lists of weights and measures (*dracma*, *untia*, *libra*, *gomor*, and others), and mathematical information about how these divide into each other. Both *De ponderibus* and *De mensuris* are rearranged and heavily paraphrased.

Lists of weights and measures are fairly common in miscellanies and in glossaries, for they frequently occur in the Bible. The names of weights and measures for liquids, volume, solids, length, distance and value, and their systems of division, were different in Hebrew, Greek, and Latin, and different again in the vernacular. The precise source and authorship of the Isidorean lists in Munich, Bayerische Staatsbibliothek, Clm 14388 is unknown—they may well have been transmitted from Isidore of Seville's *Etymologiae*, reworked by one or more of their copiers over a series of books, before they were entered into this Munich manuscript—but they were clearly useful. They were needed in order to understand the Bible fully, and, most importantly, contained information that was not otherwise provided in the preceding Eucherian glossary. The metrological texts begin with a short description and definition of weights and measures, derived from one source, and continue with their mathematical values, derived from another source—paraphrased and rearranged. The neat coherence of these themed extracts is such that it does not matter whether or not they were brought together for the first time in this manuscript: even if they are not original to Munich, Bayerische Staatsbibliothek, Clm 14388, the rationality of their pairing made them a valuable source of information.

The Munich miscellany is rounded off by two additional works. The first of these is a short paragraph about balms and spices on fol. 183v, entitled "De pigmentis nardi spicatæ" or "On the balm of the spikenard." Nard is mentioned several times in the Bible as an expensive perfume, notably in the story of Mary Magdalene drying the feet of Christ with her hair (John 12:3 and Mark 14:3) and in the Song of Songs (1:11 and 4:13–14). This paragraph is taken entirely from the Second Commentary on the Gospels from the school of Theodore and Hadrian at Canterbury, composed between the mid-seventh and mid-eighth century and preserved in its fullest form in an eleventh-century manuscript.[42] No Anglo-Saxon manuscripts of the commentaries survive, which suggests that they were transmitted to the continent by missionaries during the eighth century.

It is not surprising that, in their edition of the commentary, Michael Lapidge and Bernhard Bischoff do not list Munich, Bayerische Staatsbibliothek, Clm 14388 among its extant witnesses, since the short paragraph it contains is a patchwork of extracts from the full text that have remained unidentified until now. The paragraph is highly

**40** "Incipiunt nomina hebreorum de ponderibus." *Etymologiae* 16.25.

**41** *Etymologiae* 16.26.

**42** Bischoff and Lapidge, *Biblical Commentaries*, 1.

discriminating, choosing only those five parts of the commentary (which runs to 121 articles) on precious balms and spices, and compiling them into a single coherent entry. This demonstrates the same selectiveness that is evident in the glossaries in this manuscript. Moreover, the paragraph itself may have been intended to supplement the *Physiologus*. Like the *Physiologus* chapters, it deals with a feature of the natural world and provides an outline of its appearance and principal properties. It is copied in an otherwise empty space immediately after the *explicit* to the *Physiologus*. The paragraph also expands on the two entries for nard in the second Eucherian word-list on fol. 227r, where nard is listed as a spiky or lobed plant to be prepared by infusion.[43] As with the lists of weights and measures, the two texts supplement each other instead of providing variant readings.

The final worksin Munich, Bayerische Staatsbibliothek, Clm 14388 are Pseudo-Cicero's *Synonyma*, a straightforward list of synonyms, and Pseudo-Jerome's *Expositio quattuor evangeliorum*. The *Expositio* is thought to have been composed in the late seventh century in Ireland or an Irish foundation on the continent.[44] Although its origin is much disputed, this particular copy, together with the abbreviated commentary on nard, suggests that Munich, Bayerische Staatsbibliothek, Clm 14388 was made in a monastery with Insular connections.[45] The *Expositio* is an allegorical commentary on the four Gospels, but it reads like a glossary: "[He] blessed it, that is, the Gospel; [he] cracked it, that is, to the letter."[46] We are presented with the very vivid image of Christ cracking the holy Gospel like a loaf of bread: plumbing its mysteries to the smallest letters through his teachings. This evokes the breaking of bread at the Last Supper and the sacrament of Communion. All of the well-chosen word-lists in this manuscript are designed to help their readers to understand the Bible in the same way. They cover a broad range of subjects, are presented in a variety of arrangements—from collected glosses to the *Expositio*, which follows the Gospels in order—and range from the straightforward to the advanced (Hebrew, Greek, and allegorical).

The *Physiologus* has a rightful place among these word-lists. Within their framework, it too becomes a literal and allegorical exposition of the Bible and its contents. Thus, although not all the texts in this manuscript are glossaries, once all the contents of all the texts are considered in relation to one another, it is clear that this is a highly coherent linguistic miscellany. The natural world presented in this manuscript is very much a biblical one, containing only those items—whether ordinary or miraculous—that have some theological significance. But the inclusion of the *Physiologus* in such a

**43** "Nardum pisticum nardus fidelem idem sine inpositu [*sic*] grecum est"; and "Nardus spicatum ab e [*sic*] quod species ipsi nardi in modo spica sit quam infusa conficitur." CCSL 66.197, lines 190–93.

**44** Bischoff, "Wendepunkte," 205–73, esp. from 240; Kelly, "A Catalogue," 397–98 (no. 56A); Kavanagh, "The Ps.-Jerome's Expositio IV euangeliorum."

**45** For a summary of the historiographical disagreements about the Hiberno-Latin origin of exegetical texts, see Wright, "Bischoff's Theory of Irish Exegesis."

**46** "Benedixit, id est, Evangelium; fregit, id est, ad litteram." Pseudo-Jerome, *Expositio quattuor evangeliorum*, PL 30.572.

lexical compendium ranks its naturalistic contents as equal in importance to standard early medieval information about names, weights, measures, and Greek and Hebrew vocabulary (such as *rabbi*). What this miscellany tells us, crucially, is that the allegorical thesaurus of the natural world as represented by the *Physiologus* was, by the ninth century in Germany, part of the fundaments of biblical learning.

## Creating New Textual Contexts

Munich, Bayerische Staatsbibliothek, MS Clm 19417 dates from the first third of the ninth century and has been preserved complete and in its original quire order. Its contents are fairly limited in scope, focusing on the Creed (with two copies and two separate commentaries) and on church councils (specifically the canons of the councils of Nicaea and glosses on a large variety of other councils). The *Physiologus*, two sermons, and a question-and-answer text (the *Joca monachorum*) complete the list of texts copied during the original production stint. McKitterick suggested that this manuscript, along with a group of others, was a deliberately designed handbook "for the bishop and priest to aid him in the running of the diocese and parish."[47] It seems unlikely that Munich, Bayerische Staatsbibliothek, Clm 19417 was used personally by a bishop, who would have had little need of the material contained within it. As we shall see below, its context was probably much more modest. However, the manuscript's catechetical or pastoral use would have brought it within the bishop's mandate. It was one of the tools necessary for day-to-day spiritual work by individual priests among the local population.

The material production of this miscellany is not entirely straightforward. We know that a new scribe (scribe two) took over from fol. 28v, beginning the quire numbering anew, which implies that the two scribes had different agendas, or that the two halves of the manuscript were not at first intended to form a unit. However, the change of hand is contemporary, and it takes place within the same quire, which means that the entire manuscript was copied in the same scriptorium around the same time. The change in quire numbering may have nothing to do with a change in agenda: it may have been an oversight, or the scribe may have thought it simpler to begin again. Many of the numbers are missing, suggesting that the scribe did not pay a great deal of attention to the quires (and the ink colour throughout indicates that it was indeed the scribe who numbered the quires). The final contemporary text, the Athanasian Creed, is unfinished, and the following blank folio was filled in much later, which indicates that the miscellany was intended to continue to grow.

The presence of only two hands throughout the codex supports this possibility. The principal hand—the scribe of the codex from fol. 28v—is unpractised in its movement. From the example of fol. 52r, we see its ascenders lean both left and right (*h* in *chorus*, l. 1, and *l* in *baiolauit*, l. 13), its descenders have an irregular length, its minims descend below the baseline (*firmitates*, l. 12), the letters have an uneven height at the headline (especially obvious in the increasing height of *hoc est*, l. 4), and the horizontal

47 McKitterick, *The Frankish Church*, 41–42.

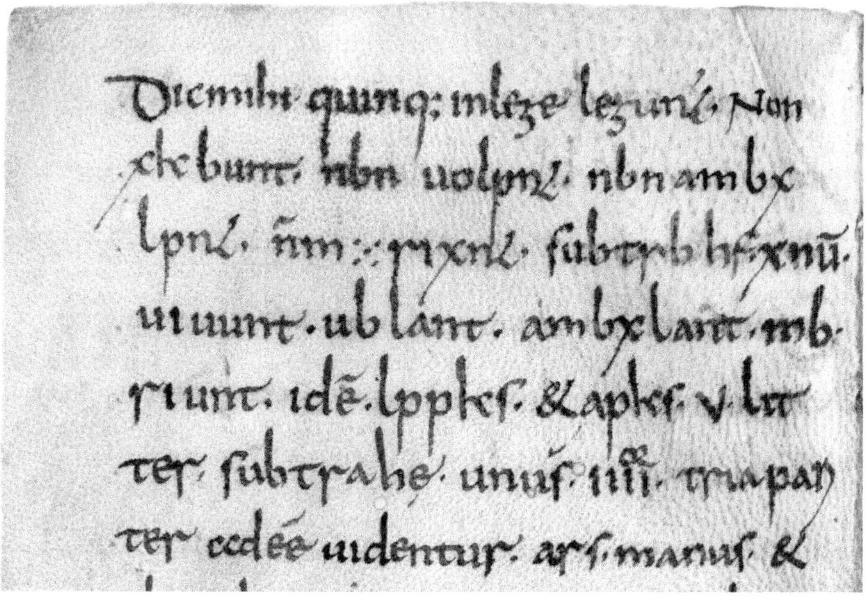

Figure 3.1. Beginning of ciphered *lapis* and *apis* riddle.
Bayerische Staatsbibliothek München, Clm 19417, fol. 100v. Used with permission.

movement is generally rounded but irregular, as seen in the contrast between the consistently left-leaning minims and right-leaning up-stroke of long s (*enim se*, l. 7). Overall, this hand is neat but not especially accomplished. Its irregularity, coupled with the very small size of Munich, Bayerische Staatsbibliothek, MS Clm 19417 (15 × 11 cm), indicates that this miscellany was most likely produced either for personal use, or for the use of a small rural community. The 180 Bavarian dialect words mixed in with the Latin glosses further characterize this miscellany as intended, from the beginning, for use only in a specific geographical zone.

Stories and story-like modes of communication, such as riddles and jokes, that touched on the natural world, were powerful tools of Christian teaching, and of encouraging cultural assimilation of the story content. The *joca*, riddles, *Physiologus*, credal texts, sermon on the Virgin and glosses on church councils in Munich, Bayerische Staatsbibliothek, MS Clm 19417 together form a collection that points towards a catechetical, pastoral, or even missionary context. The manuscript was a handbook for those working within these contexts, extending their capacity to instruct by means of story-telling and providing them with the material to explain foundational features of Christian faith and Church history.

Probably the most interesting text in this manuscript is found on fols. 100v–101r (see Figure 3.1). It begins with a (somewhat erroneously) ciphered riddle, which asks "what does not die, fly or walk" for its first component, which has five letters; and "what *does* die, fly and walk" for its second four-letter component—to get which, we are told, you need to subtract one letter from the first answer. The answer is *lapis* and *apis*: the stone and the bee. This riddle is followed seamlessly, as though they are a single text, by two very short excerpts from Augustine's first sermon on the mystery of the Trinity and of the Incarnation. These excerpts were evidently carefully selected, for they repeat the same theme: the first discusses the unity of music, composed of the trinity of art, hand, and string; and the second discusses a metaphorical nut produced from the rod of Jesse mentioned in Isaiah 11:1–2, which unites three substances: bark, shell, and kernel. The ideas in the sermon are Augustine's, but a completely original feature of the collocation of these three extracts is their clear numerological significance, which connects them and raises their joint meaning to a new level.

The riddle is a gateway text into the trinitarian extracts, explicitly teaching the reader to look at the numbers involved in the riddle's answers and in the Trinity, to pay attention to the triad of the clue and of the sermon extracts, and to understand how to read metaphorical and spiritual meanings in natural objects. Its cryptographic encoding not only makes it a game, but also slows down this learning process, making it more effective, and could even itself be regarded as a symbolic barrier between literal and spiritual knowledge. This use may have been mirrored in the *Physiologus* in this manuscript, which has a cryptographic *explicit* (see Appendix I). By repeating the same lesson in two different ways, the trinitarian extracts themselves not only demonstrate how this argument for the unity of God can be made, but also act as a meditation on the linguistic symbolism of *lapis* and *apis*. An educated reader of this riddle would have known (or been able to refer to) Isidore's etymology of *apis* as something born without feet (*pes*).[48] As the Augustinian extracts become more metaphorical, so the reader comes to understand that the bee and the stone are in binary opposition, perhaps representing the soul and the flesh (although this is not explicitly stated), and that language, particularly biblical verse, is an important way to understand such symbolism in God's Creation. Although only two pages long, therefore, these extracts were selected with a great deal of skill to create a complete lesson suitable for use in a variety of settings, including the monastic classroom or for sermon composition.

## Regional Variation

We now move on to our fourth and final miscellany. Together with seven other manuscripts, Bern, Burgerbibliothek, MS 225 + Bern, Burgerbibliothek, MS 233 + Orléans, Médiathèque, MS 313 (hereafter the Bern-Orléans codicological unit) witnesses a Carolingian corpus of texts that may have existed in upwards of fifty codices.[49] The corpus could not have been put together in its principal form before 798, assuming

**48** *Etymologiae* 12.7.8.

**49** Gorman, "The Carolingian Miscellany," 336. These manuscripts are Albi, Médiathèque Pierre-Amalric, MS RES 39 (https://cecilia.mediatheques.grand-albigeois.fr/idurl/1/114); Cologne, Erzbischöfliche Diözesan- und Dombibliothek, MS 85 (https://digital.dombibliothek-koeln.de/hs/content/titleinfo/168968); Leiden, Universiteitsbibliotheek, MS Voss. lat. Q. 122; New York, Columbia University Library, MS Plimpton 58; Paris, Bibliothèque nationale de France, MS lat. 614A (https://archivesetmanuscrits.bnf.fr/ark:/12148/cc62069f), MS lat. 2175 (https://archivesetmanuscrits.bnf.fr/ark:/12148/cc60068x), and MS lat. 10612.

this to be the correct date of Alcuin's letter—the key text of the core group of texts—and not long after the year 800, as the oldest manuscripts date from the beginning of the ninth century.[50] However, the major works of the corpus may have been compiled well before 800, "since collation reveals that the various texts seem to be distant from their archetypes and they are all rather 'old-fashioned' for the year 800."[51] The corpus may represent, therefore, an older collection expanded at the start of the ninth century.

The Bern-Orléans codicological unit contains the core corpus and a large number of other texts not listed in the two principal manuscripts used by Michael Gorman to determine the core. Isidore's *Prooemia* to the Old and New Testaments and *De ortu et obitu patrum* (on the lives and deeds of the Fathers) are added at the beginning. Further texts are inserted at the end, starting with the quires in Bern, Burgerbibliothek, MS lat. 225 and continuing with the remainder of Orléans, Médiathèque, MS 313:

1. First third of the *Liber pontificalis* (17): letters of Jerome and Damasus with lives of first 35 popes.
2. Jerome, *De uiris illustribus* (18): lives of Church Fathers.
3. Extract from Isidore, *De ecclesiasticis officiis* (19): on sin and penitence.
4. List of clerical vestments (20).
5. Isidore, *Etymologies* VI.16 (22): on church councils.
6. *De decimis offerendis in genesi* (23): on the ten tithes due to God as described in Genesis, Leviticus, and other biblical books.
7. Homily on Matthew 6:24 (25): on greed (serving two masters).
8. Lists of church council canons (27): list of priests, bishops, emperors, and popes in office, and the number of chapters in each set of canons from Nicaea onwards, followed by a list of those condemned in each council.
9. Extract from a Roman *ordo* on Septuagesima Sunday (29): general instructions on preparing for Lenten days and ultimately for Easter.
10. Three chapters from the Rule of St. Benedict (30): on observance of Lent, the proper singing of psalms and reverence in prayer.
11. Summary of the creation of the world (31).
12. Extract from Venantius Fortunatus, *Commentum in symbolum Athanasianum* (32): the Fortunatus Commentary on the Athanasian Creed.
13. Exposition of the Lord's Prayer (33).
14. *Physiologus* (35).

---

**50** Gorman noted that the astronomical reference to Mars in the invocation to Charlemagne accompanying Alcuin's letter was recorded in the *Annales regni francorum* for the year 798.

**51** Gorman, "The Carolingian Miscellany," 337.

These texts are not simply appended but are mixed in with some of the original collection: namely, the *Etymologiae*, the *Dicta Leonis* and *De diuinis scripturis*. Gorman concluded that the corpus as a whole was intended primarily for educating priests and thereafter as a reference work, containing as it does "elementary exegetical material...an introduction to the significance of names...some texts on the priesthood...a simple text on doctrinal matters...[and] a basic credal statement."[52] But this does not explain its popularity: eight extant manuscripts is a large number for a miscellaneous collection, and this one may have once existed in over fifty codices.[53] Nor does it explain why the collection was sometimes expanded, as in the Bern-Orléans unit. Gorman suggested that the corpus may have been put together by someone close to Charlemagne, which, together with its utility, guaranteed recopying for some time after the emperor's death.[54] Yet there is no reason why anyone in Charlemagne's circle would have required such an "elementary" set of texts. The day-to-day utility of the corpus was, most likely, its own impetus for wider dissemination.

The Bern-Orléans collection contains a good range of texts useful for foundational Christian learning, including exegesis of common biblical passages, Creeds, and the Lord's Prayer, reference summaries and lists (of Creation events, biblical tithes, and characters, clerical vestments and grades, councils, popes, and Church Fathers), and moral, pastoral, and spiritual texts. As in other miscellanies, an effort is made to bring together otherwise unrelated texts that form new sense-units. For example, the dangers of avarice are discussed in the texts on fol. 232, an extract from the *Dicta Leonis episcopi*, and the following exposition on the Vulgate Matthew 6:24 on fols. 233–234. The former concludes "here end the Sayings of Saint Gregory on the wickedness of money" and the exposition of the latter begins with "Mammon translates as wealth."[55] Here we find a very specific extract of the *Dicta* which was understood by the scribe or compiler to discuss the evils of greed for material wealth. The brief biblical exposition that follows on the same subject serves to complete what can be described as a lesson: an outline of the concept, followed by a biblical reference and an example interpretation. As in Munich, Bayerische Staatsbibliothek, MS Clm 19417, this would have been ideal for sermons and homilies, as well as for a teaching text.

The expansion of the core collection in three of the eight manuscripts demonstrates the significant utility of the corpus, therefore, and the extent to which regional and local needs varied in terms of such compilations. One of these three manuscripts is the Bern-Orléans codicological unit and it is related textually to one of the others—Paris, Bibliothèque nationale de France, lat. 10612, which originated at the large Benedictine abbey of St. Julien in Tours, also in the ninth century.[56] These manuscripts

**52** Gorman, "The Carolingian Miscellany," 353; p. 336 for a bibliography on the corpus. Very little study has been made of it before or after Michael Gorman, apart from brief mentions in Wright, *The Irish Tradition*, 71–72, and Reynolds, *The Ordinals of Christ*, 70.

**53** Gorman, "The Carolingian Miscellany," 336.

**54** Gorman, "The Carolingian Miscellany," 354.

**55** "Explicit dicta sancti gregori de mamona iniquitatis"; "mamona diuitio interpretantur."

**56** Gorman, "The Carolingian Miscellany," 336 and 344.

are evidence of regional links between the abbey of St. Julien and the nearby, Fleury-adjoining monastery that produced the Bern-Orléans unit. Despite their relationship, the additions to Gorman's corpus contained in each are quite different. This indicates that, though copied from the same exemplars, the two manuscripts were needed for different purposes.

Therefore, while the core corpus was suited for teaching, this was far from its only function. This supplementary function differed in different centres. In the Bern-Orléans unit, the majority of the added texts appear to be lists, with some exegetical material as well. Such texts could be used for reference purposes, whether to consult the life of a particular pope or bishop's name, or to check the correct interpretation of a common topic, perhaps while composing a sermon. Other functions are of course possible. That the miscellany was intended for use in a monastic setting is evident from the chapters from the Benedictine Rule and the question-answer version of books 7–9.3.17 of Isidore's *Etymologiae* (8), where the chapter *De monachis* is the only one to be emphasized by the inclusion of a rubric.[57] However, the monks in training at St. Mesmin, the probable place of origin of the Bern-Orléans unit, would have had recourse to literary, patristic, grammatical, and other manuscripts from which to learn. This is evident from the fact that books were exchanged for copying between the monastery near Fleury and the monastery near Tours, from their proximity to important and wealthy monastic centres associated with those cities, and from the size and quality of the Bern-Orléans collection. These details speak of resources and the availability of training. The utility of the core corpus to monastic education is clear, but for the copyists of the Bern-Orléans unit, it was more important to create a collection to which monks, priests, and lay brothers could refer as they went about their duties, rather than one which could be used to teach. As a reference work, it would therefore have been consulted often, and (perhaps in conjunction with other miscellanies) used regularly in the intellectual activity of the monastery.

## Conclusion

The Bern-Orléans collection, as well as St. Gallen, Stiftsbibliothek, Cod. Sang 230, demonstrate that miscellanies were not necessarily schoolroom books. Carolingian schools had plenty of manuscripts containing literary Latin texts read by pupils, and "textbooks" of the liberal arts.[58] While miscellanies frequently contained excerpts from such works, these were used in different contexts, or indeed included because continued learning—whether of Latin, the liberal arts, biblical content (e.g. the Psalms), or some other aspect of essential early medieval education—was a particularly important part of the community for which the miscellany was produced. As the above examples demonstrate, miscellanies were text collections for a broad range of users, who required not only basic works such as the Lord's Prayer but also complex exegetical exposition on diverse topics.

57 Gorman, "The Carolingian Miscellany," 351.

58 Grotans, *Reading in Medieval St Gall*, esp. chap. 2.

The conceptualization of miscellanies, the relationships between their texts, and especially the historical connections between these manuscripts and the communities that made them, often visible in the diplomatic evidence, require more research. Nevertheless, miscellanies were not, as this chapter has demonstrated, either poorly planned or heterogeneous. Instead, they were carefully organized, well constructed, and clearly laid out books. They were frequently used in a variety of ways, by a variety of people, and their contents varied significantly. However, viewed as a category of book, their contents indicate that they were all responding to a series of attempts at wide-ranging change. Clergy needed to learn the Roman chant thoroughly; to know and teach the Lord's Prayer and the clauses of the Creed; to be able to celebrate baptisms and masses; to know how to sing the Psalms, doxologies, and *Sanctus*; to keep the church, altar, and vessels in the proper manner; to preach against sin, especially hatred and avarice, and especially about the virtues and the resurrection of the dead; and to read only correct and canonical books. In this list of requirements—based on the *Admonitio generalis*—there is already the core outline for a miscellany. Many miscellanies include the Lord's Prayer, a Creed, liturgical texts of various kinds, Psalms, texts on sins and virtues, and excerpts from church councils, penitentials, and decrees. These texts would eventually be supplemented by suggestions taken from training manuals such as Hrabanus Maurus' *De institutione clericorum* (*On the Training of Clergy*), which he wrote at Fulda in the first half of the ninth century on the request of those of his brothers who were preparing to enter the priesthood. The first book of the treatise introduced the Church, ecclesiastical grades, sacraments, vestments, and the Mass. The second book summarized "the divine office, liturgical year, feast days, hymns, the Bible, and basic prayers and blessings, and heresies."[59]

In these miscellanies, the natural world, seen through the lens of language and grammar, came to the fore. It received pride of place in the *Physiologus*, but it was also extensively glossed and commented in a range of texts, from sermons and homilies to short extracts that were re-contextualised to create new meaning. This not only enabled early medieval scholars to explain the Bible, but also gave them the tools to point out natural objects in the real world and explain their significance to God. This was a potentially very powerful means of consolidating both the faith and Church authority among local communities. It is explored further in the following chapter.

**59** Contreni, "The Pursuit of Knowledge," 109.

## Chapter 4

# NATURE AND SALVATION

EARLY MEDIEVAL MISCELLANIES brought the natural world firmly within the province of Christian philosophical learning and co-opted it for the much larger purposes of consolidating the authority of the Church. This took a great deal of continuous recopying and re-collocation of texts, an effort that was justified by the practical utility of the material to the makers of the miscellanies. These makers—anonymous local priests tasked by their bishops with improving the faith of their flocks—recognized the utility of the natural world in accessing their audiences' imaginations as a gateway to faith.

This access was based on the fourfold way of seeing known in the later Middle Ages as the *quadriga*, first expressed by Origen and expounded by John Cassian in the fourth century (see Chapter 1 for a more detailed discussion).[1] The *quadriga* taught that the Bible has four levels of meaning: the historical, allegorical, tropological or moral, and anagogical or spiritual. The progression from the first to the second levels was the most discussed, and the full four levels were rarely expounded, but they were known to biblical exegesis from the patristic period onwards. Alcuin, for example, recognized three ways of seeing—corporeal, spiritual, and intellectual—in a letter of 798 to his pupil Fredegisus.[2] The essential point of this method was that one subject could represent several different things depending on how it was interpreted. Jerusalem could be, at the same time, the earthly city, the Church of Christ, the heavenly city, and the human soul.

The allegorical sense was of particular interest because it provided a means of linking the words of Scripture with the things of nature.[3] In other words, the allegorical hermeneutic method revealed the link between the visible physical world and the invisible world of God. It was easy to understand and easy to demonstrate. And from the allegorical sense, one could proceed to the moral and spiritual, correcting not only belief but also oneself and so coming closer to spiritual salvation. One of the stated aims of Charlemagne's reforms was, after all, the moral improvement of the empire.[4]

This chapter considers the evidence of five miscellanies containing the *Physiologus* for the application of allegorical readings both within and across texts about the natural world. The first half of the chapter investigates the different ways that these readings could be shaped and deployed. The second half is a close study of the famous *Physiologus* illustrations in Bern, Burgerbibliothek, MS 318. While these drawings have been extensively studied by art historians, they have never been considered within a whole-codex context, and their allegorical significance both for the *Physio-*

---

**1** Lubac, *Exégèse médiévale*; Lubac, *Histoire et esprit*. See also Robertson, *Lectio Divina*.

**2** "Tria sunt genera visionum: unum corporale, aliud spiritale, tertium intellectuale." MGH Epp., 203–4. See Collins, *The Carolingian Debate*, 17.

**3** Harrison, "Hermeneutics and Natural Knowledge," 345.

**4** Keefe, *Water and the Word*, 5.

*logus* and for early medieval nature within the compilatory culture discussed in this and the previous chapter, has gone largely unnoticed. They reveal the true extent to which early medieval manuscript makers innovated when working with inherited material about the natural world.

## Allegorical Interpretation

The practice of allegorical interpretation in miscellany manuscripts is especially evident in two *Physiologus* codices: Montecassino, Archivio dell'Abbazia, MS 316 and MS 323. These two manuscripts, separate volumes produced as a pair, are of Italian origin and date from the second quarter of the ninth century, according to Bischoff. If his dating is correct, the manuscripts were made not long after the composition of the letter by Jesse of Amiens on baptism in 811 or 812 and the composition of Hrabanus' *De rerum naturis* between 842 and 847, both texts that were copied into these codices. These are therefore miscellanies that were newly compiled in the mid-ninth century from available source material, rather than duplicated from available exemplars, and so can be taken to reflect a specific, contemporary set of intellectual practices and requirements.

The common factor behind the majority of texts in Montecassino, Archivio dell'Abbazia, 316 and 323 is their focus on allegorical interpretation. The Latin glossary (12), for example, prefers symbolic *interpretamenta* to literal ones: "cornum," horn or javelin, is "fortutudo uel regnum," fortitude or royal power/kingdom (p. 23, l. 19); both "dexter," "right," and "draco," dragon, are translated as the devil, "diabolus" (p. 23, lines 27 and 29); "candelabrum" is "ecclesia," the Church (p. 23, l. 12). Not only the glossary, but Eucherius' *Instructiones*, the *Physiologus*, the explanations of winds, the sun and thunder, and the texts on the mass and baptism provide allegorical explanations of their subjects. The guiding principle behind this compilation of texts is not only to assist with reading the Bible and understanding the example of biblical characters: it is also to interpret and understand the wider world as an extension of God. The sun is explained as the light of Christ, thunder is used as a way to introduce four biblical characters who rose to the sky, and the seven winds are listed so that the reader might learn, at the end, that the Saviour resides beyond the stars. These texts reveal that a knowledge of God in the world was to be gained through the study of language and of observable phenomena.

The figure at the heart of this desire for knowledge is that of the human being. Humanity was made in God's image, it was for humanity that Christ suffered on the Cross, and human history was the history of salvation. Seeking God in the natural world and in language was done so that humanity might be saved. These texts therefore reflected the human desire to come closer to God. This is evident in the anonymous explanation of the Mass on pp. 56–57. It makes use of the four-part exposition from Augustine's epistle 149, as well as the seven-part outline from Isidore of Seville's *De ecclesiasticis officiis* I.15.[5] These sources make the anonymous text a logical, author-

**5** Geiselmann, *Die Abendmahlslehre*, 104–5.

itative break-down of the Mass and its spiritual purpose. Most interesting, however, is that the explanation also contains extracts from Hrabanus Maurus' *De rerum naturis*, with their interpretations of the figure of the human being—*homo*—in Gospel parables. The vivid first interpretation is of the person with the shrivelled hand from Matthew 12:10 and Mark 3:1, "homo manu habens aridam," which is interpreted as "animam misericordiae," the soul's mercy (p. 56). The second is from Matthew 12:43, on the impure spirit leaving a person's body, "homo de quo in mundus spiritus exiens rursum," interpreted as penitence (p. 57). The third is from Matthew 13:3, on the man who sowed his field with mustard-seed, "homo qui seminauit in agro suo sinapis," interpreted as Christ (p. 57). There are two further interpretations of *homo* in Isidore of Seville's *Allegoriae quaedam sacrae scripturae* (item 3), on the man with a hundred sheep and the man who planted a vineyard (Matthew 18:11 and 21:33/Luke 15:4 and Mark 12:1, pp. 52–53). The Bible is the story of humanity, which is a subject of allegorical study in these Montecassino manuscripts. It is no accident that explanations of the sacraments of communion and of baptism, both of which are doors to salvation, were included in these books.

## Allegory for Salvation

Two other *Physiologus* miscellanies also focus very closely on human redemption and salvation. The first of these is Oxford, Bodleian Library, MS Laud. Misc. 129. It was compiled in the early ninth century in the Main river valley—that is, in the region of influence of the great Insular foundation that was Fulda, as well as many smaller monastic foundations dating back to the circle around St. Boniface. Its adherence to the same Insular traditions is evident from the calligraphic Insular minuscule used throughout, and from the regularly titled and carefully copied Insular-influenced contents. The texts are presented one after the other, without codicological breaks, in a few confident hands that at times tend to cursivity, indicating a swift and straightforward production process.

As a whole, this collection deals with penitence, but was evidently intended to be deeply spiritual, inspiring, and in places even joyful in tone. Referring to Paradise, one of the anonymous expositions describes it as a place "where there is unfailing light, everlasting joy and life eternal, where evil will not be seen and good will never fail" (23v, lines 1–3).[6] The author of these lines drew on contemporary Insular descriptions of heaven, without copying them verbatim.[7] The focus of the compilation is firmly on salvation rather than on penance, although sin, vice, and fear of God are addressed in a number of texts (e.g. item 13). In item 26, a typical example of the tone set by many of the homilies, Augustine's sermon 326, "On the birthday of the twenty martyrs," sets up the martyrs as an exalted, emotionally charged example to follow: "And indeed on earth they had nothing, but in heaven they possessed everlasting felicity. They were

**6** "Ubi lumen indeficiens, ubi gaudium sempiternum et uita perennis, ubi non uidebitur malum et non deficiet bonum."

**7** Pelle, "Source Studies," 69–70, esp. note 141.

hurrying off, fully committed, toward heaven, and running along the road of life without a worry in the world; and while still a long way off they were stretching out their hands for the palm. Run, saints; so run that you may obtain it (1 Cor. 9:24)."[8]

Allegorical interpretation is at the heart of these joyful eschatological texts. On fol. 97v, for example, the 12 apostles are interpreted as the 12 stones laid in the foundation of the heavenly city.[9] Natural imagery is used throughout the compilation to make these allegorical interpretations comprehensible. Thus, the tree in the Garden of Eden is used to explain the spiritual implications of evil in item 11; item 12 is a meditation on the sheep, wolves, serpents, and doves in Matt. 10:16, and what they represent; item 26 discusses the branches and flowers emerging from the rod of Jesse; and item 37 (the *Physiologus*) gives allegorical interpretations for each animal, plant, or stone. Other texts focus on other subjects but incorporate natural elements. This is the case with the final item, the extract from Jerome's commentary on the Song of Songs, which is concerned chiefly with the heavenly Jerusalem and with the dove; and with the brief exposition on the meaning of stars in the extract on fol. 43r. This collection therefore uses allegory of the natural world to focus closely on spiritual redemption and the hope that it offers.

The now destroyed manuscript Chartres, Médiathèque L'Apostrophe, MS 63 (125) offers a parallel compilation. It is not possible to say a great deal about it here, since all that is now available is a list of contents, together with opening lines and a few closing lines. Nevertheless, these contents do indicate that the manuscript's compilers had a similar interest in salvation history. The codex contains Ambrose's commentary on the first six days of Creation, and a text by him on the virtues and vices (2), as well as a commentary on Genesis (4) which makes use of three levels of seeing. Additional allegorical texts that would have been useful to exegetes include the commentaries on Genesis and the six days of Creation, as well as the *Physiologus* and excerpts from book 13 of the *Etymologiae* on the world and its parts. The second half of the manuscript demonstrates a significant linguistic and lexical interest. It contains the first book of Isidore of Seville's *Etymologiae* and a portion of his *Differentiae*—sets of distinctions between words which have similar meanings—as well as Jerome's epistle 29 to Marcella explaining some Hebrew words, and a work on the names of God. It seems that this miscellany was made along the lines of those compilations discussed in Chapter 3, with a focus on language as a means for understanding Scripture. Although here, too, there is a significant interest in salvation, it is approached through lexical and etymological learning: something which is entirely absent in the Oxford manuscript. This may be attributable to the difference between Insular and continental traditions.

Both these miscellanies exemplify the increased allegorical use of the natural world for stimulating thinking about moral living and salvation during the early Middle Ages. They demonstrate that those in charge of compiling books for communities were engaged in the same dialogue across different centres of production. This alle-

**8** Augustine, *The Works of Saint Augustine*, trans. Hill and Rotelle, 170.

**9** "Nunc xii lapides paradiso in fundamento ciuitatis ponuntur."

gorical hermeneutic method, with its emphasis on both language and nature, represents perhaps the highest peak of innovation for early medieval miscellanies.

## Compilation and Christian Time

As we have already seen, miscellany compilers created new contexts and meanings for texts. This was the case for Bern, Burgerbibliothek, MS 318, an apparently random compilation which, like the other *Physiologus* miscellanies, was nevertheless put together with the greatest of care. Similarly to the Montecassino, Oxford, and Chartres manuscripts, the contents of this ninth-century codex were selected based on an interest in Christian time. Here, however, the focal point was human history.

The contents are as follows:

1. Antonius, *Life of St. Simeon the Stylite*
2. Isidore of Seville, *De ortu et obitu patrum*
3. *Physiologus*
4. Fredegar, *Chronicle*
5. List of Egyptian days
6. Unknown homily on Matt. 17.1–9
7. Homily attributed to St. Ephrem
8. *Probationes pennae*, two lines of verse on the difficulty of writing, and two medical recipes for palsy and headache
9. On the seven miracles of the world

The historical work in question is a copy of Fredegar's *Chronicle*, which is also the longest text in the manuscript: it takes up 100 out of the 131 folios. The text has an uncertain authorship, problematic transmission, and extensive, frequently unique contribution to Frankish history alongside the *Annales regni francorum*.[10] It is now generally accepted that it was composed around 659 by a single author with links to Burgundy and later Austrasia, possibly Metz, who may have belonged to the Columbanian monastic milieu.[11] The *Chronicle* consists of six parts: histories by Isidore, Jerome, Hydatius, and Gregory of Tours, the *Liber generationis* (attributed to Hippolytus of Rome) and an original addition. These six parts were divided into four books, of which I to III extend from the beginning of the world to 584 CE. Book IV continues through to 642 or 658, depending on the copy.[12] The Continuations, which it seems were commissioned

10 The most recent edition of the *Chronicle* is in Wolfram, Kusternig, and Haupt, eds., *Die vier Bücher der Chroniken*; it builds on Fredegar, *Fredegarii chronicorum*. Roger Collins has compiled a detailed list of the extant manuscripts with a discussion of the text's transmission history in his *Die Fredegar-Chroniken*. The question of date and authorship is addressed in the seminal studies by Goffart, "The Fredegar Problem," and Erikson, "The Problem of Authorship." Further studies include Scheibelreiter, "Gegenwart und Vergangenheit," and Wood, "Fredegar's Fables."

11 Wood, "Fredegar's Fables," 360.

12 Goffart, "The Fredegar problem," 206.

by members of the Carolingian royal family, extend the chronicle to 768.[13] The Bern manuscript does not contain the full text of the *Chronicle*—in particular, that section of it taken from Hydatius is not present—but this is not unusual among extant copies.

The collocation of a historical text with an allegorizing text on animals, plants, and stones is unusual in the early Middle Ages. The most obvious shared feature between the *Chronicle*, the commentary on St. Matthew's Gospel, the *Life of St. Simeon the Stylite, Senior* and Isidore's *De ortu et obitu patrum* is the use of material from Isidore. The *Chronicle* gives the Augustinian six ages of the world from Isidore's history, which are divided into genealogical lists of figures from the Old Testament beginning with Adam, and lists of Egyptian and Roman rulers down to Heraclius. The *Physiologus* borrows some of its animal descriptions from the *Etymologiae*. *De ortu et obitu patrum* is itself by Isidore and in its full form contains eighty-six brief biographies of figures from the Old and New Testaments, including several prophets' lives.[14] The Bern copy begins with Adam and ends with David.

Although thematically these texts appear to be quite different, they do show by this very difference the importance of Isidore as an authority in a range of subjects. Already, perhaps, Isidore's popularity from the late seventh century onwards provides a reason for combining these texts within the same miscellany. Yet all the texts also show evidence of a certain interest in the lives—and deaths—of figures significant to Christianity. This is clear from the Isidorean genealogies in the *Chronicle* and indeed from other elements of it, such as the list of popes; from *De ortu et obitu patrum* and from the *Life of St. Simeon the Stylite*. St. Simeon is held up as an example for emulation by Gregory of Tours in the *Historia* (VIII.15), through the imitation of his lifestyle by St. Vulfolaic. The life may therefore have been selected to complement Gregory's text in Fredegar.

The early medieval focus on using the created world as part of Christian allegorical interpretation provides an even stronger reason for the collocation of these texts, however. The focus of these works is on time: not simply the historical time of the *Chronicle* but also Christian time. Indeed, history and the history of salvation are indivisible. The purpose of a list of fathers and sons, from Adam to David as in *De ortu et obitu patrum*, is to show that history is a single whole. There is an unbroken connection between the present and the biblical past. History is furthermore measurable by the lifespans of human beings: the *Chronicle* gives the age of each male biblical figure at the time he produced his heir, making it possible to calculate with relative precision the length of each historical age. The lives of human beings thus have a place in history that is commensurate with the strength of their faith.

For exegetes writing in the early Middle Ages, the biblical story of salvation was paralleled by the histories of their own peoples, and there is an echo of this in the *Chronicle*.[15] Fredegar was the first to use the fifth-century *De excidio troiae*, a history of

---

**13** Fredegar, *Fredegarii chronicorum*, xxv–xxviii.

**14** PL 83.130–156. See also Vaccari, "Una fonte."

**15** Mayeski, "Early Medieval Exegesis," 87.

the Trojan War attributed to Dares Phrygius, to stress the Trojan origin of the Franks. Fredegar of course had a political purpose: his patron was Dagobert I.[16] The claim of descent from the Trojan kings justified and asserted the right of the Franks to occupy the former territory of the Holy Roman Empire. In the early Middle Ages, this claim would furthermore have been supported by Isidore's identification of Dares Phrygius as the first secular historian (with Moses as the first ecclesiastical historian).[17]

## The Bern *Physiologus* Illustrations

The innovations of the decorative program in this manuscript have been noted by various scholars, but their astonishing significance as a whole has been entirely overlooked.[18] They are a significant supplement to this manuscript's focus on the history of salvation. The drawings are executed in an illusionistic style inherited from antique Greco-Roman pictorial traditions, like those in the fourth-century Vatican Virgil, to which the Bern *Physiologus* has been compared.[19] The effect is one of three dimensional, vivid, realistic, and skilfully executed figures. In the early Middle Ages, this was not unique to the *Physiologus* in Bern, Burgerbibliothek, MS 318: the so-called Rheims School produced several manuscripts illustrated in this style in the ninth century, including the Ebo Gospels, the Troyes and Utrecht Psalters, and the Gospels of Saint-Thierry (or the Hincmar Gospels).[20] Evidently, highly skilled artists able to work in this prestigious antique style were trained in the early medieval Frankish world. But the technical skill of these drawings is just a small part of their complexity, which is enshrined in their manuscript context and interaction with the *Physiologus* tales.

The art historian Koert van der Horst believed that the Bern *Physiologus* was a more faithful replication of late antique original illustrations than the Bibles, Gospel books, and Psalters produced at this time, as it did not undergo modifications as part of "a carefully planned reform programme."[21] However, this is not the case. The text underwent changes each time it was reproduced and at Rheims two additional illustrated chapters were added to it. The first of these is *Galli cantus*, the crowing cockerel, the text for which is taken word-for-word from sections 5, 24, and 88 of the *Hexaemeron* of Ambrose.

---

16 Young, *Troy and Her Legend*, 57. See also Yavuz, "Transmission and Adaptation."

17 *Etymologiae* 1.42.

18 Mütherich, "Carolingian Manuscript Illumination" and *Studies in Carolingian Manuscript Illumination*; Koehler and Mütherich, *Die Karolingischen Miniaturen*; Hicks, *Animals in Early Medieval Art*, among others.

19 Nees, "The Illustrated Manuscript" and *Frankish Manuscripts*. The Vatican Vergil is digitized at https://digi.vatlib.it/view/MSS_Vat.lat.3225. It is only 160 × 160 mm compared to the 255 × 180 mm of the Bern manuscript, but this does not exclude the possibility that the exemplar for the *Physiologus* was larger.

20 For a detailed discussion of these manuscripts, see Horst, Noel and Wüstefeld, eds., *The Utrecht Psalter*, 24–84 (on the Utrecht Psalter), as well as 104–19 and 168–255 (on the other manuscripts).

21 Horst, "The Utrecht Psalter," 77.

Figure 4.1. *Galli cantus*, the singing cockerel, beneath the illustration for the Indian stone. Bern, Burgerbibliothek, MS 318, fol. 21r. Used with permission.

Chapter 2

# THE EARLY LATIN *PHYSIOLOGUS*

FROM AT LEAST the early eighth century, the *Physiologus* was already well known in both Insular and continental Europe. This chapter discusses three important groups of sources: texts and images from before the eighth century that draw on the *Physiologus*; booklists from medieval libraries up to ca. 1000 that mention the *Physiologus*; and a manuscript that contains the earliest extant Latin copy of the text: Bern, Burgerbibliothek, MS 611 + Paris, Bibliothèque nationale de France, MS lat. 10756.

## The Earliest Evidence

The extant manuscripts of the *Physiologus* are listed in Table 2.1 (a more detailed description of each is provided in Appendix I, the descriptive catalogue of manuscripts, in the corresponding numbered entry).[1] In addition, nine references to what may have been the *Physiologus* are made in early medieval booklists. These refer to the following entries, listed in order by booklist date and reproduced using the original spelling and capitalization:

**a.** "Crisostomus de naturis animalium" in a ninth-century catalogue from a lost Murbach manuscript (discussed below).

**b.** "liber phisiologi" in a ninth-century catalogue from the Salvatorstift in Würzburg (from an unknown manuscript).[2]

**c.** "Liber I phisiologi" from the cathedral library catalogue Würzburg, Universitätsbibliothek, MS M.p.th.f.40, fol. 1r, compiled in Würzburg ca. 1000 with eleventh-century continuations on fols. 46r–v.[3] This may represent the same copy of the *Physiologus* as entry *b.*

**d.** "Liber bestiarum" in a list from the 863 or 864 will of Eberhard and Gisela of Friuli, bequeathed to their eldest son Unruoch from their private chapel book collection. The will survives as a copy in a cartulary from Cysoing.[4]

---

**1** This book uses the standard Latin medieval manuscript catalogue dating system (abbreviations of "saeculo ineunte/medio/exeunte" and fractions). For example, in the first entry in Table 2.1, the beginning of the ninth century indicated as s. $ix^{in}$.

**2** Becker, *Catalogi bibliothecarum antiqui*, no. 18, item 42 (hereafter Becker). This manuscript remains unidentified and may be lost. See O'Loughlin, *Adomnan and the Holy Places*, 185.

**3** Becker no. 38, item 18; MBK IV.2, 987. Digitized at http://vb.uni-wuerzburg.de/ub/mpthf40/ueber.html.

**4** Mons, Archives de l'État, Cartulaire 12 (made in 1517). The Latin text of the will can be found in Coussemaker, ed., *Cartulaire*, 1–5, no. 1, Becker no. 12, item 12; Schramm and Mütherich, *Denkmale der deutschen Könige und Kaiser*, 78–79. A partial French translation can be found in Riché and Tate, ed., *Textes et documents*, 414–15. See also La Rocca and Provero, "The Dead and Their Gifts," and Kershaw, "Eberhard of Friuli."

Table 2.1. Extant *Physiologus* manuscripts

| Manuscript | | Origin | Date |
|---|---|---|---|
| 1. | Bern, Burgerbibliothek, MS 225 + Bern, Burgerbibliothek, MS 233 + Orléans, Médiathèque, MS 313 | Saint-Mesmin, near Fleury | s. ix$^{in}$ |
| 2. | Bern, Burgerbibliothek, MS 318 | Near Rheims | s. ix$^{2/3}$ |
| 3. | Bern, Burgerbibliothek, MS 611 + Paris, Bibliothèque nationale de France, MS lat. 10756 | Bourges | ca. 727 |
| 4. | Brussels, Bibliothèque royale de Belgique, MS lat. 10066–77 | Rheims or Laon | s. x |
| 5. | Chartres, Médiathèque L'Apostrophe, MS 63 (125) | France | s. x/xi |
| 6. | Montecassino, Archivio dell'Abbazia, MS 316 + MS 323 | Cassino | s. ix$^{2/4}$ |
| 7. | Munich, Bayerische Staatsbibliothek, MS Clm 14388 | Northwest Germany | s. ix$^{med}$ |
| 8. | Munich, Bayerische Staatsbibliothek, MS Clm 19417 | Southern Bavaria | s. ix$^{1/3}$ |
| 9. | Oxford, Bodleian Library, MS Auct. T.2.23 | Eastern France (Tours?) | s. ix$^{2/3}$ |
| 10. | Oxford, Bodleian Library, MS Laud Misc. 129 | Main river valley or its environs | s. ix |
| 11. | Paris, Bibliothèque nationale de France, MS Nouv. acq. lat. 1616 + Orléans, Médiathèque, MS 18 (*olim* 15) + Paris, Bibliothèque nationale de France, MS Nouv. acq. lat. 455 | Western France | s. x$^{3/3}$–x$^{4}$ |
| 12. | St. Gallen, Stiftsbibliothek, Cod. Sang. 230 | St. Gallen | s. viii$^{2}$ |
| 13. | Vatican City, Biblioteca Apostolica Vaticana, MS Pal. lat. 1074 | Catalonia | s. x$^{ex}$–xi |
| 14. | Wolfenbüttel, Herzog August Bibliothek, MS Gud. lat. 148 | Eastern France | s. ix$^{4/4}$ |

**e.** "Liber Esopi de natura animalium" in a fragment of a Fulda library catalogue, now lost. It was partially transcribed in the eighteenth century by Johann Friedrich Schannat in his history of the monastery.[5] Though we do not know the precise date of this catalogue, it may have been created before 830 during the abbacy of Hrabanus Maurus, who had various lists made when he took up office in 822. Other extant booklist fragments are thought to have been part of the same long catalogue.[6]

**f.** "et libros bestiarum Ysiodori" in a Passau charter of property exchange dated 8 September 903, extant as a copy in Munich, Bayerisches Hauptstaatsarchiv, Hochstift Passau, Inneres Archiv 5, fols. 124v–125v (the *Codex Lonsdorfianus* or *Lonsdorfer Codex*).[7]

**5** Schannat, *Historia Fuldensis*, 63.

**6** Raaijmakers, *The Making of the Monastic Community*, 194–95.

**7** Digitized at https://bavarikon.de/object/bav:GDA-OBJ-00000BAV80016806. Transcribed at Becker no. 28, item 37; MBK IV.1:142–49.

Figure 4.2. *Caballus*, the horse, with its rider. Bern, Burgerbibliothek, MS 318, fol. 22r. Used with permission.

The second is *Caballus*, the horse, taken from Isidore's *Etymologiae* XII.1.42–8.[22] The *Galli cantus* image shows three birds, two of which may be singing, perched on a beam, with three archways in the background (Figure 4.1). *Caballus* depicts a man on a galloping horse (Figure 4.2). It is possible that they represent an original ninth-century addition to this manuscript.

The text of *Galli cantus* may have been added to the *Physiologus* at some stage long before Bern, Burgerbibliothek, MS 318 was copied, particularly since it is also found in Brussels, Bibliothèque royale de Belgique, MS lat. 10066–77, which is not related pictorially to Bern, Burgerbibliothek, MS 318. The two manuscripts may therefore depend on an exemplar or set of exemplars that already contained *Galli cantus* and dated to before the ninth century. However, the text of *Caballus* is not found in any other Greek or Latin *Physiologus* and relies on Isidore's seventh-century encyclopedia. It cannot therefore have been present in a Greek exemplar, nor in the fourth-century Latin translation which served as, or lay behind, the exemplar for the Bern manuscript. The art historical evidence suggests that the exemplar from which the *Physiologus* in Bern, Burgerbibliothek, MS 318 was copied may have been made no earlier than the seventh century.[23] The existence of a *Physiologus* with a new chapter as early as the seventh century indicates an early medieval interest in and desire to expand the text. Similarly, the presence of the additional drawings in the Bern *Physiologus* reveals both the text's significance, and the willingness of early medieval compilers to expand and innovate. This is especially clear from the example of *Galli cantus*, as we shall see below.

**22** Henkel, *Studien zum Physiologus*, 26.

**23** Tselos, "A Greco-Italian School," 5–13.

Figure 4.3. The bathing devil in Bern, Burgerbibliothek, MS 318, fol. 17v. Used with permission.

The Bern *Physiologus* also contains one other drawing that distinguishes it from other manuscripts and may not have been copied from a pre-existing model: the illustration that seems to accompany the chapter on the salamander on fol. 17v. This chapter is entitled *De natura animalis qui dicitur salamandra* and lists the characteristics of the animal: if it enters a furnace it extinguishes the flame; if it enters a warm bath, the entire bathroom grows cold. The *Physiologus* compares this behaviour with the biblical parable of the three boys who were thrown into a furnace, where the power of Christ protected them and induced the flames to attack their enemies instead. Just before the chapter's *incipit* is the small painting in question, without an enclosing frame, depicting a small, hairy, man-like figure with slightly curving horns, sitting in a round wooden bathtub surrounded by grass and plants (Figure 4.3). His prominent lower lip gives him a comically mournful air.

This drawing floats ambiguously between the chapter on the salamander and the preceding chapter on the stag.[24] The stag is said to spew water at its enemy, the dragon, which it then swallows, and this action is interpreted in the text as the drowning of the devil's works in the bath of rebirth or baptism. The figure in the bath has all the features of a devil rather than a salamander, but it could be sitting either in a cold bath, or in a bath of baptism. The ambiguity of its interpretation was almost certainly a deliberate choice made by the Frankish artist (and we have no reason to suppose that the illustrator of this devil was not also the manuscript's main illustrator).[25]

---

**24** Woodruff, "The *Physiologus* of Bern," 250, believed it was part of the salamander chapter.

**25** Nees, "The Illustrated Manuscript," studied this illustration. Some similar motifs, of both baptisms and demons, are present in other early medieval art. A few examples include Musée de Picardie, Amiens, M.P.1875.61: an ivory book plaque from Rheims, dated to the last quarter of the ninth century with three scenes from the life of St. Remigius, whose central scene shows the miraculous baptism of a dying man; British Museum, OA.3065: an ivory from the Tournai

The carefully curated placement of the bathing devil illustration is only one of a series within this manuscript. Like the added drawings, the positioning of the entire series of paired illustrations is also meticulously thought-out. On fol. 8v, the miniatures for the story of the lizard and the charadrius bird appear next to each other. Each of these stories describe the loss of eyesight and the importance of looking at the right object in the right way or at the right time. The person responsible for the organization of the manuscript made an effort to link these related stories together visually through the illustrations—consciously intending them to be discussed as part of the subject of vision—even though this meant that the following page was a plain block of text. In general, the illustrator avoided leaving pages unadorned: only four of the thirty-two folia containing the *Physiologus* are plain.

The same paired layout is repeated on fol. 21r, with the miniatures for the stories of the Indian stone at the top and *Galli cantus* at the bottom. The link between these stories is less obvious, since *Galli cantus* is not a canonical *Physiologus* story, but the text is essentially a long list of the virtues of the cockerel's song, which—like the Indian stone—possesses miraculous restorative properties. The Indian stone heals sick people by pulling foul water from their bodies into itself and releasing it again after three hours in the sun. The miniatures are linked through the image of three drops of water falling from the stone and the image of three cockerels under three arches. By adding the *Galli cantus* miniature to the miniature of the Indian stone, the illustrator created a mirrored set of relevant symbolic materials for meditating on the miracle of the Resurrection. The discovery of Christ's empty tomb at dawn after three days—one of the best-known Christian stories—is symbolised by the number three, the sun (whose rise is heralded by the cockerel's song), the cockerel himself, and the salvific power of water.

Finally, the placement of the illustration for the story of the elephant was also a highly creative and deliberate choice. This was first observed by Bent Gebert. In this part of its story, the elephant, which cannot lie down and must sleep leaning against a tree, falls over when a hunter cuts down the tree. Other adult elephants come to help it stand, but fail (until a young elephant, which is not depicted here, succeeds). The miniature is presented vertically, at 90 degrees to the reader (Figure 4.4). The art historian Otto Homburger thought this was because the illustrator didn't have enough horizontal space to represent the full range of figures from left to right, but as Gebert noted, a variety of solutions was available to resolve this potential issue in other miniatures (such as the one on fol. 7r, where the figures are arranged vertically in varying

School showing a reed basket baptism of Christ in the top scene, dated ca. 900; Walters Museum, 71.305: a similar tenth-century baptism ivory; Hannover, Niedersächsische Landesbibliothek, MS I 189, fol. 5r (http://digitale-sammlungen.gwlb.de/index.php?id=6&tx_dlf%5Bid%5D=9809&tx_dlf%5Bpage%5D=1): a tenth-century manuscript from Fulda containing the *vita* of St. Kilian which shows Kilian baptizing man in a wooden bath made of planks with two horizontal metal stripes; St. Peter fends off a devil: from the New Minster *Liber Vitae*, England (Winchester), ca. 1031, London, British Library, MS Stowe 944, fol. 7r (www.bl.uk/manuscripts/FullDisplay.aspx?ref=Stowe_MS_944).

femina & concipit. Cum ergo tempus uene
rit ut generet intrat in stagnum aquae & sit
aqua ad mamillas eius & dimittit natum ut na
uigando sup aquas proximum habeat natem ma
tris suae. Serpens autem inimicus est aelifanto quia
pedibus suis interficit eum.

Figure 4.4. The falling elephant in Bern, Burgerbibliothek, MS 318, fol. 19v. Used with permission.

sizes).[26] The decision to turn the illustration 90 degrees and thus compel the reader to turn the book is a physical reminder of the elephant's fall. It creates haptic involvement in the story, providing an opportunity for active reading.

For all its simplicity, this is an astonishingly imaginative technique, still used today in printed books for creative audiences (such as storytellers and children). At the same time, there is a great deal of spatial ambiguity within the image: the tree is the central pictorial axis, where the fallen elephant is clearly struggling to rise with the help of a second, standing elephant. This was perhaps intended as a moralizing reminder of the tree in Paradise, which was often the central axis of pictorial depictions of the Fall. Yet, looking from the bottom of the page upwards, there is empty space behind the fallen elephant's elongated body, suggesting the movement of the fall into the bent—not yet broken—tree. This is also the direction from which the third elephant approaches. It is on higher ground and its head is hidden: it has not yet arrived, or it is mid-movement.[27] These details mean that the painting as a whole does not satisfy even as it is flipped. It demands not only movement in space to make sense, but also explanation of its movement in time. This elevates it from a clever device in an educational book to an exegetical tool, wielded by an artist aware of the moral value of Christian time.

The ease with which the *Physiologus* could be used to move from literal to allegorical to moral interpretations, so clear in this and the other images, must have been one of the principal reasons for its popularity, since it enabled teachers of all kinds to make that crucial connection between the visible, immediate world, and the insubstantial spiritual realm. The Bern *Physiologus* is a masterpiece that enables even modern eyes to see these associations. But glorious illustrations, while surely always welcome, would not have been essential in the early Middle Ages. All that was needed was a moderately competent guide to the material. Part of the genius of the *Physiologus* was that it was adaptable to any level, of both teacher and student: the story of Christ's baptism and its associated symbolism, for example, could be told simply, or it could be made rich and complex. The *Physiologus* provided the necessary material for both ends of the scale. Its different textual families almost certainly played a role in this, as well: each new copy and arrangement of the text provided new opportunities for juxtaposing the stories and creating new contextual links between them. Textual variance is not a unique feature of *Physiologus* manuscripts, of course, but it might, in this case, have contributed to the adaptability and therefore to the spread of the *Physiologus* as a work in miscellanies.

The arrangement of the miniatures in the Bern *Physiologus* wasn't simply a matter of having the intelligence to spot the iconography: it involved, firstly, reading the text at the planning stage to decide where both the old and the new drawings would be placed; and secondly, arranging the spacing of the text-blocks against the space the miniatures would take up, to ensure that the execution would match the planning. Both these matters required significant experience both of writing Caroline minuscule

**26** Gebert, "Der Satyr im Bad," 29.

**27** I am very grateful to Tina Bawden for her many insights into this illustration, which are reflected in this paragraph.

Figure 4.5. Haecpertus completes the text by writing around the miniature. Bern, Burgerbibliothek, MS 318, fol. 8v. Used with permission.

and of painting framed miniatures. (It is clear from occasional overlaps that the drawings were added after the pages were ruled but before the text. One such overlap is evident in the last line on fol. 8v, where the loop of the g in "intellegibilis" extends into the red border of a miniature.) Moreover, the creator of the Bern *Physiologus* needed to be a master manuscript-maker, not only skilled in the arts of brush and pen, but also trained in the kind of project management ability demanded by the complex *mise-en-page* evident here. The manuscript is principally the work of a single scribe, probably the Haecpertus named in the colophon on fol. 130r.[28] I suggest that Haecpertus must have been not only the scribe but also the principal artist and master-maker of the *Physiologus.*

Haecpertus' skills would have been far from unusual among early medieval manuscript-makers across Europe. Eadfrith both copied and painted the seventh- or eighth-century Lindisfarne Gospels, and innovated in the depiction of the Evangelists, deviating from his exemplar.[29] The Irish monk Macregol left a colophon in the eighth- or ninth-century Gospels on which he worked, in which he described himself as a scribe who had painted ("dipinxit") the manuscript.[30] The scribe of the eighth-century Trier Gospels from Echternach, Thomas, signed his name on the pages containing miniatures, which suggests that he was also the artist.[31] Similarly, Eadwig Basan is known

**28** "Haecpertus me fecit" ("Haecpertus made me").

**29** Netzer, "The Design and Decoration of Insular Gospel-Books," 235.

**30** "Macregol dipinxit hoc Evangelium. Quicumque legerit vel intellegerit istam narrationem orat pro macreguil scriptori" ("Macregol painted this Gospel. Whoever should read or hear these words, pray for Macregol the scribe"); Oxford, Bodleian Library, MS Auct. D.2.19 (Rushworth Gospels or Macregol Gospels), fol. 169v: https://digital.bodleian.ox.ac.uk/inquire/p/4aee97d4-0845-44fa-8dd7-8c4902090db2.

**31** Trier, Domschatz, MS 61, fols. 5v, 11r, 125v. See Netzer, *Cultural Interplay*.

to have contributed to multiple manuscripts as a scribe and artist at Christ Church Canterbury.[32] Haecpertus' execution was not flawless—he ran out of space at the top of fol. 8v and had to finish writing around the right-hand side of the miniature, for example—but his overall achievement is nevertheless impressive (Figure 4.5).

## Conclusion

In the Middle Ages, memory was a sense-image inseparable from learning.[33] It was through the things one learned and memorized that one shaped one's inner self. This idea was reflected in monastic reading or *lectio divina*, which prescribed hours of silent meditation on texts. It also applied to lay people, who were no less in need of correction and assistance than those in the service of the Church in finding the path to salvation. The *Physiologus* was a means by which images of God's work could be imprinted in the memory—whether though silent reading or through hearing the text read aloud. This may be one reason for the many images in the copies of the *Physiologus* across the different linguistic traditions of this text. But even without illustrations, allegorical interpretations of the tales deepened their colours, imbued every object with significance and lent the actions of their characters special interest. In the *Physiologus*, allegory made the Created world of God part of the story of human salvation.

The manuscripts explored in this chapter highlight that the *Physiologus* was an extremely malleable text which could be adapted here to a very specific context. Through the miscellanies in which it was included, with their narrow focus on virtue and salvation, and allegory as a method for exploring and achieving these, the *Physiologus* itself became an eschatological work. By refining on the kind of textual collocation discussed in Chapter 3, these compilations had become not only even more focused, but able to move smoothly between an exploration of language, human history, and the wider world as an extension of God. It is in these and similar manuscripts that we see the peak of Carolingian innovation as regards both miscellanies and the natural world.

---

**32** As a scribe, he worked on Florence, Biblioteca Medicea Laurenziana, MS Plut. XVII 20 (a Gospel lectionary) and three British Library manuscripts: Add. 34890 (the Grimbald Gospels, www.bl.uk/manuscripts/FullDisplay.aspx?ref=Add_MS_34890); Cotton Vespasian A.I (www.bl.uk/manuscripts/FullDisplay.aspx?ref=Cotton_MS_Vespasian_A_I); and Harley 603 (www.bl.uk/manuscripts/FullDisplay.aspx?ref=Harley_MS_603). As an artist, he is believed to have worked on Hannover, Museum August Kestner, MS W. M. XXIa 36; London, British Library, MS Arundel 155 (www.bl.uk/manuscripts/FullDisplay.aspx?ref=Arundel_MS_155), which contains his portrait (fol. 133); and York, Minster Library, MS Add. 1. See Gameson, "The Colophon" and *The Scribe Speaks?*, and Karkov, "Writing and Having Written." My grateful thanks to Tina Bawden for drawing my attention to these scribes and references.

**33** Carruthers, *The Book of Memory*, 19.

Chapter 5

# NATURE AND LEARNING IN THE TENTH CENTURY

THE RECEPTION OF the *Physiologus* in the early Middle Ages was bound up with the political and cultural changes associated with the invention of the miscellany manuscript in Western monastic circles and its success as a book format. This context prioritized flexibility of selection in the compilation and use of texts, in order to meet a range of new educational, political, and pastoral challenges. With the emergence of new modes of rule from the late ninth century onwards, the large-scale cultural patronage supported by the stability and political programs instituted the Carolingians gradually waned.[1] It continued in some places, especially East Francia under the Ottonian kings, a period with such sumptuous book art that it has sometimes been called the Ottonian renaissance.[2] But new political priorities also meant new priorities in book production. Some of these will be brought out in this chapter. Since the early medieval perception of the natural world was closely bound up with the written word, these changes had an impact on the cultural history of nature in this period as well.

The full extent of the impact of wider tenth-century political and economic realities on the mobility of scribes, and ecclesiastical agendas as regards educational goals and pastoral care, is still being explored.[3] Part of the debate concerns the feudal revolution, and the influence of Carolingian intellectual and conceptual innovations on later developments in the way that society became structured. There are practical questions to be asked about the fragmentation of the Carolingian empire, its effect on book patronage as well as trade and intellectual networks, the consequences of extreme violence and the new idea of the Peace of God, and the impact of the displacement and movement of people.

The findings of this chapter, which examines a small and specific set of sources, should be seen within our rapidly changing understanding of this context, an analysis of which is beyond the scope of this volume. The copies of the *Physiologus* discussed below do, however, provide evidence of two emerging features that point to wider tenth-century tendencies relating to book culture: firstly, a new trend towards systematic collocation of texts on the natural world; and secondly, confirmation that the *Physiologus* was beginning to be used more systematically as a classroom text. These features indicate that both the interpretation of the *Physiologus* and the perception of nature were different in the tenth century to what they had been previously. These changes were associated with developments in educational practices that led to the

1 A useful introduction to the historiography of political decline in this period is Jong, "The Empire."

2 Schenkluhn, "Ottonische Renaissance."

3 Patzold, *Presbyter*; Greer, Hicklin, and Esders, eds., *Using and Not Using the Past*; Vanderputten, *Dark Age Nunneries*, chap. 4; West, *Reframing the Feudal Revolution*; Patzold, *Das Lehnswesen*; Head and Landes, eds., *The Peace of God*.

decline of the miscellany and the rise of the thematic compilation. This was, in turn, related to the gradual emergence of a new phase in the early medieval reception of the natural world, which prioritized the consolidation of the allegorical and moral learning that has been developed in past centuries. This chapter explores the encoding of these changes in the four extant tenth-century witnesses of the *Physiologus*: Vatican City, Biblioteca Apostolica Vaticana, MS Pal. lat. 1074; Wolfenbüttel, Herzog August Bibliothek, MS 148; Brussels, Bibliothèque royale de Belgique, MS lat. 10066–77; and Paris, Bibliothèque nationale de France, MS Nouv. acq. lat. 455.

## Nature and Orthodoxy

The tenth-century manuscript of the *Physiologus* now in the Vatican Library (Vatican City, Biblioteca Apostolica Vaticana, MS Pal. lat. 1074) is small, both in size and in the number of texts it contains. In this discussion, the manuscript classmark will be used only to refer to the first part of this composite codex, which is also its earliest. It contains four quires, numbered I–IV by a contemporary hand, of which the fourth is now missing the final folio. This indicates that the manuscript was designed as an independent booklet, or even as part of a complete codex. Its Catalan or Iberian origin, suggested in the catalogue by Dorothea Walz,[4] is indicated by the recurrent addition of an initial aspirate *h* to Latin words such as "heleuatur," "hodorem," and "hélephans."[5] The southern European origin of the manuscript is confirmed by the common mark of abbreviation, which has evolved from a horizontal stroke into a tall and narrow 2-shaped mark, though the more standard form also sometimes appears. An example is the phrase "de aquila talem," where the *e* in *de* is suspended using a horizontal stroke across the shaft of the *d* and the *m* in "talem" is suspended using the 2-shaped mark.[6] This mark is not dissimilar to the 2-shaped mark of abbreviation used in Montecassino in the early to mid-Middle Ages, though it is not part of the Beneventan system used there.[7]

Unusually, some of the uses of this manuscript are preserved in its palaeographical features. The text is written in multiple hands with a clear system of *positurae* (punctuation marks), primarily the *punctus versus* and the *punctus elevatus* for the final and medial pauses respectively, but also occasionally the *punctus interrogativus*.[8] The acute accent is also used frequently, apparently over stressed vowels—not long vowels in the manner of the Roman *apex*—though this is debatable in words such as "ínuicem."[9] At the beginning of the text this accent can be found in almost every word; it becomes less frequent—though still on every folio—by the middle, and increases again towards

4 Walz, *Die historischen und philosophischen Handschriften*, 255–7.

5 "Is elevated," "odour," "elephant." Respectively, on fol. 3v, three lines from the bottom; fol. 12r, l. 15; and fol. 17v, l. 13.

6 Fol. 3v, l. 17.

7 Newton, *The Scriptorium and Library*, 182.

8 E.g. fol. 5r, l. 8.

9 "By turns"; fol. 17v, l. 19.

the end. It is not used by every scribe, nor does it appear to have been used systematically for certain words like "animal."[10] The word "Phisiólogus," too, sometimes has an accent on the first *o*, and sometimes doesn't.[11] In some instances a consonant is marked with an accent as if it were a vowel, as with "aúes";[12] but this is rare, and may simply be due to displacement of the accent to the right during writing. These inconsistencies are not associated with a change in scribe; in fact, a single hand can copy the same word with and without an accent in quick succession, as with four instances of the word "necticorax" on fol. 3v.[13] This means that these accents are not systematic rhythmic or stress marks inserted specifically to prepare the text for public reading. Neither are they corrections or later additions, since the shades of ink are evidently a match in accents and letters.[14] Rather, they are guides to pronunciation inserted as and when the scribe remembered to do so, or when they felt it necessary for a particular word. This shows that the manuscript was produced in a milieu in which Latin pronunciation was becoming (or was already) a particular concern, perhaps because knowledge of the language was declining, or in efforts to distinguish it from the vernacular. As Roger Wright has shown, late Latin and early Romance were not entirely separate languages in early medieval Spain and France.[15] Wright suggested that the accents in this particular manuscript were made by scribes trying to help themselves in copying by "thinking aloud."[16] For example, double *i* was usually copied with accents on both letters in words such as "periít" and "páriít": perhaps a reminder that in Latin, doubled vowels are pronounced separately.[17] The presence of these accent and punctuation marks does not exclude the possibility that the text was used for *lectio publica*, but their features indicate that in the first instance, they were writing aids, and perhaps later also functioned as reading aids.

This, then, was the educational and linguistic context of the manuscript. Its extant texts are:

1. *Physiologus*.
2. Descriptions of the pearl and three birds from Isidore's *Etymologiae*.
3. Anonymous commentary on two passages from Daniel.
4. Profession of faith sent to Iberian bishops in 794 after the Council of Frankfurt in response to the Adoptionist controversy.
5. Commentary on the Creed by Venantius Fortunatus.

---

**10** With the accent: fol. 1v, 5 lines from the bottom; fol. 8r, four lines from the bottom; and fol. 9r, l. 16. Without the accent: fol. 9v, l. 22.

**11** Fol. 9v, l. 23; fol. 10r, l. 19.

**12** Fol. 21v, final line.

**13** Lines 3, 4 and 14.

**14** See especially fol. 17v with its different gradations of ink colour.

**15** Wright, *Late Latin and Early Romance*.

**16** In private correspondence of 13 June 2013.

**17** Fol. 17v, l. 12 and l. 21.

6. Smaragdus of Saint-Mihiel, letter on the procession of the Holy Spirit.
7. Charlemagne, letter to Alcuin from 798 on Septuagesima, Sexagesima, Quinquagesima, and Quadragesima Sundays.

These texts can be divided into two groups: the first focused on the natural world, and the second on orthodox faith, with a strong emphasis on Carolingian authority. The arrangement of the first group is focused on the *Physiologus*. Each of its chapters has been modified by additional sentences from the *Etymologiae* on the same animal, stone, or plant, titled *De ethymologiarum*. These additions are present in every chapter, but they are not visually distinguished: the titles are on the same line and in the same colour as the main text. The only exception is the first uncial title, "De libro etimologiarum sancti hysidori." Because of this and several other contemporary additions (described below), this copy of the *Physiologus* is sometimes considered to be the first bestiary, of the B-Is version. Florence McCulloch described it as follows:

> A point of particular interest in this manuscript is that it shows how the large Second Family bestiary could have been composed. At the end of the entire *Physiologus* with its fixed passages from Isidore, the scribe has continued (f. 21v) *De etimologiarun libro*, and has apparently aimlessly copied and somewhat elaborated on Isidore's descriptions of the Psitacus, Ercine, and Coturnix—birds totally unrelated to the old *Physiologus* contents. All that was then needed was a more systematic borrowing from Isidore, which would result in the highly organized Second Family bestiary.[18]

However, there are two issues with this description. Firstly, the scribe made a clear distinction between the *Physiologus* and the appended sections from Isidore. The final chapter, on the *mirmicoleon* or ant-lion, ends on fol. 21r with the phrase, copied in uncial: "Here ends the book. The Physiologus argued well. Amen. From the etymologies."[19] The words "From the etymologies" introduce an extract on the pearl. Another uncial title, "From the book of the etymologies," then introduces the extracts on the parrot, Harz forest bird (*hercinia*) and quail from fol. 21v.[20] While there is clearly a thematic relationship between the *Physiologus*, the pearl, and the three bird-chapters, the latter were not incorporated into the main *Physiologus* text. This copy is not, therefore, an early, as yet disorganized bestiary. Explaining these additions as early attempts to create a bestiary pre-empts the causes that led to the systematic re-organization and expansion of the *Physiologus* into the bestiary in the eleventh century.

Secondly, expansions of and additions to the *Physiologus* were not uncommon, as can be seen in the ninth-century examples of Bern, Burgerbibliothek, MS 318—which contains additional chapters on the cockerel and the horse—and Munich, Bayerische Staatsbibliothek, Clm 19417, which arguably contains an appended chapter on nard. Should we consider these to be proto-bestiaries? The usefulness of labels at this early

---

**18** McCulloch, *Mediaeval Latin and French Bestiaries*, 29n29. Also listed as a bestiary in Henkel, *Studien zum Physiologus*, 28, and Clark, *A Medieval Book of Beasts*, 116.

**19** "Explicit liber bene phisiologus. arguit. amen. de ethimologiarum."

**20** "De etimologiarum libro." *Etymologiae* 16.8.10; 12.7.24, 31 and 64. None of these subjects are included in the B Family *Physiologus*, to which this Vatican manuscript belongs.

stage in the history of the text's gradual alterations is limited. It is more interesting instead to consider this manuscript on its own terms, in order to explore how the textual context of this particular copy affected its function.

Perhaps more importantly, a focus on text-internal additions ignores the wider manuscript context. In this case, it is just as relevant, since a commentary from the Book of Daniel follows the *Physiologus* and contains the same discussions of natural objects. Commencing without a title or other notice of its contents immediately after the added extract on the *coturnix* (quail), the text explains the Latin meanings of words associated with the appearance, rich with gold and precious stones, of the man from Uphaz in Daniel's vision on the bank of the Tigris (Dn. 10.4–5); and of Daniel's dream of four beasts (Dn. 7): a winged lion, a bear, a four-headed winged leopard, and a huge beast with ten horns. The passage on the beasts is significantly longer and is covered in much greater depth in this commentary extract: only the first seventeen or so lines are concerned with the first passage on the man from Uphaz. The anonymous author purposefully linked Daniel with Jacob's son Dan in Genesis 49:17, in which Dan is called a serpent who bites the horse's heels so that the rider falls backwards.[21] The appearance and behaviour of the four beasts are explained and linked with the Second Coming.[22] This commentary therefore adds to the reader's understanding of the allegorical treatment of animals in the Bible, and their relevance to the moral and spiritual well-being of humanity.

The *Physiologus*, the four added extracts from the *Etymologiae* and this partial commentary form a reference collection on biblical beasts, birds, and stones. These are not simply described and explained, but also linked with human beings. The quail, for example, suffers from "falling sickness," like humans: "sicut et omo." Nature is shown to consist of signs that have a significance in human—moral or eschatological—terms. In the exposition, priority is given to those animals, plants, and stones that are mentioned in the Bible. The reader is taught to see an impression of the Bible overlaid on the physical world, highlighting certain features. At the same time, this impression emphasizes the illusoriness of nature, since not all the things mentioned in the Bible can be found in life, and those that can are conduits to a more profound reality. The resulting landscape, which exists largely in the imagination, is powerfully spiritual and memorable.

The other extant texts comprise a Creed, a credal commentary, a letter on the procession of the Holy Spirit and a letter from Charlemagne to Alcuin requesting information about the liturgical function of the named Sundays in Lent. Three of these texts can be linked directly to the circle around Charlemagne at the end of the eighth century and beginning of the ninth. The first of these is the *Epistola de processione Spiritus Sancti* by Smaragdus of Saint-Mihiel. It was one of several works sent by Charlemagne

21 Fol. 23r, l. 14: "et fiat inquid dán serpens in uia" ("And it is said that Dan shall be a snake by the roadside").

22 Fol. 23v, lines 16 and 21–2: "Post hec adueniet dominus...post hec erit celum nouum et terra noua" ("Thereafter the Lord shall come...thereafter there shall be a new sky and a new earth"). This is a reference to Rev. 21:1, "Then I saw a new heaven and a new earth."

to Pope Leo III in response to the *filioque* controversy at the end of 809, to explain the Frankish position.[23] The copy of the *Epistola* in Vatican City, Biblioteca Apostolica Vaticana, MS Pal. lat. 1074 is the earliest that survives.[24] It is supplemented by the credal commentary of Venantius Fortunatus, whose works were well known in the early Middle Ages. The second Carolingian work is the Creed sent to Iberian bishops in 794 after the Council of Frankfurt in response to the Adoptionist controversy. It is a "mosaic" based on the Niceno-Constantinopolitan as well as other credal statements, and may have been composed by Alcuin.[25] The detailed letter by Charlemagne to Alcuin on the Sundays in Lent completes this small grouping. Though the quire has missing leaves, cutting short the letter and any other texts that may have followed, this collection has a clear focus on the correct interpretation of liturgical matters, in particular the fundamental Christian statement of faith, and the observance of the most significant period in the Christian calendar (Lent and Easter). Evidently the creators of Vatican City, Biblioteca Apostolica Vaticana, MS Pal. lat. 1074 had access to a set of texts that represented both Charlemagne's court and its concern with the orthodox faith as expressed in the *filioque* and Adoptionist controversies, which are among the major ninth-century religious disputes.

In an Iberian manuscript, such a collection might have functioned as a re-affirmation that the Iberian Church had renounced Arianism. The adoption of the *filioque* wording in the Creed had been an important anti-Arian event at the Third Council of Toledo in 589. It may also have helped to explain the inclusion of the *filioque* in the Latin rite, spread from Francia during Charlemagne's reign. If this manuscript was indeed produced in Catalonia, such a Carolingian collection of texts can also be explained by Catalonia's positioning of itself as a Frankish domain in the tenth and eleventh centuries. Catalan charters in this period were dated by the reigns of Frankish kings and take as their example Frankish rather than Visigothic documents.[26] The *Physiologus*, and the texts that supplement it, modify this interpretation of the Carolingian collection. Read together, both sets of texts function as a commentary on orthodox belief, with particular attention paid to the avoidance of heresy, and to the Bible as a means of exploring the moral and eschatological significance of the created world.

This manuscript witness of the *Physiologus* demonstrates strong continuity in the text's use during the tenth century. The original codicological whole is incomplete, making it impossible to tell whether its careful arrangement was not already made in some earlier exemplar, and so whether it simply transmits pre-existing textual traditions that reflected earlier concerns. But the act of recopying such a collection itself

**23** See Haugh, *Photius and the Carolingians*. The manuscript from which the work has become known today—Vatican City, Biblioteca Apostolica Vaticana, MS Reg. lat. 190—is now lost: Willjung, "Zur Überlieferung der Epistola."

**24** The next-oldest manuscript is the eleventh-century Vatican City, Biblioteca Apostolica Vaticana, MS Ottob. lat. 339.

**25** Bullough, *Carolingian Renewal*, 191–92. The text of the creed has been analyzed by Wallach, *Alcuin and Charlemagne*, 152–54.

**26** Jarrett, *Rulers and Ruled*, 6. See also Chandler, *Carolingian Catalonia*.

indicates its continued relevance. The combination of texts such as explanations of the Creed, that clarify the basic foundation of the Christian faith and emphasize the orthodox path, and of texts that explain the natural world through allegory and various forms of linguistic interpretation, is one that is repeated over and over in the *Physiologus* manuscripts. The continued collocation of these texts ranks its allegorical and etymological interpretation of the natural world alongside the central tenets of Christianity in terms of knowledge that was thought to be essential. For the first time, however, there is also a consolidation of this knowledge. The texts are not brought together in a convenient volume to fulfill a diversity of practical needs, but rather to form a compilation of existing knowledge on the created world and its relevance to the Christian faith. This difference is small but important, because it indicates that the justification of the natural world as a landscape of signs from God was complete. In the tenth century, it was now time to begin a new phase: collecting these signs together as a body of evidence. In Vatican City, Biblioteca Apostolica Vaticana, MS Pal. lat. 1074, therefore, the importance of the expansion of the *Physiologus* from the *Etymologiae* lies not in its status as an early bestiary, but as perhaps the earliest symptom of eleventh- and twelfth-century encyclopaedic culture. This is expressed even more clearly in another tenth-century *Physiologus* manuscript now in Wolfenbüttel.

## Nature and Virtue

Wolfenbüttel, Herzog August Bibliothek, MS Gud. lat. 148 was copied by a single scribe. A number of features show that this scribe was somewhat inexperienced, though the hand is not unskilled. Ascenders and descenders are generally of a regular height, and the letters are well-formed and regularly placed. However, the ductus is not always entirely certain. There is a marked tendency to slant to the right.[27] This slant is often corrected at the beginning of a sentence with the formation of a *littera notabilior*, but sometimes returns after a few words. It improves significantly later in the manuscript. On some folios (e.g. 14r) the ink varies significantly in colour every few words, emphasizing the places where the scribe dipped the quill. He or she was either not yet adept enough at making the ink flow smoothly from the quill to the page, or had not quite learnt the optimal amount of ink to collect on the nib. Inexpert ink or quill preparation may also have played a role.

The manuscript's contents are as follows:

1. Julian of Toledo, *Prognosticum futuri saeculi*: question-and-answer text based on patristic sentences, mainly from Augustine, Gregory, and Isidore, on life after death and especially the state of the soul before the Resurrection.
2. Short Latin glossary, mainly on Greek words relating to different kinds of discourse, from Isidore of Seville, *Etymologiae* 1.37.22–34.
3. Isidore of Seville, *In Deuteronomium* 16.3, on the eight vices.
4. Phaedrian fables.

27 As on fol. 11v, l. 12, "in sermonibus."

5. Alcuin of York, commentary on the Song of Songs.
6. Note in upper margin recording the appearance of a new star and of a rainbow.
7. *Physiologus.*
8. *Liber monstrorum*, on monsters, and two short texts on whales and serpents.

This is the only early medieval Latin *Physiologus* manuscript whose entire contents appear to have been collocated because they share the same theme: the natural world. The fables, the *Physiologus*, and the *Liber monstrorum*, which make up around half of the total number of folios, contain allegorical and moral interpretations of animal behaviour and appearance, stones, plants, and geographical features. The remaining texts shift the focus of the collection even further towards the meaning of the natural world in terms of its moral value for the life of the soul.

This is especially evident in Julian of Toledo's *Prognosticum futuri saeculi* and in Alcuin of York's commentary on the Song of Songs. The commentary preserves the theme of nature, but in the context of the heavenly Paradise. The *Prognosticum* also discusses Paradise as part of a set of questions and answers on life after death: "How death entered the world, the types of paradise, our age and sex after death, whether the disembodied soul can sense pain, the resurrection of those consumed by animals, whether what the soul senses after death is any more vivid than that which it once sensed in dreams, and whether our corporeal eyes, that now see the sun and moon, will also be the eyes that see God."[28] The text aims to answer practical and philosophical questions and comfort those who fear death, but also to encourage its readers to lead a more moral life, as Julian wrote in his prologue: "May this ordered collection of chapters be in its combined wisdom a mirror wherein our spirit may recognize its very self. For if we consider in careful meditation what we will become in the future, I believe that we would rarely or never sin."[29] The topic of moral living is explored in the *Physiologus*, with its dogmatic tales of the wickedness of the devil and of evil men; in the list of vices; in the fables; and in the *Prognosticum*. A clear link is made between the good life and the afterlife.

The natural world is presented as an essential component of salvation, through the idea that every physical object has a meaning. The short list of Greek words for different kinds of interpretative text from Isidore's *Etymologiae* (allegory, enigma, tropology, parable, paradigm, prose, dialogue, apologetic text) emphasizes the importance of biblical exegesis as the working principle behind every written work. Both the fables and the *Liber monstrorum* transmit important information about the names, history, appearance, location, behaviour, and other characteristics of animate and inanimate things in the world. These required both explanation and interpretation, which this collection as a whole provided. It taught that behind the visible physical world was the invisible, intangible world of God, and understanding the mysteries of the connection between them brought Christians closer to the Creator of both. The advan-

**28** Stork, "A Spanish Bishop," 44.

**29** Trans. in Stork, "A Spanish Bishop," 48.

tage of this was evident, since the Creator was the source of all that is morally good. This manuscript represents the first collection containing the *Physiologus* to systematize received thought about the natural world, and its relationship with allegory and morality, from previous centuries. By doing so, it also consolidated these ideas. Like Vatican City, Biblioteca Apostolica Vaticana, MS Pal. lat. 1074, because of its emphasis on nature as a subject rather than on a method such as allegorical exegesis (as had been the case in earlier *Physiologus* copies), the Wolfenbüttel codex represented a move from practice to preservation. It was a compendium of knowledge, perhaps copied for reference purposes.

## The Natural World in the Classroom

Unlike the Vatican and Wolfenbüttel manuscripts, the tenth-century copy of the *Physiologus* now in Brussels (Brussels, Bibliothèque royale de Belgique, MS lat. 10066–77) is much more difficult to categorize. Part of the reason for this has been its composite structure and unclear origin. Hubert Silvestre, following François Masai, argued that it was made in Rheims or Laon. He showed that the Benedictine theologian Rupert of Deutz used the manuscript during his time at the Abbey of Saint Laurent in Liège (1085–1119), which means that it arrived there before the end of the eleventh century. But Silvestre pointed out that there was no community to speak of at Saint Laurent before the eleventh century, and that the manuscript was not included in the list recording books donated to the Abbey by Prince-Bishop Reginard of Liège between its foundation in 1026, and until 1034.[30] Consequently, the Abbey only acquired this tenth-century manuscript between ca. 1034 and 1085.[31] Its Rheims or Laon origin is made plausible by the fact that links existed between these cities, the Abbey of Saint Hubert near Liège and the Abbey of Saint Laurent, from the end of the eleventh century. Silvestre, too, believed that "close contacts existed at the end of the eleventh century and in the first years of the twelfth century between Saint Laurent, Saint Hubert and other centres associated with Laon and Rheims."[32] These links were expressed principally in the circulation of monks between these communities.

However, Robert Babcock showed that this manuscript was associated with the flourishing of Liègeois schools in the tenth and eleventh centuries, and with the intellectual circles of Liège more generally.[33] This is supported by the manuscript's codicology. Almost all its six independent parts either originated, or were put together, in the Meuse valley. The first three parts were assembled at the Abbey of Saint-Laurent in Liège. The last two parts date from the second half of the tenth century and so cannot

30 Silvestre, "À propos du 'Bruxellensis'"; Masai, "Le manuscrit à miniatures."

31 Silvestre, "À propos du 'Bruxellensis'," 156, suggested that this occurred during the temporary exile of Abbot Berenger of Saint Laurent to the nearby priory of Evergnicourt in 1092–95. He was reinstated at Saint Laurent with the assistance of ecclesiastics from Rheims and Laon.

32 "Entre Saint-Laurent, Saint-Hubert et les centres rémois et laonnois des contacts intimes existaient à la fin du XIe et dans les premières années du XIIe s."

33 Babcock, *The Psychomachia Codex*, esp. 28.

originate at the Abbey of Saint-Laurent (as noted by Silvestre).[34] They are linked by the illustrations in the *Psychomachia* and the *Physiologus*, which were executed by the same artist and traced to the Meuse valley by several art historians.[35]

The eleven quires (fols. 80–162) of the fourth, fifth, and sixth parts of this manuscript were, as Babcock showed, produced in the same context, and gradually filled with text over the following centuries. In the following discussion, the manuscript's classmark shall refer only to these eleven quires. Their contents are as follows:

1. Alphabetized list of Biblical Hebrew names
2. Greek-Latin glossary, letters A–C
3. Biblical glossary (*Expositio dicionum difficilium bibliae*)
4. Prudentius, *Psychomachia*
5. *Physiologus*
6. Gerbert of Aurillac, scholia on Boethius' arithmetic (excerpt)
7. Anonymous commentary on Boethius' *De consolatione Philosophiae*
8. Micon of Saint-Riquier, *Opus prosodiacum*
9. Boethius, commentary on Cicero's *Topica* (excerpts from books 1, 2, 4, and 6) and mythological notes
10. Calcidius, excerpt from *Commentarius in Timaeum*

In addition, *glossae collectae* (also known as the *Glossae Bruxellensis*) to works by Horace, Juvenal, Lucan, and Vergil, with some definitions in Old High German and Old French, are scattered throughout the quires. They provide the most secure evidence that these quires had been brought together by the early eleventh century: they can be dated palaeographically to this period, and their contents are related to the biblical glossary that covers quires two to four.[36]

Babcock has demonstrated that the above texts were copied into this Brussels manuscript with gaps, which were later filled in with other texts; that none of the quires were necessarily planned to be joined into a volume together, since all three of the longer texts (the biblical glossary, the *Psychomachia*, and the *Physiologus*) begin on the recto of a new quire and since the latter two end with blank pages (the end leaf or leaves of the glossary are lost); that the first quire was once the fourth; and that the final quire (containing Micon's *Opus*, the extracts from Boethius, and the excerpts from Calcidius as well as later texts) was probably not originally planned to be included in the volume.

As Babcock concluded, the codicology indicates that the last eleven quires of this Brussels manuscript were used within the same context in the late tenth and early

**34** Fraeys de Veubeke, "Un catalogue de bibliothèque scolaire"; Silvestre, "À propos du 'Bruxellensis'"; Gheyn, *Catalogue des manuscrits*, 2 (1902).

**35** Gaspar and Lyna, eds., *Les principaux manuscrits*, 1:25; Nilgen, "Der Codex Douce 292," 204–7; Swarzenski, *Monuments of Romanesque Art*, 180; Euw, *Rhin-Meuse*, 225.

**36** See De Cesare, "Su di un gruppo di glosse," 439.

eleventh century, but were not planned as a coherent whole, instead circulating separately—perhaps as single booklets. But although the biblical glossary, the *Psychomachia*, and the *Physiologus* were not copied together, they form the earliest core of the manuscript. This core was used for allegorical exegesis, both in novice education and in more advanced scholarship. From the beginning of the eleventh century, additions to these texts had moved the manuscript more clearly into a general schoolroom context, which nevertheless remained geared towards allegorical interpretation.[37] These additions indicate that its compilers were interested in prosody, arithmetic, dialectics, philosophy, and Platonic physics, topics that sit comfortably within the framework of the seven liberal arts as defined by Martianus Capella: grammar, rhetoric, dialectic, arithmetic, geometry, astronomy, and music.[38] In Brussels, Bibliothèque royale de Belgique, MS lat. 10066–77, as in other manuscripts from this period, a late antique, Capellian attitude to education is mingled with Christian texts and concerns. All the manuscript's texts, original and added, were associated with Heriger, schoolmaster and abbot of Lobbes, and Notger, bishop of Liège, both active in the second half of the tenth century.[39] The compilation of Brussels, Bibliothèque royale de Belgique, MS lat. 10066–77 was probably undertaken by one or more of their students and successors.

This manuscript suggests, therefore, that the *Physiologus* was gradually becoming a classroom text by the end of the first millennium. The association of the *Physiologus* with the *Psychomachia* in the tenth century also supports this, since Prudentius' poem is known to have been used in schools even in the Carolingian period. Babcock argued that the *Physiologus* and the *Psychomachia* were not paired because they were useful school-texts, but because of the strength of their allegorical readings. The drawing of the temple of Sapientia in the *Psychomachia* (fol. 137r), for example, probably depicted Solomon's temple instead, providing a teacher of allegory with an image to illustrate the struggle of human souls to enter the temple of Heaven.[40] The utility of the *Physiologus* for allegorical exegesis, especially in conjunction with other texts, is undeniable. As a group, the core of tenth-century texts in this manuscript—the *glossae collectae*, the *Psychomachia* and the *Physiologus*, together with their illustrations—not only supplied readers with literal and allegorical levels of interpretation, but also used allegory to link the natural world with morality. The allegorical hermeneutic method revealed the link between the visible physical world and the invisible world of God. From the allegorical sense, one could proceed to the moral and spiritual, correcting not only belief but also oneself and so coming closer to spiritual salvation. Nevertheless, the texts in this manuscript show a clear shift of this kind of allegoresis from miscellanies into school-room books. Miscellanies had been community books and, most likely, a product of a kind of informal on-the-job training for monks and priests: a practical

37 On this, see also Baxter, "Learning from Nature."

38 Greenfield, *Humanist and Scholastic Poetics*, 20. For an introduction to and bibliography for Martianus Capella, see Teeuwen and O'Sullivan, eds., *Carolingian Scholarship*.

39 Babcock, *The Psychomachia Codex*, 42 and 101.

40 Babcock, *The Psychomachia Codex*, 170–71.

instrument and, at their most innovative, a mirror of the early medieval world. With the gradual formalization of training during the tenth and eleventh centuries, exegesis became more clearly part of the medieval curriculum.

## The Siren and the Centaur

In practice, advanced reading—both the texts in Brussels, Bibliothèque royale de Belgique, MS lat. 10066–77, and of the physical world—could sometimes produce apparently contradictory statements or descriptions. Yet they were all considered to be meaningful. This can be demonstrated using the example of the siren in the *Physiologus*. Its story reads as follows:

> Formerly, Isaiah the Prophet pointed out that the sirens and ass-centaurs and hedgehogs will come into Babylon and dance [cf. Is. 13:21 and 34:14]. Physiologus treated the nature of each one, saying of the sirens that they are deadly animals living in the sea which cry out with odd voices, for the half of them down to the navel bears the figure of a man, while the other half is that of a bird. [They sing a most pleasant song so that through the sweetness of the voice they charm the hearing of men sailing far away and draw them to themselves. By the great sweetness of their extended song they charm the ears and senses of the sailors and put them to sleep. When they see the men lulled by most heavy sleep, they attack them and tear them to pieces. Thus, they deceive men unacquainted with the persuasion of their voices and kill them. Just so are those men deceived who delight in the charms of the world, in games and the pleasures of the theater. Dissipated by tragedies and various melodies and lulled to sleep, these men become the prey of their enemies.]
>
> Likewise, the ass-centaurs from their breasts up bear the figure of a man and that of an ass from there down. "Thus the man of deceitful heart is confused in all his ways" [Jas. 1:8]. Such are the impulses of the souls of wicked merchants; they even sin secretly while gathered together in church. As the Apostle said, "Holding the form of piety, they deny its virtue" [II Tim. 3:5]. And in church their souls are like sheep, yet when they are released from the congregation they become like the herd. "They are like brutish beasts" [Ps. 49:20].
>
> Such beasts, sirens or ass-centaurs, represent the figures of devils.[41]

The siren appears in both the *Physiologus* and in the *glossae collectae* on various books of the Bible which precede the *Psychomachia* and *Physiologus*, and which contain several mentions of animals also found in the *Physiologus*. They are given below, with their chapter number in the Brussels *Physiologus* indicated in brackets (see Appendix II):

91r: *Caradrion* (5)

92v and 95v: *Sirena* (11)

95v: *Onocentaurus* (11)

99v: *Onager* (17)

102v: *Pelicanus* (6), *Nocticorax* (7)

103r: *Aquila* (8)

---

**41** Curley, trans., *Physiologus*, 23.

These unalphabetized *glossae collectae* do not rely on the *Physiologus* for the above creatures. The list derives from other sources. However, the fact that these creatures appear both in the *Physiologus* and in the *glossae collectae* underscores the early medieval desire for information about natural and biblical creatures. The two descriptions of the siren in the *glossae collectae* are of particular interest for the allegorical interpretation of the natural world, because they derive from two different traditions: one that depicts sirens as bird-women and one that depicts them as winged serpents. Despite these contradictory portrayals, they are brought together within the same codicological and textual context. How were these different readings reconciled? To understand this, we must look at their sources and background.

The first and longest gloss is as follows:

> It is said that there were three sirens in the sea, in one part maidens and in one part birds having wings and claws. One made music with her voice, one with a flute, one with a lyre. They drew sailors, enticed by the song, into shipwreck. But in truth they were prostitutes, who, because they seduced passers-by into destitution, were imagined as bringing shipwreck upon them.[42]

This description derives from the *Etymologiae* 11.3.30–31, from which it differs only in minor ways:

> People imagine three Sirens who were part maidens, part birds, having wings and talons; one of them would make music with her voice, the second with a flute, and the third with a lyre. They would draw sailors, enticed by the song, into shipwreck. In truth, however, they were harlots, who, because they would seduce passers-by into destitution, were imagined as bringing shipwreck upon them.[43]

Isidore himself took the above information directly from Servius' *Commentary on Virgil*:

> According to legend the three sirens were part maidens, part birds, the daughters of Achelous the river and Calliope the muse. Of these sirens one made music using her voice, another using the flute, and another using a lyre. At first they dwelt near Faro Point, and after in the islands of Capri. They drew those seduced by their songs into shipwreck. But in fact they were prostitutes, who, because they seduced passers-by into destitution, were imagined as bringing shipwreck upon them.[44]

---

**42** "Sirenę tres finguntur fuisse. in fluctibus ex parte uirgines et ex parte uolucres habentes alas et ungulas quarum una uoce altera tibia tertia lira canebat. quae illectos cantu nauigantes. in naufragium trahebat. Secundum ueritatem autem meretrices fuerunt, quę transeuntes quoniam deducebant ad egestatem. his fictę sunt inferre naufragia."

**43** Isidore of Seville, *The Etymologies*, trans. Barney et al., 245. Original: "Sirenas tres fingunt fuisse ex parte virgines, ex parte volucres, habentes alas et ungulas: quarum una voce, altera tibiis, tertia lyra canebant. Quae inlectos navigantes sub cantu in naufragium trahebant. Secundum veritatem autem meretrices fuerunt, quae transeuntes quoniam deducebant ad egestatem, his fictae sunt inferre naufragia."

**44** "Sirenes secundum fabulam tres, parte virgines fuerunt, parte volucres, Acheloi fluminis et Calliopes musae filiae. Harum una voce, altera tibiis, alia lyra canebat: et primo iuxta Pelorum, post in Capreis insulis habitaverunt, quae inlectos suo cantu in naufragia deducebant. Secundum veritatem meretrices fuerunt, quae transeuntes quoniam deducebant ad egestatem, his fictae sunt inferre naufragia." Servius, *Servii grammatici*, ed. Thilo and Hagen, 2.2:654–55.

Servius' commentary may have been based on an earlier fourth-century commentary by Aelius Donatus, St. Jerome's teacher. The Servius text exists in two versions, long and short; the long version (known as *Servius auctus* or *Servius Danielis*, after its editor Pierre Daniel) is based on a seventh-century expansion of the short version, using material from Donatus that Servius himself originally omitted.[45] It is this longer version that Isidore used.[46] It's apparent that he drew on late antique sources for his information—ones, like Servius' *Commentary*, that were in active circulation and use in Isidore's own time. This, then, is the source of the description of sirens as bird-women.

The second gloss on sirens in Brussels, Bibliothèque royale de Belgique, MS lat. 10066–77 reads: "Sirens. Demons or a kind of monster, or large, crested, and flying dragons."[47] This description derives from Jerome's *Commentary on Isaiah*, in which he described sirens as demons or monsters or large, flying dragons: "Moreover, sirens are called THENNIM (תנים) which we interpret as either demons, or some kind of monsters, or indeed great dragons, who are crested and fly."[48] This description may have come to the Brussels manuscript via two authors who are likely to have known Jerome's text: Isidore, who stated that "moreover in Arabia there are snakes with white wings, called sirens";[49] or, more probably, Eucherius, whose *Instructiones* mention that sirens appear in Isaiah: "Sirens are, in Isaiah, either demons or great dragons, crested as well as flying, as is thought by some."[50] There exist, therefore, two separate traditions of the siren's appearance: classical and Christian. Both these different descriptions, of bird-women and of winged serpents, made it into the *glossae collectae* of Brussels, Bibliothèque royale de Belgique, MS lat. 10066–77.

The iconography of the Brussels *Physiologus* also suggests that the illustrator was aware of both the siren forms. The chapter on the siren (146v), in accordance with the text, depicts three bird-women, one playing a lyre and two others tearing a man apart. The illustration to the chapter on the saw-fish, several folios earlier (142r), depicts a snake or fish-woman, swimming in water, with several sets of wings growing out of her human arms. The remaining details of the scene, however, reveal it as an illustration not to the saw-fish but to the siren: some sailors sleep in a boat next to the winged woman. The double illustration may have been a result of the production process, which was apparently somewhat chaotic, and in which not all the planned images

**45** Martindale, ed., *The Cambridge Companion to Virgil*, 73. This idea was originally suggested by Rand, "Is Donatus's Commentary on Vergil Lost?"

**46** Isidore of Seville, *The Etymologies*, trans. Barney et al., 12.

**47** "Syrenę. demones. uel monstra. quedam uel dracones magni atque cristati. ac uolantes."

**48** Jerome, *In Esaiam*: "Sirenae autem thennim (תנים) uocantur, quas nos aut daemones, aut monstra quaedam, uel certe dracones magnos interpretabimur, qui cristati sunt et uolantes." CCSL 73.

**49** "In Arabia autem serpentes albi sunt cum alis, quae sirenae vocantur." *Etymologiae* 12.4.29.

**50** "Syrenae, in Isaia, daemones, aut dracones magni, cristati pariter ac volantes, ut a quibusdam putatur." Eucherius, *Instructiones*, CCSL 66.

Figure 5.1. Siren. Paris, Bibliothèque nationale de France, MS lat. 12048 (the Gellone Sacramentary), fol. 1v. Courtesy of gallica.bnf.fr / Bibliothèque nationale de France.

were later executed.[51] In such an unrestricted context, could the illustrator have been responding not only to the *Physiologus*, but also to the *glossae collectae*?

Certainly, for the compiler of the *glossae collectae*, these twin descriptions were neither contradictory nor redundant. The two illustrations in Brussels, Bibliothèque royale de Belgique, lat. 10066–77 show that the early medieval makers of the manuscript were aware of two traditions for the siren, and their appearance in the same codicological unit indicates they were not mutually exclusive. The two manifestations of the siren are not explicitly linked with the fourfold method of exegesis anywhere in the manuscript, but they are part of the same mode of thinking: one that accepts a plurality of forms or interpretations (an "emblematic worldview"). The *Physiologus* itself contains several different interpretations of the same animal, such as the three natures of the ant. Supporting this idea is the suggestion made by Jacqueline Leclerq-Marx that the translation of the *Physiologus* into Latin may have been the catalyst for the development of an illustrative tradition in the West—absent in the East—that emphasized allegorical interpretation.[52]

The Brussels sirens and those in Bern, Burgerbibliothek, MS 318 (fol. 13v) are also the first known depictions of fish-tailed women, rather than the bird-women of antiquity. The origins of the fish-woman remain unclear. The Gellone Sacramentary (Figure 5.1, made ca. 780 in Meaux) contains an illustration depicting a woman's head with a fish-tail, but it is decorative and has no connection to the story of the siren.

**51** Babcock, *The Psychomachia Codex*, chap. 5.

**52** Leclercq-Marx, "L'illustration du Physiologus," 151.

Figure 5.2. Zoomorphic initial D. Paris, Bibliothèque nationale de France, MS lat. 13159, fol. 13v. Courtesy of gallica.bnf.fr / Bibliothèque nationale de France.

A similar mermaid appears in a late eighth-century psalter as a zoomorphic initial D (Figure 5.2). The earliest known fish-siren is described in the seventh-century *Liber monstrorum de diuersis generibus,* in which the storyteller paints "a little picture of a sea-girl or siren, which if it has a head of reason is followed by all kinds of shaggy and scaly tales."[53] Fish-tailed beings proliferated in antiquity, but it remains unclear how the early medieval siren went from bird to fish.[54]

The *Physiologus* illustrations could also represent snake or dragon-women, however, following Jerome's description. This possibility has rarely been recognized, though it is a plausible outcome of the illustration of a classical text by Christian art-

**53** "Marinae puellae quandam formulam sirenae depingam, ut sit capite rationis quod tamen diuersorum generum hispidae squamosaeque sequuntur fabulae"; Orchard, *Pride and Prodigies,* 88. Lapidge, "Beowulf," gave the *terminus post quem* for the composition of the *Liber monstrorum* as 636, based on the fact that it borrows extensively from the *Etymologiae,* which were published after Isidore's death in 636. He believed that it can therefore be dated to ca. 650–750. Two copies of the *Liber* were made at Rheims, and one at Fleury, in the ninth and tenth centuries. Of the others, St. Gallen, Stiftsbibliothek, Cod. Sang. 237 is contemporaneous with St. Gallen, Stiftsbibliothek, Cod. Sang. 230, while the *Liber monstrorum* in Wolfenbüttel, Herzog August Bibliothek, MS Gud. lat. 148 is itself bound together with a *Physiologus.* See Leclercq, "De l'art antique à l'art médiéval," 62. The possible origin of the sirens in the Bern *Physiologus* is also discussed in Vieillard-Troiekouroff, "Sirènes-poissons carolingiennes"; Pakis, "Sirens and their Victims"; and Holford-Strevens, "Sirens."

**54** Tselos, "A Greco-Italian School," 8, suggested that the fish-tailed siren in Bern, Burgerbibliothek, MS 318 results from a series of errors by the miniaturist, but this is unlikely. See also Leclercq-Marx, *La sirène dans la pensée* and "La sirène et l'(ono)centaure," as well as Pakis, "Contextual Duplicity" and "Sirens and Their Victims."

Figure 5.3. The viper in Bern, Burgerbibliothek, MS 318, fol. 11r. Used with permission.

ists. A similar depiction of the viper, a creature that is half woman and half serpent, in Bern, Burgerbibliothek, MS 318 (fol. 11r; Figure 5.3), for example, has similar tail-coils to the Brussels siren. Dragons and serpents were often used interchangeably in Christian writing, though the dragon was also seen as a particularly unpleasant and ferocious member of the serpent family (to which Isidore assigned it).[55] Jerome associated sirens with dragons and so by extension with Satan, who as the snake in the Garden of Eden tricked Eve with his sweet voice and words into eating the forbidden fruit.[56] This idea is reflected in another of Jerome's exegetical works, the commentary on Micah: "And they shall weep like the daughters of sirens, for sweet are the songs of the heretics, and with their pleasant voice they deceive the people. Nor can any pass by their singing, but he who has stopped his ear and as it were gone deaf."[57] The association of sirens and onocentaurs with heretics is new to the *Physiologus* in Bern, Burgerbibliothek, MS 318, and may derive from Jerome's commentary on Micah.[58] This suggests that the fish-tailed siren in Bern, Burgerbibliothek, MS 318 could also derive

**55** *Etymologiae* 12.4.4.

**56** This re-interpretation of classical mythology in Christian terms is well-attested; as Holford-Strevens ("Sirens," 21–22) noted, Ambrose in his commentary on Luke (*Expositio in euangelium secundum Lucam* 4.2) interpreted Odysseus bound to the mast as a metaphor for Christ on the Cross, an idea also used by Clement of Alexandria some centuries earlier.

**57** Jerome, *In Michaeam*: "Et lugebunt quasi filiae Sirenarum, dulcia enim sunt haereticorum carmina, et suaui uoce populos decipientia. Nec potest eorum cantica praeterire, nisi qui obturauerit aurem suam, et quasi surdus euaserit." CCSL 76.

**58** Travis, "Of Sirens and Onocentaurs," 34. Version B of the *Physiologus*, for example, describes those lulled by the voices of the sirens as deceived by worldly pleasures; see Carmody, *Physiologus latinus: éditions préliminaires, versio B*.

from Jerome's descriptions, and that the contradiction with the classical description of the siren as a bird-woman in the passage accompanying the illustration arises from the ninth-century miniaturist's knowledge of Jerome's exegetical work, as well as the Vulgate as the standard Bible text.

For the early medieval cultural context of the *Physiologus*, the origins of the siren's different forms are less important than their existence. The illustrations diverge from their text in both manuscripts. The Brussels *glossae collectae*, too, list the siren as both bird and dragon-woman. These glaring divergences are not the result of inattention. They have not been changed or removed because one of the key messages transmitted by the *Physiologus* is that an idea can have more than one meaning. The siren may have been human, fish, bird, dragon or snake, or a combination of these, but what mattered most was that this was recorded: not only so that audiences might learn the siren's different guises and be able to understand biblical passages in which it appears, but also so that they might comprehend its allegorical, moral, and spiritual significance in each incarnation.

## The New Cosmography Full-Circle

One other tenth-century *Physiologus*, generally overlooked in scholarship on this text, also contains evidence of a schoolroom context. This is Paris, Bibliothèque nationale de France, MS Nouv. acq. lat. 455, copied in western France. Extant as an incomplete single quaternion, Paris, Bibliothèque nationale de France, MS Nouv. acq. lat. 455 begins with the chapter on stars from Isidore of Seville's *De naturis rerum*, followed by an abridged Y recension of the *Physiologus*.[59] The quaternion is part of a longer early medieval compilation (the Paris–Orléans collection), which contains computistical and biblical texts useful for the classroom. This collection has a strong thematic emphasis on the natural world, with most of the individual texts exploring animals, the seasons, astronomy and the human lifespan from different viewpoints, but its primary purpose is clearly didactic. This is particularly evident from the script in Paris, MS Nouv. acq. 455. This quaternion will be the primary focus here, since its place in the Paris–Orléans collection is somewhat doubtful (see entry 11 in Appendix I).

The manuscript was clearly written by three student hands learning Caroline minuscule. It has many of the features of Visigothic minuscule, including but not limited to: very tall ascenders; very tall and narrow initials; an *nt* ligature where the *t* looks like it has been flipped upside-down; the second vowel in a diphthong placed above the first vowel; the letter *e* with an open lobe and rising above the x-height (the height of those letters without ascenders) when in ligature; Visigothic minuscule *g*; *f* with a descender and a top stroke that rises above the x-height; a distinctive majuscule *N* resembling an open box or square *u* shape with a long descender on the first vertical stroke; very tall *i* at the beginnings of words; *Q* and *O* often flattened on the top right; and a horizontal stroke added on the left side of majuscule *L*. There is a further range of transitional features. These include three allographs of *a*—u-shaped, single-lobed,

**59** Digitized at https://archivesetmanuscrits.bnf.fr/ark:/12148/cc71179s.

and Caroline (an oblique stroke with a belly on the left); both Visigothic and Caroline *g*; both open and closed lobe of *e*; and an unusual *ur* abbreviation that looks like a wavy macron with a dot underneath (in Visigothic script the dot is more commonly on top). All three scribes are, however, clearly attempting to conform to Caroline minuscule, which is generally evident in the letter morphology and system of abbreviation (including the Caroline 4-shaped abbreviation for *um*).

The manuscript was dated by Delisle to the tenth or eleventh century. The third part of the Paris–Orléans collection in Paris, Bibliothèque nationale de France, MS Nouv. acq. lat. 161, was dated by Bischoff to the ninth or tenth century, or to the first half of the tenth century.[60] Given the transitional nature of the script used, a more precise date is not easy to establish. The letters do not observe the lateral compression of pre-Gothic script, and the accompanying division of the page into two or three columns is absent, as is the reduction in ascender and descender height. However, the script also has a range of eleventh-century features. For example, the down-stroke of ascenders is begun with a wedge shape to the left, though this is irregularly applied here—straight or clubbed ascenders are just as common. This may be a result of the shift from Visigothic, which has very clear left-leaning wedge-shaped ends on ascenders and minims. As a result, the finishing of ascenders and minims is an unreliable guide to the date of the manuscript. Other eleventh-century features include a contemporary—to judge by the ink colour—manicule on fol. 6r (manicules are almost entirely unattested in Western manuscripts before the eleventh century);[61] an oval-shaped *o*; shading, which begins to acquire the contrast observed in pre-Gothic script; and strokes on double *i*, to distinguish them from *u*.[62] But the general scarcity of pre-Gothic features, combined with the presence of oval *o*, suggests that this manuscript dates to the last third or last quarter of the tenth century.

The three scribes—A, B, and C—took turns copying the manuscript, but scribe A's hand is by far the most common. Scribes A and C are significantly less experienced with writing than scribe B, who has a better grasp of letter-shape balance: letters are generally evenly spaced and formed, conform uniformly to the x-height as well as the baseline, and have the same length of ascender and descender. Visigothic influence is nevertheless clear in all three hands. The differences in the hands are most clearly visible on fol. 3v (Figure 5.4), where they took turns: scribe A to midway through line 17; scribe B to midway through line 23; and scribe C to the end of the page.

This quaternion is the earliest firm evidence we have that the Latin *Physiologus* was read by students rather than masters. The palaeographical evidence is supported by the collocation with another text often used in schools, Isidore of Seville's *De naturis rerum*, which in this manuscript also has some interlinear annotations with linguistic and interpretative glosses (top third of fol. 1r)—another feature pointing to a school context. Classroom use need not mean that the *Physiologus* was a rudimentary

60 Delisle, *Catalogue des manuscrits*, 80; Bischoff, *Katalog* 3:242.

61 Steinová, *Notam superponere studui*, chap. 6.

62 E.g. fol. 2r, l. 11. Derolez, *The Palaeography of Gothic Manuscript Books*, 59.

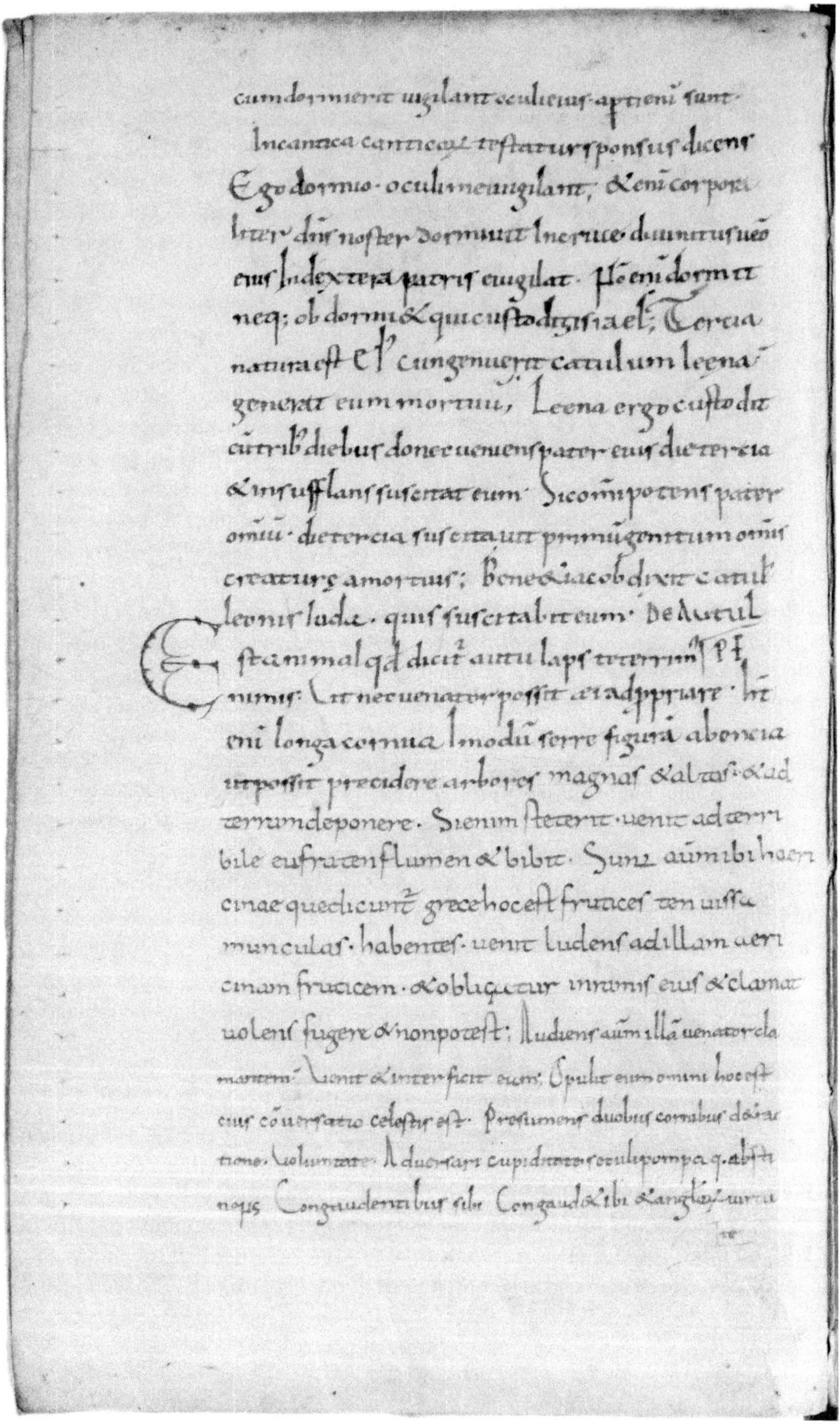

Figure 5.4. All three scribal hands in Paris, Bibliothèque nationale de France, MS nouv. acq. lat. 455, 3v. Reproduced by permission of gallica.bnf.fr / Bibliothèque nationale de France.

text, but rather that it was foundational reading that paved the way for more complex material. The copyists of this quaternion had already been trained to write Visigothic minuscule, which suggests that they were not children; but the difficulties that all three had in maintaining an even ductus indicates that they were still in training. Although poorly developed hands could belong to fully trained priests and even bishops, the fact that these scribes took turns to work on the same quire in a script new to them makes it probable that they were relatively young—perhaps participating in a monastic exchange to learn new skills, as was common in the early Middle Ages. If this quaternion was indeed part of the Paris–Orléans collection, the scribes may have worked in a large centre—such as Fleury, which possessed the collection by the twelfth century—that also had a Breton scribe, and perhaps others from elsewhere in the Carolingian world (see entry 11 in Appendix I). At the very least, the presence of these Visigothic students in a monastic centre in west France associated with Caroline minuscule is a sign of the continued lively intellectual exchange across cultural frontiers that characterized the transmission of the Latin *Physiologus*.

In schoolroom manuscripts like Paris, Bibliothèque nationale de France, MS Nouv. acq. lat. 455, and in thematic compilations like Vatican City, Biblioteca Apostolica Vaticana, MS Pal. lat. 1074, and Wolfenbüttel, Herzog-August-Bibliothek, MS Gud. lat. 148, the concept of subject-based knowledge paved the way for the development of the encyclopedia: the all-encompassing compilation, derived from the Greek idea of the circle of knowledge or *enkýklios paideía* (ἐγκύκλιος παιδεία). Such compilations also make clear what "the natural world" had become: a Christian mirror of the world, combining both observation from nature and imagination about nature. Like the *Physiologus*, the chapter on stars from Isidore's *De natura rerum* in Paris, Bibliothèque nationale de France, MS Nouv. acq. lat. 455 presents verifiable facts alongside Christian morality and cultural literacy. The rise of Orion, for example, signals the beginning of winter and symbolises martyrdom, "for just as they [the Orions] arise in the heavens in wintertime, so martyrs appear in the Church in time of persecution."[63] Thematic miscellanies such as these demonstrate that by the late tenth century, a specific natural semiotics had been fully developed that would continue to exert substantial influence throughout the Middle Ages.

In these texts and their manuscript contexts, the new cosmography discussed in Chapter 1 also came full-circle. It had begun as a diverse range of new texts and objects that amalgamated the Christian worldview with explanations of the structure and function of the physical world. Here, at the turn of the first millennium, the physical world had become fully integrated into the miscellaneous manuscript codex, recognized through thematic compilation as a distinct category of basic knowledge, originally arising out of the liberal arts—especially grammar—and now taught as an important component of those liberal arts, as well as being capable of filling an entire book. The creation of multi-text manuscript volumes on a single theme anticipated the much more ambitious twelfth and thirteenth-century compilations of scholars such as Lambert of Saint-Omer, Bartholomeus Anglicus and Thomas of Cantimpré. This was

**63** Isidore of Seville, *On the Nature of Things*, trans. Kendall and Wallis, 153.

itself a natural development from the fusion of natural allegory, biblical exegesis, and the study of *grammatica* during the preceding centuries.

## The Natural World and Thematic Compilation

Within the monastic book production context of the early Middle Ages, the new tendency towards thematic compilation is a visible sign of a fundamental change in education, and in intellectual culture more generally. From around the tenth century, the wane in Carolingian cultural dominance resulted in a decrease in the emphasis on the trivium, especially grammar, in schools. The subjects of the quadrivium—arithmetic, astronomy, geometry, and music—gained more prominence. Perhaps most importantly, as Claudio Leonardi noted, "schools ceased to be totally identified with culture and began to take on a preparatory and introductory role which continued through the eleventh and the beginning of the twelfth centuries."[64] The close associations of the monastic and early cathedral schools with the court, and with the court-driven rhetoric of scholarly endeavour for the glory of God, had been a Carolingian phenomenon. With its disappearance, the emphasis on the written word as the principal pathway to divine revelation was also set aside to make room for new ideas.

Crucially, the rhetoric around the role of teachers and scholars was no longer primarily to serve their communities by advancing knowledge of the Bible and God's Creation. Instead, they deliberately aimed to reach for high political and ecclesiastical office as the leading thinkers of their day. These tenth-century *magistri* have been seen as the forerunners of modern intellectuals.[65] And this changing role was reflected in their books. While the early miscellanies (discussed in Chapter 3) had been linguistically focused, practical community books, covering a wide range of subjects useful to a wide range of people, including school-masters, the thematic compilations that began to emerge in the tenth century were a symptom of knowledge accumulation. They were no longer pragmatic tools. This does not mean that they were superfluous, but rather that they signalled a shift away from earlier knowledge selection practices. In the Carolingian period, both inherited and new texts had been mined for their practical utility. Conversely, from the tenth century, the available knowledge began to be codified in more set forms. This is the context in which the *Physiologus* appeared at this time.

By the end of the century, its practical application had moved wholly into the schoolroom, as evidenced by Paris, Bibliothèque nationale de France, MS Nouv. acq. lat. 455 and the eleventh-century additions to Brussels, Bibliothèque royale de Belgique, MS lat. 10066–77. The miscellaneous handbooks in which it had found its place in earlier centuries gradually fell out of use or were transformed into subject-based collections. The natural world, too, came to be seen through the lens of these collections, recognizable to us as early natural encyclopedias—a somewhat anachronistic description that nevertheless conveys the general function of these manuscripts.

---

**64** Leonardi, "Intellectual Life," 188.

**65** Vocino, "Migrant Masters."

In this changing context, it is no surprise that the *Physiologus* text itself began to be adapted to suit new purposes. Both the *Dicta Chrysostomi*, which omits stones and plants as well as dividing the animals into beasts and birds, and its shorter descendant the metrical *Physiologus Theobaldi*, were novel eleventh-century versions of the text. They (particularly the *Physiologus Theobaldi*, which was glossed and commented) were popular schoolroom reading.[66] *Physiologus* material continued to be used in schools throughout the later Middle Ages, as attested by the *Aviarium* or *Book of Birds*, composed by Hugh of Fouilloy ca. 1132–52 to teach illiterate lay brothers.[67] From the twelfth century, the *Dicta Chrysostomi* became the foundation of numerous redactions and expansions of the text, whose recensions—which continued to develop until the sixteenth century—are collectively called the bestiary.[68] As noted in the introduction to this book, bestiary material derived largely from Isidore of Seville but also from a wide range of other sources (including the *Aviarium*), and contained as many as 150 chapters, in contrast to the maximum 40–48 chapters of the Greek and Latin *Physiologus*. These bestiaries were encyclopaedic, or else highly decorated texts for courtly entertainment or moral instruction. They did not replace the *Physiologus*, which continued to be copied in the Latin and in various vernaculars; but in the second millennium these *Physiologus* copies were often based on the versions of the Latin text adapted from the eleventh century onwards. The *Dicta Chrysostomi*, for example, was the foundation of the vernacular German *Physiologus*.[69] The first vernacular French version was the *Bestiaire*, a rhymed translation of the *Physiologus* made by Philippe de Thaon ca. 1121.

Other authors writing in French, among them Gervaise (using the *Dicta*), Guillaume le Clerc, and Pierre de Beauvais, continued to popularize *Physiologus* material into the central and late Middle Ages. Further vernacular translations, among them Old Norse and Middle English, also appeared in the period from the eleventh to the fifteenth centuries. These numerous adaptations, reworkings and translations indicate that the cultural context of the *Physiologus* had thoroughly changed from what it had been in the eighth and ninth centuries. The text was no longer read as a companion to the Bible, or as an introduction to natural allegoresis. While it continued to be copied alongside bestiaries even into the early modern period, the later medieval *Physiologus* was a diminished and entirely different beast.

---

**66** The *Physiologus Theobaldi* was printed in Morris, *An Old English Miscellany* (in Bibliography under *Physiologus*). See also Curley, trans., *Physiologus*, xxviii.

**67** Clark, *The Medieval Book of Birds*, 2.

**68** A wide range of scholarship on the bestiary has been published, but there is still much disagreement about the relationship of the bestiary families to each other and to the *Physiologus*. The introductory literature includes Dines, "The Problem"; Stewart, "The Mediaeval Bestiary"; Clark, *A Medieval Book of Beasts*; Faraci, "The Bestiary"; Clark and McMunn, *Beasts and Birds*.

**69** Henkel, *Studien zum Physiologus*.

# CONCLUSION

THE PROCESS BY which Christian learning could be achieved was described by Augustine in *De doctrina christiana* 2.16.23–4:

> As for metaphorical signs, any unfamiliar ones which puzzle the reader must be investigated partly through a knowledge of languages, and partly through a knowledge of things...Ignorance of things makes figurative expressions unclear when we are ignorant of the qualities of animals or stones or plants or other things mentioned in scripture for the sake of some analogy...Just as a knowledge of the habits of the snake clarifies the many analogies involving this animal regularly given in scripture, so too an ignorance of the numerous animals mentioned no less frequently in analogies is a great hindrance to understanding. The same is true of stones, herbs, and anything that has roots. Even a knowledge of the carbuncle, a stone which shines in the dark, explains many obscure passages in scripture where it is used in an analogy; and ignorance of the beryl and of the adamant often closes the door to understanding.[1]

Augustine formulated the notion that language is a system of signs for the Western tradition. His thought about the inherent semiotic richness of the natural world had a deep impact on early medieval ideas—particularly through Isidore of Seville, who cited Augustine more than any other author in his works.[2] Augustine's philosophy of language has been criticized, by Wittgenstein and others, as being simplistic, though he never attempted to put forward a systematic semantic theory in the modern sense.[3] Augustine's idea that things were "learned by signs" was one of the foundations of early medieval ideas about language.[4] As noted in Chapter 1, it was Isidore in the seventh century who recognized the increasing importance of the written word, modifying Augustine's original thought: "Letters are the tokens of things, the signs of words."[5] Writing was not only the record of the spoken word, but also had an intrinsic value or power, which could be obtained by understanding their origin or etymology.[6] The subsequent reception of this idea, up to the tenth century, embedded the concept of intrinsic, hidden meanings into the very idea of text, from the letter to the codex. This

---

1 "In translatis vero signis si qua forte ignota cogunt haerere lectorem, partim linguarum notitia, partim rerum investiganda sunt...Rerum autem ignorantia facit obscuras figuratas locutiones, cum ignoramus vel animantium vel lapidum vel herbarum naturas aliarumve rerum, quae plerumque in Scripturis similitudinis alicuius gratia ponuntur...Ut ergo notitia naturae serpentis illustrat multas similitudines quas de hoc animante Scriptura dare consuevit, sic ignorantia nonnullorum animalium, quae non minus per similitudines commemorat, impedit plurimum intellectorem. Sic lapidum, sic herbarum, vel quaeque tenentur radicibus. Nam et carbunculi notitia, quod lucet in tenebris, multa illuminat etiam obscura librorum, ubicumque propter similitudinem ponitur; et ignorantia berylli vel adamantis claudit plerumque intellegentiae fores." PL 34.46–7 and St. Augustine, *On Christian Teaching*, 43–44.

2 Elfassi, "Presence of Augustine of Hippo in Isidore of Seville."

3 King, "Augustine on Language."

4 "Res per signa discuntur." Augustine, *De doctrina christiana* 1.2.2 (CCSL 32).

5 "Litterae autem sunt indices rerum, signa verborum." *Etymologiae* 1.3.1.

6 Parkes, *Pause and Effect*, 21.

approach ultimately blurred the line between reality and its interpretation. Just as texts, and indeed language itself, were a tapestry of meaning waiting to be read, so the physical world was a symbolic landscape that could be decoded in much the same way.

This symbolic landscape was developed as part of the compilation and re-compilation of texts that took place within miscellanies. The early medieval concept of the natural world had its roots in the Christian philosophical concepts of *natura* and *physica*, but it did not rely on them. As the manuscripts discussed above have shown, its features were quite different. Firstly, it was an intellectual concept which developed gradually out of both inherited and newly composed texts. In addition, its conceptualization was always very local in nature, dependent as it was on the requirements and the intellectual standards of the miscellaneous compilations which carried texts about the natural world. Moreover, when texts about the natural world, such as the *Physiologus*, were included in miscellanies, they were associated with salvation or morality as the essential religious goal. And finally, readers were taught practical ways to interpret these texts, based on allegory and grammar: texts had a very specific application that was bound up with the cultural role of miscellanies.

Early medieval miscellanies therefore offer a fascinating glimpse into a Church-wide effort to make the natural world intelligible as a part of God, and as a gateway to salvation, undertaken on a broad scale across the Frankish empire. The people who produced these miscellanies were highly educated monks, priests, and scholars, who, as we saw in Chapter 1, understood the realities of the environment but who were largely removed from its practicalities. The majority of the population was focused on "reading" nature in the ways that were necessary to ensure the successful growth of plentiful and various enough foods to provide for immediate needs, to pay tax, and to feed animals. People needed to do this continuously through changing seasons, while contending with both plentiful and scarce yields. In comparison to this life, miscellanies reveal a deeply unrealistic approach to the environment. Yet this approach mattered a great deal, as it taught that there was a more important future existence for humans not on this earth. Miscellanies didn't attempt to offer an alternative "reading" of nature to early medieval farmers and craftspeople, but they did provide training to clergy in co-opting the natural world for promoting a strong ideological message.

Throughout these miscellanies, we have also seen a coherent set of intellectual practices being developed, refined, and changed to meet shifting social and political practices. The preceding chapters have offered a close investigation of how this happened in practice during the early Middle Ages. The ways the compilations were brought together and used reflect many of the aims of the Carolingian project: the idea, rooted in the Bible, of betterment for both society and the self; the centrality of education and language-learning; patronage as well as the fostering of intellectual networks as a way of consolidating power. These miscellaneous compilations also reflect one of the major strengths of the early medieval period: horizontal knowledge exchange, embodied in the many texts, and re-contextualised texts, without a named author, which reflect the flowering of interest in knowledge even among those with modest training or skill. Transformative change could be, and was, effected through relatively humble manuscript books, which remain relatively poorly understood.

However, despite the small size of the manuscript corpus investigated in this study, it has been possible to make a number of observations about the organizational structures underpinning early medieval miscellanies. They demonstrate an evolution, from the eighth to the tenth centuries, in practices of compilation, that were closely linked to social and political changes. The precise connections between these changes and the associated compilation practices require further research. Certainly the ideologies of the Carolingian period come through clearly in the miscellanies produced in the late eighth and during the ninth centuries. During this period, as many scholars have already observed, book-making centres cultivated "house" styles that were reflected in books and script. By doing so, they made manuscripts, and writing, into one of the ways it was possible to make statements about institutional identity, as well as faith. Miscellanies fit perfectly within this context. They were well-planned and clearly laid-out books, carefully organized to respond to specific intellectual requirements. As discussed in Chapter 3, this, together with their participation in the conversation about salvation and reform, made them a very precise and powerful tool, and an important category of book in a culture which had raised books to an unprecedented status.

In the tenth century, knowledge about the natural world began to be recorded in different kinds of books. With the fragmentation of the Carolingian empire, the production of manuscript miscellanies declined. Thematic compilations—forerunners of eleventh and twelfth-century encyclopedias—began to emerge. Miscellanies were no longer practical tools, used in the active service of those working on behalf of ecclesiastical rulers, but rather repositories of knowledge. Their makers seem to have aimed to consolidate the learning of previous centuries. This, too, was a gradual process, which has not yet been fully investigated, and which took place at least in part in tenth-century schools. This not only demonstrates a shift in book production practices, but also a shift in how the natural world was understood. It had, by the early Middle Ages, become a fully Christian concept, combining both observation from nature and imagination about nature. Scholars and exegetes had succeeded in conceptualizing the physical world as part of a Christian cosmography. From the tenth century onwards, the natural world became solidified as part of this cosmography. Knowledge about the natural world became its own category of basic knowledge and was now taught in schools as a component of the liberal arts.

Although this was quite a different context from the preceding two centuries, and although the Carolingian promotion of knowledge appears to have been brief, the early medieval period did make a crucial contribution. During this time, inherited knowledge about the natural world was reworked and digested, while new ideas and new textual modes for those ideas were developed. Much of this processing and "re-tooling" for a new age was done, as the *Physiologus* manuscript corpus suggests, in miscellanies. The implications of this rich body of evidence for eleventh and twelfth-century knowledge building remains to be explored.

The miscellanies within which the *Physiologus* is found shed new light on the text itself as well. As its codicological context shows, it was not just a stand-alone work.

It can therefore only be comprehended as part of the manuscript collections in which it was copied, which represented deliberate compilations used for promoting an allegorical understanding of nature as God's moral creation. Although the manuscript evidence suggests that it was a relatively popular text, it was not among the most widely read works in circulation at this time. Nevertheless, the *Physiologus* had the important function, within the context of its manuscript miscellanies, of unfolding Christian natural allegory, morality, and eschatology to its readers. Although its format was relatively simple, this meant that it was easily adaptable and flexible: an important characteristic for a successful text during the Carolingian period.

Although the *Physiologus* circulated in miscellanies, it was never broken up into extracts like the other works with which it was collocated: it seems always to have been copied whole from the exemplar. Its "bite-sized" chapters rendered further breakdown of the text unnecessary. It was either a sufficiently large commentary on the natural world, or an excellent locus for further expansion from other sources, such as the *Etymologiae*, within the miscellany tradition. An implication of this is that extracts were more important than whole texts for miscellanies. By prioritizing extracts, compilations continuously reconstituted early medieval Latin Frankish culture along new lines, re-drawing the boundaries of knowledge anew with each codex. At the same time, the compilers and scholars engaged in this activity did not deny the authority of antique and patristic authors, and their works. They simply sought to interpret their texts better through new contexts.

The manuscripts of the *Physiologus* reveal that the early medieval natural world was interpreted as an allegorical, moral, and spiritual work, which could be read and understood as part of the human journey towards salvation. The *Physiologus* manuscripts played an important role in this discourse, by making nature intelligible in literal, allegorical and grammatical terms. They, and the other manuscripts in the multilingual *Physiologus* traditions, remain a rich source of complex and surprising interpretations about the world.

Dulcis amice, gravem scribendi adtende laborem,
Tolle, aperi, recita, ne laedas, claude, repone.[7]

7 "Dear friend, think of the great labour of writing; take up [this book], open, read, take no offense, close, and put it back." From a poem composed by Walahfrid Strabo ca. 846. Berschin, "Vier karolingische Exlibris," 170.

Appendix I

# DESCRIPTIVE CATALOGUE OF MANUSCRIPTS

THE CATALOGUE PRESENTED here describes the fourteen known manuscript collections containing the Latin *Physiologus* (full or fragmentary) that have been dated to the beginning of the eleventh century or earlier. They are:[1]

1. Bern, Burgerbibliothek, MS 225 + Bern, Burgerbibliothek, MS 233 + Orléans, Médiathèque, MS 313 (266)*
2. Bern, Burgerbibliothek, MS 318*
3. Bern, Burgerbibliothek, MS 611* + Paris, Bibliothèque nationale de France, MS lat. 10756*
4. Brussels, Bibliothèque royale de Belgique, MS lat. 10066–77*
5. Chartres, Médiathèque L'Apostrophe, MS 63 (125) [now lost]
6. Montecassino, Archivio dell'Abbazia, MS 316 and MS 323
7. Munich, Bayerische Staatsbibliothek, MS Clm 14388*
8. Munich, Bayerische Staatsbibliothek, MS Clm 19417*
9. Oxford, Bodleian Library, MS Auct. T.2.23
10. Oxford, Bodleian Library, MS Laud Misc. 129*
11. Paris, Bibliothèque nationale de France, MS Nouv. acq. lat. 1616* + Orléans, Médiathèque, MS 18 (*olim* 15)* + Paris, Bibliothèque nationale de France, MS Nouv. acq. lat. 455*
12. St. Gallen, Stiftsbibliothek, Cod. Sang. 230*
13. Vatican, Biblioteca Apostolica Vaticana, MS Pal. lat. 1074*
14. Wolfenbüttel, Herzog August Bibliothek, MS Gud. lat. 148*

Two other manuscripts deserve a special mention. The first is Cambridge, Corpus Christi College, Parker Library, MS 448.[2] It is an insular manuscript, of probable Worcester origin and twelfth-century provenance in Winchester.[3]

---

1 An asterisk (*) indicates that a full digital surrogate is available.

2 https://parker.stanford.edu/parker/catalog/kk233ff6415. A facsimile and description based on James, *A Descriptive Catalogue*, 2, 360–61 can be found at https://parker.stanford.edu/parker/catalog/kk233ff6415. The manuscript is also described in Budny, *Insular, Anglo-Saxon, and Early Anglo-Norman Manuscript Art*, 219–23; and Gneuss and Lapidge, *Handlist of Anglo-Saxon Manuscripts*, 38. A description based on James and Gneuss/Lapidge can also be found on the DigiPal website at www.digipal.eu/digipal/manuscripts/572. See also Conrad-O'Briain, D'Arcy, and Scattergood, ed., *Text and Gloss*, 173.

3 Dumville, *English Caroline Script*, 56n245. He had previously questioned the generally accepted date and origin of this manuscript in Dumville, "English Square Minuscule," 176.

It is divided into two parts as follows:[4]

> **Part I:** Fols. 1–86, s. x$^{1}$ or x$^{med}$. Written in Anglo-Saxon square minuscule.
>
> **Part II:** Fols. 87–103, s. xi or s. xi$^{2}$. Written in Anglo-Caroline minuscule by a single round hand. Rubrics in rustic capitals.

Part II contains the *Physiologus*. Its dating has caused some disagreement. Gneuss and Lapidge assigned a date of s. xi/xii, but Dumville characterized its script as Style I Anglo-Caroline of s. x/xi.[5] Dumville noted his opinion in a short footnote and did not include a palaeographical discussion. He seems to be alone in suggesting a date as early as the tenth century: the principal scribe of part II also wrote fol. 41v onwards of London, British Library, MS Royal 13.C.V, which T. A. M. Bishop saw as an example of eleventh-century Worcester script.[6] In a personal communication, Julia Crick noted the angled descender of the letter *g* (45 degrees from the baseline), found in post-Conquest manuscripts; and the double height of the *l* in the rustic capitals used as section headings, a feature of post-Conquest production.[7] In her opinion, moreover, the use of the *ra*-ligature in this manuscript is not a twelfth-century feature, but the remaining evidence indicates that the manuscript originated at the end of the eleventh century. On the basis of these observations, Cambridge, Corpus Christi College, Parker Library, MS 448 has been excluded from this study, as it cannot be taken as evidence of an early medieval, pre-bestiary manuscript production context.

The second manuscript is a fragment *sine nomine* which was recovered from a book binding and sold as a fragment of the *Physiologus* in lot 2 of the Christie's valuable books and manuscripts sale on 12 July 2017 (sale no. 14299).[8] The fragment is 22 × 109 mm in size, showing three partial lines of text and part of the margin. The Christie's cataloguer thought it to be of tenth-century origin: "The script can be dated to the tenth century: the *s* is in the long form, the *t* is short with a wide crossbar, the *l* has a clubbed and notched ascender, and the *g* has the form of two closed loops." The sale catalogue description also indicates German origin.

The visible text on the recto is from the B version of the *Physiologus*, chapter XXXIII on the first nature of the elephant: "...[concu]piscentiam uero cogitur insemine / [me fetus habet]. Tempore suo cum uoluerit filios / [procrea]re vadit in orientem cum femina [...]." The previously unidentified text on the verso has been largely rubbed away and is difficult to decipher: "[...]atum / eutro[...] E[...] [...]ti [...] qui nunc seq[...]." Line two is most likely the phrase "Eutropii: Eleati nomen gentis qui nunc Sequani dicuntur" from the *Liber glossarum*. This gloss gives two alternative names for the peo-

4 Gneuss and Lapidge, *Handlist of Anglo-Saxon Manuscripts*, 120.

5 Dumville, *English Caroline Script*, 56n245.

6 Bishop, *English Caroline Minuscule*, 20n1.

7 Personal correspondence.

8 www.christies.com/lotfinder/Lot/an-early-fragment-from-the-physiologus-in-6089213-details.aspx.

ple of the Helvetii ("Eleati"). This is therefore a fragment of the *Liber glossarum* and not the *Physiologus*, and is for this reason also excluded from the below catalogue.[9]

Each entry in the catalogue consists of a description of a putative medieval collection or miscellany. The entries provide the codicological properties and list of contents for each manuscript, followed by an account of script, decoration, and provenance. The word "manuscript" refers to a bound volume held in a modern library. In cases where one or more manuscripts were originally bound together, they are listed as one entry and their relationship indicated by means of a "+."

The principles used to present the manuscripts are as follows:

**1. Titles, classmarks and abbreviations:** The Latin titles used here are those used in manuscript catalogues; where the title is not available because the text has not been identified, an English description of the text has been provided. All abbreviations in *incipits* and other quotations from the text have been fully expanded.

Manuscripts are identified by their classmarks, which consist of city, library, collection (where applicable) and number (*olim* numbers in brackets where applicable). Cities are listed in English, and libraries in one of the principal languages of the countries in which they are located.

**2. Dating:** The date of each manuscript or manuscript section is given in quarter or half-century intervals, or indicating the beginning, middle or end of the century. Bern, Burgerbibliothek, MS 611 + Paris, Bibliothèque nationale de France, MS lat. 10756 is the sole exception, since its (more or less) precise date is known. The standard Latin medieval manuscript catalogue dating system (abbreviations of "saeculo ineunte/medio/exeunte" and fractions) is used to indicate these intervals. For example, the beginning of the ninth century would be indicated as s. $ix^{in}$. All dates are taken from the relevant manuscript catalogues as well as secondary scholarship. Where these disagree, a discussion and date range are provided in the manuscript description elsewhere in this book together with an examination of the palaeography, where relevant. Please note that there exists no consensus as to what precise dates these intervals indicate. They may have meant different things to different cataloguers and should be taken as a rough guide only.

**3. Contents:** Each individual text in the manuscript, whether original or a later addition, is numbered using Arabic numerals. Blank pages or folios are also numbered separately in order to avoid confusing the foliation or pagination. Codicological units broken up across different manuscripts have been listed as one entry, with the texts itemised in the original order as determined from quire numbers. The precise chapter or paragraph number of extracts have been identified where possible. When a text is listed as incomplete rather than as an extract, this indicates that a folio or folios are missing from the manuscript.

**9** See entries EL32 (Helvetii) and EL97 (elephant) in the *Liber glossarum* online: http://liber-glossarum.huma-num.fr/exist/apps/libgloss/index.html.

**4. Collation:** Each manuscript entry begins with an indication of foliation or pagination, origin, dimensions and dating, and a short description of the manuscript. A collation formula is then provided in a separate section. In the formula, quires are listed in Roman numerals, in groups by quire type (ternio, quaternio, etc.). The number of leaves in each is given in an exponent, missing leaves are indicated in brackets, and additions are expressed in a bracket with a + sign. For example: $I^{6}$ + 1 after 6 (wants 1, 5), II–$VI^{8}$. As the example indicates, quires of the same type are grouped together, and the exponent is added only to the final quire of the group. Singleton parchment additions to a codex are shown by means of + I; single paper sheet additions are shown by means of + i. This applies equally to the formula and to the foliation/pagination.

**5. Number of leaves, foliation, pagination:** End-leaves, paste-downs, and additional leaves are indicated both in the collation and in the lists of contents. For foliated manuscripts, the lists of contents indicate recto and verso; for paginated manuscripts, the lists of contents indicate page numbers only.

**6. Material:** The primary support is parchment; any paper leaves are indicated in the foliation/pagination and collation formulae.

**7. Dimensions:** Page dimensions have been supplied for each manuscript. All text is in long lines unless otherwise stated in the list of contents.

**8. Pricking and ruling:** This is noted only in the case of empty but ruled folios. All ruling is drypoint unless otherwise noted.

**9. Script, punctuation, and transcription:** The principal script is Caroline minuscule, unless otherwise noted. Individual texts or titles in a different script are noted in the contents rather than in the manuscript description. A general description of the script of each manuscript is provided after the contents; it excludes textual additions made after the primary production stint of the early medieval codicological unit(s). The letters *u* and *i* are used to represent *v* and *j*, regardless of the form used by the various scripts and hands. These letters are not, however, replaced in transcriptions of post-medieval notes. Spacing has been inserted in *incipits* and other Latin quotations, but the original punctuation and capitalization of the manuscript have been preserved (this will be particularly evident in early modern inscriptions made in some of the manuscripts). Hierarchy of script has not been indicated in transcriptions. *Incipits* generally begin with the title or rubric of a text, where it exists, and continue with its first line. Lacunae are shown by means of a dot within angle brackets. The same brackets are used to supply missing letters where these can be deduced. The word *sic* in square brackets is used to show that the preceding word is quoted exactly as it stands in the original. Corrections are supplied in square brackets within or next to the word being corrected.

**10. Decoration:** The general decorative program and appearance of each manuscript are described in a separate section. Illustrations are noted but discussed only when they are relevant to the manuscript's localization or production circumstances.

**11. Binding:** In every case, bindings are modern or early modern. They are therefore not described or mentioned in this study.

**12. Referencing:** Where a reference to the *Codices Latini Antiquiores* (*CLA*) exists, it is indicated in a footnote. The *CLA* entry, together with the relevant catalogue or catalogues, form the basis of each manuscript description. All information provided has been checked against the physical manuscript unless otherwise noted; minor divergences from printed catalogues of spelling, dimensions, text titles etc. are not indicated. N. R. Ker's system has been followed in showing the relationship between the manuscript and an edition of a text it contains: in the relevant footnote, the word "pr." (printed) indicates that the edition is based on the manuscript, and the word "as" indicates that the manuscript was not used at all.[10] Links have been provided to the online facsimiles of those manuscripts that have been digitised.

It should be noted that many of the catalogues in which the manuscripts below are listed date from the nineteenth or early twentieth century, and that their descriptions were therefore in need of amplification and revision. Occasionally the information they contain has been superseded by the scholarship presented in more recent editions and studies, which contradict the older catalogues or omit certain information depending on their aims. These contradictions have been noted or resolved insofar as this has been possible.

For reasons of space, secondary scholarship has been referenced sparingly, principally in those cases where it is available for the more obscure works.

In very many cases, this catalogue identifies texts and extracts for the first time. These new identifications are too many to note individually. The catalogue especially prioritizes the contents and form of anonymous texts, by describing them rather than simply providing a Latin title. Moreover, by placing these texts alongside smaller excerpts and marginal notes, the catalogue re-focuses attention on the whole manuscript as a textual collection in its own right, and as an important framework for the *Physiologus*.

10 Ker, *Catalogue of Manuscripts Containing Anglo-Saxon*, xxii. See also Ker, *Medieval Manuscripts in British Libraries*.

## 1. Bern, Burgerbibliothek, MS 225 + Bern, Burgerbibliothek, MS 233 + Orléans, Médiathèque, MS 313 (266)[11]

Bern, Burgerbibliothek, MS 225[12]: i + 103 fols. + i | 29 × 18.5–19 cm | 26–32 lines.

Bern, Burgerbibliothek, MS 233[13]: i + i + 13 fols. + i + i | 31 × 19.5 cm | 29 lines.

Orléans, Médiathèque, MS 313[14]: i + i + i + i + 255 pp. + i + i + i + i | 26.7 × 17 cm | 29 lines.

All parts from s. ix$^{in}$, Loire region, possibly the monastery of Saint-Mesmin at Micy near Fleury. Bischoff dated Orléans, Médiathèque, MS 313 more narrowly to s. ix ¼ , while Pellegrin gave a broader date of s. ix.

Bischoff thought it possible that Bern, Burgerbibliothek, MS 233 and Orléans, Médiathèque, MS 313 were part of the same manuscript.[15] In-house codicological notes from the Burgerbibliothek confirm this and add fols. 88–103 of Bern, Burgerbibliothek, MS 225, a composite codex containing ninth, eleventh and twelfth-century quires (parts A (fols. 1–31), B (fols. 32–87), and C (fols. 88–103)).[16]

The composition of the original medieval codex can be reconstructed from the Roman numerals that appear on each quire:

1. Orléans, Médiathèque, MS 313 (pp. 1–223): quires 1–14.
2. Bern, Burgerbibliothek, MS 225 (fols. 88–103): quires 15–16.
3. Orléans, Médiathèque, MS 313 (pp. 224–55): quires 17–18.
4. Bern, Burgerbibliothek, MS 233 (fols. 1–13): quires 19–20.

Quire 19 in Bern, Burgerbibliothek, MS 233 was initially marked XII in brown ink but this was corrected to XVIIII by the addition of rubricated V and II, as noted by Bischoff. Quire 20 lacks the final page and so a quire number. The unit is well-decorated, the writing neat, and the text well laid-out.

Apart from the quire numbers, the evidence that these three manuscripts were initially one lies in their distinctive and colourful decoration, and their similar materi-

---

11 Only the Orléans manuscript is digitized: https://mediatheques.orleans-metropole.fr/ark:/77916/FRCGMBPF-452346101-01A/D18012396.locale=fr.

12 Hagen, *Catalogus codicum Bernensium*; Bischoff, *Katalog der festländischen Handschriften*, 2:351.

13 Bischoff, *Katalog der festländischen Handschriften*, 2:351.

14 Bischoff, *Katalog der festländischen Handschriften*, 2:351; Pellegrin and Bouhot, *Catalogue des manuscrits*, 435–444; Pellegrin, "Membra disiecta Floriacensia (II)," 83. See also Everett, "The Interrogationes de littera," 253. Images of the decorative elements also available online at http://bvmm.irht.cnrs.fr/consult/consult.php?COMPOSITION_ID=3079&corpus=decor.

15 Bischoff, *Katalog der festländischen Handschriften*, 2:351.

16 These notes were generously provided by Florian Mittenhuber of the Burgerbibliothek. See also Mittenhuber, "Die Berner Physiologus-Handschriften." Bern, Burgerbibliothek, MS 225 is also listed as part of Orléans, Médiathèque, MS 313 in Mostert, *The Library of Fleury*, 66.

ality.[17] Moreover, all show water stains and tears to top outer page corners, most obviously on pp. 213–50 in Orléans, Médiathèque, MS 313, fols. 88–94 in Bern, Burgerbibliothek, MS 225 and fols. 3–13 in Bern, Burgerbibliothek, MS 233. This means that quires 14 and 15 (Orléans, Médiathèque, MS 313, p. 223; Bern, Burgerbibliothek, MS 225, fol. 88) have matching torn corners; while quires 16 and 17 (Bern, Burgerbibliothek, MS 225, fol. 1; Orléans, Médiathèque, MS 313, p. 255) have matching water-stained corners. This further suggests that the manuscripts were once bound together.

## COLLATION

Bern, Burgerbibliothek, MS 225: I–III$^{8}$, IV$^{8}$ (8 cut away), V–X$^{8}$, XI$^{6}$ + 2 singletons before 1 and after 6, XII–XIII$^{8}$.

Bern, Burgerbibliothek, MS 233: I$^{8}$, II$^{4}$ + singleton before 1.

Orléans, Médiathèque, MS 313[18]: I–XVI$^{8}$.

## CONTENTS

### Orléans, Médiathèque, MS 313

1. 1–24: Isidore of Seville, *Prooemia in libros ueteris ac noui testamenti.*
   *Inc.* "Incipit liber praemiorum de libris noui et ueteri testamenti. Plenitudo noui et ueteris…"

2. 24–59: Isidore of Seville, *De ortu et obitu patrum.*
   *Inc.* "Incipit uita uel obitus sanctorum patrum qui in domino praecesserunt. Item praefatio. Quorundam sanctorum…"

3. 59–93: Isidore of Seville, *Allegoriae quaedam sacrae scripturae.*
   *Inc.* "Domino sancto ac reuerentissimo fratri orosio esidorus. Quaedam notissima…"

4. 93–180: Pseudo-Jerome (Pseudo-Gregory), *Expositio sancti euangelii.*
   *Inc.* "In christi nomine incipit expositio sancti euangelii edita gregorio papa urbis romae. Matheus sicut in ordine primus ponitur…"

5. 180–8: Alcuin of York, *Epistula* 136 *De gladio.* Missing prologue but with invocation on p. 188.[19]
   *Inc.* "Item de gladio secundum lucam. Est enim locus euangelii…"

---

**17** The arguments are summarised in Pellegrin and Bouhot, *Catalogue des manuscrits*, 442. The authors themselves argue that the contents of Bern, Burgerbibliothek, MS 225 do not fit those of the Orléans manuscript, but this is not a convincing argument for miscellanies.

**18** This manuscript is paginated; it has 256 pages, corresponding to 16 regular quaternions, but the pagination begins on the verso of fol. 1, since the recto is blank.

**19** This invocation is missing from Dümmler's MGH Capit. edition, but is present in several copies of the letter and was thought by Dolbeau to have been its genuine part. See Dolbeau, "Du nouveau sur un sermonnaire de Cambridge," 256. The prologue, in which Alcuin responds to Charlemagne's letter, did not normally circulate with *De gladio* in the ninth century. See Lauwers, "Le glaive et la parole," 4nn16–17.

6. 188–90: Anonymous commentary on Isaiah 11:1. "De septiformis spiritu sancti."[20]
   *Inc.* "Egredietur uirga de radice iesse..."

7. 191–5: Anonymous sermon on Luke 11.5–13.[21]
   *Inc.* "Primo enim sciendum est quod omnes similitudines euangelicae..."

8. 195–204: Isidore of Seville, *Sententiae*, with a condensed question-answer version of books 7–9.3.17 of Isidore's *Etymologiae.*
   *Inc.* "Summum bonum deus est. Quid est deus. Deus igitur spiritus est..."

9. 204–6: First set of questions on "letters" from a northeastern Italian set of *interrogationes et responsiones* written in or before the eighth century.[22]
   *Inc.* "Incipiunt quaestiones de litteris uel singulis causis. Quia uideo te..."

10. 206–7: Second set, based on Isidore's *Prooemia* with details from the *Etymologiae.*
    *Inc.* "Incipiamus de sanctam scripturam. Et a [*sic*] sacrorum numero..."

11. 207–12: Third set, on the books of the New Testament.[23]
    *Inc.* "Item noui testamenti libri isti sunt. Primum euangeliorum..."

12. 212–13: Fourth set, initial section of the questions on Genesis.
    *Inc.* "Incipit questio de libro genesis. Ubi primum in sacris..."

13. 213–18: Pseudo-Jerome, *Chronica sancti Hieronimi.*
    *Inc.* "Hyeronimus ait. De principio caeli..."

14. 218–22: Fifth set, on Exodus and other Old Testament books.
    *Inc.* "Item de exodo. Quare moyses non alium signum..."

15. 222: Ordinal of Christ in the Hibernian Chronological version.[24]
    *Inc.* "Hii sunt grados septem in quibus christus aduenit. Primus lector fuit..."

---

**20** Mistakenly printed as Isaiah 1:11 in Gorman, "The Carolingian Miscellany," 338 (but given correctly on p. 350).

**21** Also known as the anonymous *Omelia in laetanias.* "Laetania" derives from Greek λιτανεία: "supplication," "litany" (in the sense of "litany sung during a procession"), and "in laetanias" refers to the Rogation days, which involved public procession and the singing of litanies. This may have been a sermon intended for delivery during, or associated with, this period in the liturgical year. The sermon also appears in a ninth-century homily collection in Wolfenbüttel, Herzog August Bibliothek, MS Guelf. 102 Weiss. It was inspired by Augustine, *Quaestiones euangeliorum* 2.22 and *Sermo* 105.5–7. For more details see Gorman, "The Carolingian Miscellany," 350.

**22** Gorman, "The Carolingian Miscellany," 351.

**23** Not listed by Gorman as one of the texts in his corpus of exegetical miscellanies. However, it appears in the tenth-century manuscript Cologne, Erzbischöfliche Diözesan- und Dombibliothek, MS 85 (https://digital.dombibliothek-koeln.de/hs/content/titleinfo/168968), which is one of the codices in Gorman's corpus, and it has the appropriate erotematic form. Gorman, "The Carolingian Miscellany," 353.

**24** As Reynolds, *The Ordinals of Christ*, 58.

16. 222–3: *Chronica sancti Hieronymi*, conclusion. Based on Isidore's *Differentiae*, followed by definitions of *liber*, *mundus* and four kinds of *dilectio*. Remainder of page blank.
*Inc.* "Dicamus de sacerdote. Sacerdos christus est..."

### Bern, Burgerbibliothek, MS 225

17. 88r–97r: First third of the *Liber pontificalis*, with the letters of Jerome and Damasus followed by the lives of 35 popes (Linus to Liberius).[25]
*Inc.* "Beatissimo pape damaso hieronimus gloria..."

18. 97r–103v: Jerome, *De uiris illustribus*.
*Inc.* "Hortaris dexter ut tranquillum seque[n]s..."

### Orléans, Médiathèque, MS 313

19. 224: Isidore of Seville, *De ecclesiasticis officiis* chapters 80–1, 84.
*Inc.* "Poenitentiam abolere peccata indubitanter credimus..."

20. 224: Anonymous text on clerical vestments.
*Inc.* "De uestimenta sacerdotale. Quod nostris temporibus dalmatica dicitur..."

21. 226–8: Theodulf of Orléans, *De diuinis scripturis* (extract).
*Inc.* "Dilectissimi fratres, sanctissimi consacerdotes..."

22. 229–30: Isidore, *Etymologiae* 6.16. Preceded by a list of the four ecumenical councils and followed by a list of a further six councils.
*Inc.* "De canonibus conciliorum ex libro ethymologi[arum esidori]. Canon autem græcæ..."

23. 230–2: Anonymous text on the one-tenth tithe in Genesis (Gen. 14:18–20).
*Inc.* "De decimis offerendis in genesi. Et dedit ei decimas..."

24. 232–3: Leo I, *Dicta Leonis episcopi*.
*Inc.* "Dicta leonis episcopi. Credo in deum..."

25. 233–4: Anonymous commentary on Vulgate Matthew 6:24.
*Inc.* "Nemo potest duobus dominis seruire..."

26. 234–6: The same text as on pp. 229–30.
*Inc.* "De canonibus conciliorum. Canon autem grece..."

27. 236: List of church council canons, giving the number of bishops and priests named, and the number of chapters in each.
*Inc.* "Sub constantino augusto imperatore. Canon nicenis..."

---

**25** According to the Felician epitome, but omitting Pope Eusebius. The letters are known to be forgeries made by one or more of the sixth-century compilers of the *Liber*. See Davis, trans., *The Book of Pontiffs*, and McKitterick, *Rome and the Invention of the Papacy*.

28. 236–7: List of the emperors and popes in office during each council, and the number of Roman bishops who were present, followed by a list of those condemned in each council. With rubricated title in an irregular substitution cipher ("ef npnkoicxs dpndipsvm"), which may be read as "de nominibus conciorum [*sic*]."
    *Inc.* "Primum concilium nicenum…"

29. 237–8: Extract from a Roman *ordo* on Septuagesima Sunday.[26]
    *Inc.* "Item de septuagesimo die. Deinde septuagesimo die ante pascha dominica tamen ingredient…"

30. 238–9: Three chapters from the Rule of St. Benedict.

    49: On the Observance of Lent.
    *Inc.* "Item de quadragensimi obseruatione. Licet omni…"

    19: On the Proper Manner of Singing the Psalms.
    *Inc.* "De disciplina psallendi. Ubi cumque…"

    20: On Reverence in Prayer.
    *Inc.* "De reuerentia orationis. Si cum hominibus…"

31. 239: Anonymous summary of the creation of the world.
    *Inc.* "Dogmatum caelesticorum. Ante exordium creaturarum fundauit…"

32. 240–1: Extract from Venantius Fortunatus, *Commentum in symbolum Athanasianum.*[27]
    *Inc.* "Qui si sol aut ignis aliquid inmundum tetigerit…"

33. 242–3: Anonymous commentary on the Lord's Prayer.[28]
    *Inc.* "Oratio dominica propria dicitur quia christus filius dei de inpenetrabile sapientia sua docuit discipulis suis dicens. Pater noster…"

34. 243–55: Gennadius of Marseille, *De ecclesiasticis dogmatibus.*
    *Inc.* "Incipit doctrina dogma ecclesiasticorum secundum niceum concilium. Credimus unum esse deum…"

---

**26** As Batiffol, *History of the Roman Breviary*, 359–60. This Roman *ordo* can be found in full in Montpellier, Bibliothèque de l'École de Médecine, MS H 412 (s. ix). The Montpellier *ordo* is given in *Les "Ordines romani,"* ed. Andrieu, 1 (1931), 211–13, but the Bern-Orléans *ordo* is not.

**27** Pr. Burn, *The Athanasian Creed*, 36.

**28** Listed, together with the other manuscripts that contain this commentary, in Bloomfield, et al., eds., *Incipits of Latin Works*, 633.8699. This is the same commentary as St. Gallen, Stiftsbibliothek, Cod. Sang. 230, item 20.

### Bern, Burgerbibliothek, MS 233

35. 1r: *Physiologus* in 37 chapters (B family).
 *Inc.* "Incipit liber fisioloto exposito de natura animalium uel auium seu bestiarum. i. De natura leonis..."

## SCRIPT

Many different hands writing Caroline minuscule. Bischoff noted that the script of the principal hand is generally very flat and inclined, with frequent use of cc–*a*. This hand is broken up with one that is "brittle" (*spröd*). Quire signatures visible at the end of every quire. Clear hierarchy of script, with use of uncials and rubrication.

## DECORATION

Decorated initials throughout in yellow and purple.

## PROVENANCE

The collection was in Fleury until 1562, when it was taken away by Pierre Daniel along with many other books after part of the monastic library was burned by Calvinist protestants.[29] On Daniel's death in 1603 most of his library was bought at auction by Jacques Bongars (d. 1612) and Paul Pétau (d. 1614), while the rest went to private collectors. Bongars' heir Jakob Graviseth donated his collection to the Bern Burgerbibliothek in 1632, while Pétau's son sold his father's library, including his share of the Fleury books, to Queen Christina of Sweden, via Isaac Vossius, who also kept some volumes for himself. Vossius' manuscripts are now in Leiden and Queen Christina's have been at the Vatican since 1690. Pierre Daniel's Fleury manuscripts are therefore spread between Bern, Leiden, and the Vatican.[30] This history, and the fact that one part of the present collection is now at Orléans, means that it must have been taken apart soon after leaving Fleury—probably by Daniel, who is known to have disassembled some of the manuscripts in his possession—and certainly prior to the donation made by Graviseth to the Bern Burgerbibliothek.

**29** Carey, "The Vatican Fragment," 97.

**30** Pellegrin and Bouhot, *Catalogue des manuscrits*, xxiv.

## 2. Bern, Burgerbibliothek, MS 318[31]

I + 131 fols. + I | 25.5 × 18 cm | 23–4 lines.

Made near Rheims (possibly in Hautvillers) in s. ix$^{2/3}$. With the exception of the additions, corrections and annotations, the manuscript is the work of a single scribe, presumably the Haecpertus named in a colophon on fol. 130r: "Haecpertus me fecit." In the last full quire (fols. 123–130), the written space changes to 190 × 120 mm and the number of lines increases to 24, which, together with a darker shade of ink, might suggest a change of hand. As Roger Collins pointed out, however, the difference is possibly due to the same scribe refining and narrowing their writing in order to fit in more text.[32] A finer pen/nib would have been used for this purpose.[33] The final folio is an added singleton which has a written space of 210 × 130/40 mm and 26 lines, different from the rest of the manuscript, and is written in a pre-Gothic minuscule of the eleventh or early twelfth century.[34]

### COLLATION

$I^2$, $II–XVI^8$, $XVII^8$ + 1 after 8.

### CONTENTS

1. 1r–5r: Antonius, *Life of St. Simeon the Stylite.*
   *Inc.* "Incipit uita sancti symeonis syre serui dei qui in coluna stetit. Sanctus symeon ex utero matris..."
2. 5r–6v: Isidore of Seville, *De ortu et obitu patrum.*[35]
   *Inc.* "Incipit de ortu et obitu patrum. A. Adam pater generis humani..."
3. 7r–22v: *Physiologus*, illustrated, in 24 chapters (C family).[36]
   *Inc.* "Est leo regalis omnium animalium et bestiarum..."
4. 23r–125r: Fredegar, *Chronicle.*
   *Inc.* "In nomine domini nostri iesu christi incipiunt capitula chronici libri primi. i. De inicium mundi..."
5. 41r: List of Egyptian days (s. ix addition).
   *Inc.* "Incipiunt dies egyptiaci quos obseruare oportet. mens ianuarii..."

---

**31** www.e-codices.unifr.ch/en/list/one/bbb/0318. Bischoff, *Katalog der festländischen Handschriften*, 1:122; Homburger, *Die illustrierten Handschriften*, 101–17; Mostert, *The Library of Fleury*, 71; Nees, *Frankish Manuscripts*, cat. no. 55.

**32** Collins, *Die Fredegar-Chroniken*, 66.

**33** Woodruff, "The *Physiologus* of Bern," 229.

**34** Collins, *Die Fredegar-Chroniken*, 66.

**35** Homburger, *Die illustrierten Handschriften*, 57, noted that this text is numbered using Greek numerals.

**36** First listed, with notes, in Sinner, *Catalogus codicus*, 130–36.

6. 125r–130v: Unknown homily on Matt. 17.1–9.
   *Inc.* "Sequuntur sancti euangelii secundi mattheum. In illo tempore...."

7. 125v–130r: Homily attributed to St. Ephrem, known as *In transfigurationem Domini*. Homily on the same passage as in item 6.[37]
   *Inc.* "Incipit tractus eiusdem lectionis effrem. De regionibus messis gaudii..."

8. 130v: *Probationes pennae*, two added lines of verse on the difficulty of writing (s. x or ix) and two medical recipes for palsy and headache (s. x).
   *Inc.* "Ad capitis dolorem. ruta manipulus i; ueronica manipulus i..."

9. 131r–v: On the seven miracles of the world (s. xi–xii$^{in}$). Addition in pre-Gothic script.
   *Inc.* "De septem miraculis mundi. Primum miraculum..."

## SCRIPT

The script used by the scribe of the main text is Caroline minuscule, characterized by lightly clubbed ascenders and a slight slant to the right. According to Bischoff, it is "not typical of Rheims" ("nicht typisch reimsisch").[38] Most chapter and section headings and numbers are written in uncial and rubricated. Brown ink is otherwise used throughout, including *litterae notabiliores* in uncial at the beginning of sentences and large initials in rustic capitals beginning new sections or paragraphs. The punctuation consists of the *distinctio* and the *punctus interrogativus*, used irregularly.

## DECORATION

Skilfully executed illusionistic illustrations accompany most *Physiologus* chapters, showing figures of humans and animals in a landscape usually but not always framed by a thick band of red with a narrower black line on the inside. Apart from the *Physiologus* illustrations, there is no decoration, except an initial S on fol. 1r in light green, yellow and purple wash, with ornament in these colours as well as light red. Bischoff described the style as Franco-Saxon.[39]

## PROVENANCE

A note in an early fifteenth-century hand on fol. 131v indicates the codex then belonged to Ragonde Bachellier, a member of a prominent Rheims family.[40] The same hand has also made notes throughout the codex, particularly in Fredegar's *Chronicle*. By 1603 it had come into the hands of Pierre Daniel; on his death it was acquired by Jacques Bongars, whose heir, Jakob Graviseth, donated it to the Bern Burgerbibliothek along with the rest of Bongars' collection in 1632.

---

**37** Listed as an unusual ("merkwürdig") homily by Ephrem in Siegmund, *Die Überlieferung*, 70.

**38** Bischoff, *Katalog der festländischen Handschriften*, 1:122. Homburger however noted that the writing of the scribe of Bern, Burgerbibliothek, MS 318 is not in conflict, and in many cases coincides, with a list of characteristics of the Rheims script made by Frederick Carey. See Carey, "De scriptura Floriacensi," and Steiger and Homburger, *Physiologus Bernensis*, 20.

**39** Bischoff, *Katalog der festländischen Handschriften*, 1:122.

**40** Homburger, *Die illustrierten Handschriften*, 102.

## 3. Bern, Burgerbibliothek, MS 611 + Paris, Bibliothèque nationale de France, MS lat. 10756[41]

Bern, Burgerbibliothek, MS 611: 153 fols. | ca. 727 | 18–19 × 14–14.5 cm | 16–27 lines.

Paris, Bibliothèque nationale de France, MS lat. 10756: i + i + i + i + 71 fols. + i + i + i | ca. 727 | 20 × 15 cm | 16–27 lines.[42]

This collection has been localized to east Francia with Corbie influence.[43] The Paris codex was identified by Zeumer as one of three manuscripts containing formularies associated with Bourges (text 12b).[44] The date of the collection was established by Lowe from the computus on fols. 94r–96v, which states that 5928 years have passed since the beginning of the world to the present day, bringing it up to 727. Lowe believed this date to be "not incompatible with the palaeography of the manuscript," whose Merovingian minuscule places it within the first three quarters of the eighth century.[45] Lindsay stated that the collection was "written, in part at least, before 721,"[46] although he did not specify why he thought this. In addition, formula no. 6 (part of item 11) is dated to the fourteenth year of the reign of an unnamed king, but this is vague and may have been taken directly from the exemplar text.[47] The Easter table on fol. 96r covers the years 727–48.[48]

A quire of eight leaves has been removed after fol. 72 in Bern, Burgerbibliothek, MS 611 and currently forms fols. 62–69 in Paris, Bibliothèque nationale de France, MS lat. 10756, which also contains other, unrelated early medieval texts.
The collection consists of six separate parts:

I: Fols. 1–86 (quires 1–14, where quire 12 is in Paris, Bibliothèque nationale de France, MS lat. 10756).

II: Fols. 87–93 (quire 15).

III: Fols. 94–113 (quires 16–18).

IV: Fols. 114–115 (quire 19).

---

**41** Bern, Burgerbibliothek, MS 611: CLA VII.604a–e and VII.866–7, with a bibliography on p. 55; digitized at www.e-codices.unifr.ch/en/list/one/bbb/0611. Paris, Bibliothèque nationale de France, MS lat. 10756: CLA V.604; digitized at https://archivesetmanuscrits.bnf.fr/ark:/12148/cc724569. See also Bischoff, *Katalog der festländischen Handschriften*, 1:131, and Mostert, *The Library of Fleury*, no. 1167. I am grateful to David Ganz for generously sharing his comments and notes on this manuscript.

**42** Delisle, *Inventaire des manuscrits latins*, 93.

**43** CLA VII.604e.

**44** MGH Form., 166; Rio, *Legal Practice*. The other two manuscripts are Paris, Bibliothèque nationale de France, MS lat. 4629 and Leiden, Universiteitsbibliotheek, MS BPL 114.

**45** CLA VII.604e.

**46** Lindsay, *Palaeographia Latina* 1:22.

**47** Rio, *Legal Practice*, 111.

**48** Ganz, "In the Circle of the Bishop of Bourges," 274.

V: Fols. 116–145 (quires 20–24).

VI: Fols. 146–153 (quire 25).

A full codicological discussion is provided in the section on Bern, Burgerbibliothek, MS 611 + Paris, Bibliothèque nationale de France, MS lat. 10756 in Chapter 2.[49]

The first four folios are badly damaged with up to three-quarters of the page lost. Modern white parchment has been glued to their outer edges. On some pages, especially in the first quires, the text has faded to near-illegibility. Fols. 116–142 and 145 are palimpsest, as are fols. 143–144. The former contained the Life of St. Sebastian (s. vii) and the latter part of the Vulgate Gospel of Mark (s. $v^2$). Both were probably written in Italy. Traces of damage from attempts to reveal palimpsest text chemically are evident throughout these folios.

## COLLATION

Bern, Burgerbibliothek, MS 611: $I^{10}$ (wants 3, 7), $II^{10}$ (wants 4, 8), $III^2$ (wants 2), $IV^2$, $V^{10}$ (wants 3, 6, 9), $VI^8$, $VII^8$ (wants 8), $VIII–IX^8$, $X^{10}$ (wants 1), $XI^6$, $XII^8$, $XIII^6$, $XIV^8$, $XV^8$ (wants 3), $XVI^6$, $XVII^8$, $XVIII^6$, $XIX^2$, $XX^{10}$ (wants 2, 10), $XXI–XXII^8$, $XXIII^4$ (wants 1, 3), $XXIV^4$, $XXV^8$.

Paris, Bibliothèque nationale de France, MS lat. 10756: Collation unavailable. Based on the digital facsimile, the quires may be as follows: $I–II^8$, $III–IV^6$, $V^8$, $VI^4$, $VII–IX^8$, $X^2$, $XI^6$.

## CONTENTS

### Bern, Burgerbibliothek, MS 611

#### Part I

1. 1r–18v: Latin glossary "Abba" (D–Z). The start of the quire is severely damaged, suggesting that letters A–C and perhaps other texts have been lost.
2. 18v–19r: Latin glossary (A–H).
   *Inc.* "Agellus modicus ager..."
3. 19r: Notes (some upside down) in a darker ink and pre-Caroline hand, listing named creditors or book-borrowers.
   *Inc.* "Deodici <.>..."
4. 19v–20r: Birth lunarium.[50] Contemporary addition.
   *Inc.* "L .i. qui natus fuerit uitalis erit..."

---

49 See also Dorofeeva, "Visualizing Codicologically and Textually Complex Manuscripts."

50 This appears to be the same text as that in the following British Library manuscripts: Cotton Tiberius A.iii (s. $ix^{med}$); Harley 3017 (s. $ix^2$); Cotton Titus D.xxvi, xxvii (s. $xi^1$); and in Biblioteca Apostolica Vaticana, Pal. lat. 235 (s. xi); edited as part of a group of similar lunariums in Liuzza, ed., *Anglo-Saxon Prognostics*, 158–63.

**Part II (from 20r)**

5. 20v–26r: Isidore of Seville, *Etymologiae* 5:39, *De descriptione temporum*.
*Inc.* "Prima aetas mundi in exordio..."

6. 26r–40v: Isidore of Seville, *Etymologiae* 9:2 (sections 2–135), *De gentium uocabulis*.
*Inc.* "De diuisione gencium pauca dicamus..."

7. 40v–41v: Palladius of Hellenopolis (Paradisus), *Historia Lausiaca*, chap. 38 (life of Pachomius).[51] This and the following text copied in a pre-Caroline hand. Contemporary addition.
*Inc.* "De libro qui apellatur paradisus. Quomodo beatus pagomeus..."

**Part III**

8. 42r: Poem on the winds (*Carmen de uentis*).[52] Contemporary addition.
*Inc.* "Quattuor a quadro consurgunt limite uenti..."

9. 42v–72v: Asper, *Ars Asporii*.[53] Pen trials on 57v.
*Inc.* "Partes orationis quot sunt octo..."

10. 72v: Isidore of Seville, *Etymologiae* 1.22 (incomplete). On Tironian notes.
*Inc.* "De notis uulgaribus. Uulgares notas sennius..."

**Paris, Bibliothèque nationale de France, MS lat. 10756**

11. 62r–64r: Five formulae: "Charter recording an exchange of property, the renewal of a precarial grant, a *mandatum* asking someone to record a donation to a monastery in the *gesta municipalia*, a *securitas* guaranteeing a grant to a bishop and a document asking an archdeacon to support a newly ordained priest."[54] Tironian notes in the lower margins of fols. 62r–63r.[55]
*Inc.* "Carta conmutationis. Quod conuenientibus partibus placida..."

---

**51** From text family 1a, according to Wellhausen, *Die lateinische Übersetzung*, 158–63.

**52** Pr. in Alberto, "The Textual Tradition." This poem is similar to Isidore's description of the winds (*De natura rerum* 37, PL 83.1006–8), but it is not clear if it was his source or if it post-dates him. See Obrist, "Wind Diagrams," 49n82.

**53** The section of the text from fol. 71r appears to be unique to Bern, Burgerbibliothek, MS 611. Asper's text is edited in Hagen, *Anecdota helvetica*, 39–61. Homburger, *Die illustrierten Handschriften*, 21–23, perhaps following Lowe, believed that the hand of this text was influenced by Corbie or Luxeuil.

**54** Ganz, "In the Circle of the Bishop of Bourges," 272.

**55** Text edited in full from this manuscript in MGH Form., 169–71; and in Pardessus, "Notice sur les manuscrits," 10–11, 20–22.

12. 64r: a. Jerome, *Contra uigilantium*, chapter XIV in Tironian notes (5 lines).[56] On the rightness of sending alms to monks in the Holy Land. Contemporary addition.
    *Inc.* "Nec nos necamus cunctis pauperibus..."

    b. Formula of a mandate to register a donation in Bourges by the *ordo curiae*. In a pre-Carolingian hand over four drawings of circles.[57] Addition.
    *Inc.* "Anno xiii regni domini nostri illius gloriosissimi regis..."

13. 64v–67r: Table of a 19-year lunar cycle.
    *Inc.* "Kal. ian. luna..."

14. 67r: Section from a paschal cycle, attributed in the manuscript to Victorius of Aquitaine.
    *Inc.* "Parte quaedam de cyclo uicturii..."

15. 67v–68r: Verses on the creation and end of the world, mainly in Tironian notes.[58] Contemporary addition.
    *Inc.* "Cantenus domino christo cantenus honorem..."

16. 68v: Birth lunarium of 30 days. First line partially trimmed away. Contemporary addition.
    *Inc.* "Incipit <.> quis qua<.> luna <.> uerum uitulis..."

17. 68v–69r: Dionysius Exiguus, *Argumenta paschalia* (argument 16 on the rationale of the leap day). Ends with two lines on 69r, the remainder of which is blank and unruled.[59] Contemporary addition.
    *Inc.* "Incipit ratio bisexti. Non illum diem..."

18. 69v: Gregory I, *Regula pastoralis* III:12, on the admonition of the healthy and the sick. In Tironian notes.[60] Contemporary addition.
    *Inc.* "Aliter admonendi sunt incolumes..."

**56** PL 23.350. Edited from Bern, Burgerbibliothek, MS 611 and discussed in Schmitz, "Tironianum."

**57** The identification of this short text and of its script is that of Pardessus. See his "Notice sur les manuscrits," 21, 22.

**58** Edited from Bern, Burgerbibliothek, MS 611 in Chatelain, *Introduction*, 226–29. Thought by Schmitz to be a mixture of extracts from Christian and classical poets, put together by the compiler: "Tironianum," 77.

**59** Cuppo believed Felix of Squillace to be the author of this text. Felix extended the Dionysian cycle by ninety-five years. See Cuppo, "Felix of Squillace."

**60** Discussed in Schmitz, "Tironische Noten."

## Bern, Burgerbibliothek, MS 611

19. 73r–80v: Latin *aenigmata*, known as *Aenigmata Bernensia*.[61] A Greek alphabet with corresponding letter-names and Latin phonemes has been written into the bottom margins of fols. 77v–78r in a contemporary hand; underneath it is a Latin alphabet with numbers assigned to each letter, for the most part in accordance with the Roman numeral system.
    *Inc.* "...[osten]dire possunt patulo sed flammas..."

20. 80v–81r: Metrical sentences, arranged alphabetically (A–T) and written partly in Tironian notes.[62] Ruling partially ignored. Remainder of second page blank. Addition.
    *Inc.* "Corpus ad longe positus..."

21. 81v: List of measurements for Noah's Ark. In uncial.
    *Inc.* "xviiii. Arca noe dicitur habuisse..."

22. 81v–82r: List of various measurements and how they fit into the *stade* (stadium). In uncial.
    *Inc.* "Stadium unum habet atropennes duo semis..."

23. 82r: Excerpt from Jerome, *In Danielem* 10:12–14. Contemporary addition.
    *Inc.* "Noli meaiere [*sic* for *metuere*] danihel qia ex die primo quo posuisti..."

24. 82v–85v: Pseudo-Galen, *Epistula de febribus*.[63]
    *Inc.* "Epistula gallieni de febribus. Multa genera febrium..."

25. 86r: Jerome, *In Danielem* III (extracts). Contemporary addition.
    *Inc.* "Si autem nabuchodonosor quus uolebat..."

26. 86v: Three epitaph formulae for abbesses. Bischoff believed they were based on real epitaphs.[64] Traces of two lines of Tironian notes in a darker ink visible in bottom margin. Addition.
    *Inc.* "Hic tumulat artus hic..."

---

**61** Ed. CCSL 133A.541–610. The Bern manuscript is the earliest witness of this text, extant in at least ten manuscripts, over half of which are from the early Middle Ages. The first riddle, *De lucerna*, begins in the middle of a word, indicating that it continues from a previous page. The missing page is no longer found in Bern, Burgerbibliothek, MS 611 or Paris, Bibliothèque nationale de France, MS lat. 10756. Leipzig, Deutsche Nationalbibliothek, Deutsches Buch- und Schriftmuseum, Klemm-Sammlung, MS fragm. 1966/Bl/95 (s. xi$^{1/4}$) represents a hitherto unknown fragment of the *Aenigmata Bernensia*.

**62** Edited in full from Bern, Burgerbibliothek, MS 611 in MGH Poet. 4:1, 648–51.

**63** Flammini, "L''Epistula' pseudogalenica 'de febribus'"; pr. there and in Hagen, *Zur Geschichte der Philologie*.

**64** Bischoff, "Epitaphienformeln," 150. Bischoff also edited these formulae. The texts are partially translated in Effros, *Caring for Body and Soul*, 128.

27. 87r: Prognostics for a good or bad summer or winter (7 lines) in Tironian notes. The remainder of the page left blank.[65] Contemporary addition.
*Inc.* "Kalendas ianuarias si fuerint die dominico erit annus pessimus..."

28. 87v: Jerome, extracts (*In Matthaeum* V.12, 18, 29, 30; *In Jonam* II.2), of which a few words are written in Tironian notes. Contemporary addition.
*Inc.* "Non quaeras gloriam et non dolebis..."

29. 88r: a. Anonymous compilation of questions and answers on grammar, in 18 lines of Tironian notes. Contemporary addition.
*Inc.* "Ars quid est? Rei cuiusque scientia..."

b. Verse epigram by Taio, Bishop of Saragossa from the beginning of his *Sententiae*.[66] Contemporary addition.
*Inc.* "Epigramma operis subsequencis. Quisquis amas sacram..."

30. 88v–89r: Part of a Roman mass *ordo* containing an exposition of the Lord's Prayer.[67] Ends in Tironian notes. Copied in a pre-Caroline hand. Contemporary addition.
*Inc.* "Dominus et saluator noster discipulis suis..."

31. 89v: *In aurium apertione* Lent service for the induction of catechumens. Slightly adridged. Mostly in Tironian notes.[68] Contemporary addition.
*Inc.* "Euangelium dicitur proprie bona adnunciatio..."

32. 89v: Gregory I, *Moralia in Iob* XII.36 (4 lines) followed by his *Regula pastoralis* III.11 (2.5 lines). Contemporary addition.
*Inc.* "Praefixi dies singulis ab interna Dei presciencia..." and "admonendi sunt igitur simplices, ut sicut fallaciam..."

33. 90r: a. Gregory I, *Regula pastoralis* III.3 (7 lines). Contemporary addition.
*Inc.* "Dicendum est clericis ne prepositorum..."

---

**65** Legendre believed the texts in Tironian notes on fols. 87r-v, 88r, 90v, 91v, 91r and 92r to have been anonymous homilies, but these passages have been identified and transcribed differently by Schmitz. See the relevant folio entry for details, as well as Legendre, *Études tironiennes*, 55, and Schmitz, "Schnellschriftliches," 360–67.

**66** PL 80.730. Edited from Bern, Burgerbibliothek, MS 611 in Hagen, *Carmina medii aevi*, 12. The name "Taio" is sometimes given in its Latin form, "Taius."

**67** PL 74.1091–3. David Ganz noted that this text also appears in the Gelasian Sacramentary (personal correspondence). A detailed study and edition of this *ordo*, though without reference to Bern, Burgerbibliothek, MS 611, is to be found in Odermatt, *Ein Rituale*, 264–66. The *ordo* is listed, though without reference to Bern, Burgerbibliothek, MS 611, in Bloomfield, et al., eds., *Incipits of Latin Works*, 590.8245.

**68** Discussed and edited from other manuscripts in Brown, *The Book of Cerne*, 110–12.

b. Isidore, *Sententiae* III.57. *De oppressoribus pauperum* (10.5 lines). Contemporary addition.
*Inc.* "De oppressoribus pauperum. Pauperum oppressores tunc se sciant grauiori dignos..."

c. Gregory I, *Moralia in Iob* XX.21 (9 lines).
*Inc.* "Magis mala facientibus quam mala patientibus..."

34. 90v–91r: Brief sentences from Jerome (*In Isaiam* 58, 66), Augustine and Isidore, almost entirely in Tironian notes.[69] Contemporary addition.
*Inc.* "Si abstuleris de medio tui catenam..."

35. 91v: Unidentified homily on penitence. Begins with text from Julianus Pomerius, *De uita contemplatiua* II.7. Largely in Tironian notes.[70] Contemporary addition.
*Inc.* "Ab ipsis confessa, nec ab aliis publicata..."

36. 92r: Unidentified homily in Tironian notes. Quotations from Amos and Zacharias; questions on penitence. Contemporary addition.
*Inc.* "In amos propheta..."

37. 92v–93r: Planned program of contents for unit III in uncial. Six lines of text have been added to the bottom of 92v from the *Moralia in Iob* V.22 of Gregory I, on the ant-lion. A gap has been left on 93r between item 15 and item 16.
*Inc.* "I. Ars donati exposita ab aspero..."

38. 93v: Isidore of Seville, *De natura rerum*, wind diagram. Probably intended to accompany the poem on the winds on fol. 42r and perhaps a Carolingian addition. Remainder of page blank except a quire signature, later medieval or modern dry-point doodles (a four-square grid and the word "ggehana" (?)) and an extremely faded line of writing on the bottom.

**Part IV**

39. 94r–96v: *Computus*, given in dialogue form, based on the table of Victorius of Aquitaine.[71]
*Inc.* "Alius interrogare uolo de racione conpoti..."

40. 97r–v: Excerpt from Isidore, *De ecclesiasticis officiis* XI.4–7, XII.1–7.
*Inc.* "Sanctis isidoris in [illegible] officiis..."

---

**69** Pr. in Schmitz, "Notenschriftliches."

**70** Pr. in Legendre, *Études tironiennes*, 48–50 (with notes on the Tironian elements on pp. 85–87).

**71** The literature on this computus is rather extensive and is not listed in full here. The principal works include Borst, *Computus: Zeit und Zahl*, 15, 42, 152; Borst, *Der Streit um den karolingischen Kalender*, 84, 168; Borst, *Schriften zur Komputistik*, 29:348–74; Krusch, "Das älteste fränkische Lehrbuch," 241; Krusch, *Studien zur christlich-mittelalterlichen Chronologie*; Schmid, *Die Osterfestberechnung*, 82. For a fuller bibliography, see Warntjes, *The Munich Computus*, xxiv–xxv.

41. 98r–99r: Gregory I, *epistola* IX.213 to Brunhilde on the ordination of laymen as bishops (same shortened form as in the *Collectio Vetus Gallica*).[72]
*Inc.* "Dominae gloriosissemi adque precellentissimae filiae Bruniheldam…"

42. 99r–100r: Gregory I, *Dialogues* IV.25.2–41.3. On purgatory.
*Inc.* "De anime iustorum si anti restitucione corporum in celo recipiuntur…"

43. 100r: Gregory I, *Moralia in iob* XVIII.54 (from line 5).
*Inc.* "Iob dicit abscondita est…"

44. 100v: Blank, ruled.

45. 101r–113r: Pseudo-Methodius, *De initio et fine saeculi* (*The Apocalypse*).[73] Title in uncial and text in two Merovingian and pre-Caroline hands. Last page ends with three lines of text and one line of pen trials, and is otherwise blank and ruled.
*Inc.* "Incipit facciuncola uel serm [*sic*] sancti methodii…"

46. 113v: Blank, ruled.

47. 114r–115r: Excerpt from Jerome, *Epistula de uirginitate seruanda ad Eustochium* (*epistola* 22.30 on Jerome's dream about the temptation of classical literature). In a distinctive Merovingian chancery minuscule.
*Inc.* "In epistula sancti hyeronimi presbyteri ad eustochium…"

**Part V**

48. 115v: Blank, unruled. Traces of two lines of text visible.

49. 116r: Blank, ruled.

50. 116v–138v: *Physiologus* in 40 chapters (Y family). Heading, table of contents, and first part of chapter on lion in uncial.
*Inc.* "Incipit tractatus episcopi ortodoxi de natura animalium. Incipimu [*sic*] loqui de leone rege bestiarum…"

51. 138v–145v: Twenty-two Greek and African church canons on the election of bishops (the *Collectio Bernensis*).[74] Headings given in uncial for each extract.[75]
*Inc.* "Excarpsum de canonis in canonem serdicensis…"

---

**72** MGH Epp., 199. The version of the letter in Bern, Burgerbibliothek, MS 611 is discussed in Mordek, *Kirchenrecht und Reform*, 108–9.

**73** As Aerts and Kortekaas, ed., *Die Apokalypse des Pseudo-Methodius*. Bern, Burgerbibliothek, MS 611 is considered on pp. 50–4.

**74** See Ganz, "Bureaucratic Shorthand," 69.

**75** Mordek, *Kirchenrecht und Reform*, 107–9. See also Mordek, "Bischofsabsetzungen."

**Part VI**

52. 146r–v: Gargilius Martialis, *Medicinae ex oleribus et pomis* 72–97. On the dietary effects of plum, medlar, mulberry or blackberry, hazelnut, chestnut, almond, and pine nut.
    *Inc.* "Incipit de arte medica ad stomachum…"

53. 146v–147r: Pseudo-Hippocrates, *Dynamidia* II.9–10. On the dietary effect of wild garlic and henbane.
    *Inc.* "Et stringus hoc est uva lopina…"

54. 147r–148v: Unidentified recipes, including those against headache and gout.
    *Inc.* "De capitis dolorem aniti floris…" and "Incipiunt curas podagriorum…"

55. 148v: Pseudo-Oribasius, *Commentaria in aphorismus Hippocratis.*
    *Inc.* "In aforismum. Medicina autem partitur…"

56. 148v–153v: Unidentified recipes, including those against gout and vertigo.
    *Inc.* "Item ad cures podagrorum…" and "Antidotum qui sacr<.> ad virtigi…"

## SCRIPT

Pre-Caroline and Merovingian minuscules (see the palaeographical discussion in Chapter 2).

## DECORATION

With the exception of small flourishes on initial letters, the diagram on fol. 93v, and occasional rubrics, this manuscript is entirely plain.

## PROVENANCE

Bern, Burgerbibliothek, MS 611:

> Owned by Pierre Daniel (1530–1603), whose name is mentioned on a document dated 30.7.1571, formerly attached to the binding and now kept separately as Bern, Burgerbibliothek, MS 847.9.[76] Later owned by Jacques Bongars (1554–1612), whose name is entered in the margins of fols. 5r and 153v. Gifted to the Bern Burgerbibliothek by Jakob Graviseth in 1632.

Paris, Bibliothèque nationale de France, MS lat. 10756:

> A Fleury provenance is probable. Previous owners include Pierre Pithou and possibly Louis Le Peletier de Rosambo.[77] Eventually acquired by the Bibliothèque de Rosny, as evidenced by the ex-libris on the front pastedown with the arms of Marie-Caroline de Bourbon-Sicile, Duchess of Berry (1798–1870). Acquired by the Bibliothèque nationale de France on 14 December 1844 from Simonin de Nancy.[78]

---

**76** www.e-codices.unifr.ch/en/list/one/bbb/0847-09.

**77** On the history of Pithou's library, see Bibolet, "Bibliotheca Pithoeana."

**78** Details from the online catalogue of the Bibliothèque nationale de France at http://gallica.bnf.fr/ark:/12148/btv1b9065920c, and CLA V.604.

## 4. Brussels, Bibliothèque royale de Belgique, MS lat. 10066–77[79]

I + I + I + I + 162 + I + i + i | 25.5 × 17.5 cm | 27 lines |
2 columns (5r–8r, 89v–111v, 157v–160v), 3 columns (80r–85r).

The manuscript was made in northeastern France (fols. 5r–65r) and Belgium (probably the Meuse valley). The binding was restored in 1958, and the nature of this restoration has since made it impossible to determine the quire structure from the physical manuscript.[80] The structure given below is therefore based on the analysis of Hubert Silvestre, who examined the manuscript prior to its restoration and wrote a thorough codicological description.[81]

I was unable to examine this manuscript physically as it was being restored at the time. Unless otherwise stated, all details are therefore taken from the study made of it by Robert Babcock.[82] Additional details are taken from the MA thesis of Valérie Mareschal, who examined the codex in 1997, and from the study made by Silvestre, on which Mareschal's thesis is partly based.[83]

Brussels, Bibliothèque royale de Belgique, MS lat. 10066–77 is a composite made up of six independent codicological units, not counting fol. 88, which is an independent singleton:

1. 4v–65r: Solinus, *Collectanea rerum memorabilium*. Twelfth-century, five hands. Seven quaternios and a final ternio.
2. 66r–70r and 72r–79v: Richard of Saint-Victor, *Liber exceptionum*. Fifteenth-century. Ternio in one hand, quaternio in three. Text parts separated by blank folios.
3. 80r–87v: Glossaries; glosses; antiphons. Tenth–eleventh centuries (see list of contents).
4. 89r–111v: *Glossae collectae*. Fol. 88 was attached to these three quaternios. Tenth-century.
5. 112r–139r: Prudentius, *Psychomachia*. Tenth-century.
6. 140r–162v: *Physiologus*; Gerbert of Aurillac, *Scholium*; commentaries; glossary; *probationes pennae*; fragments. Tenth–eleventh centuries (see list of contents).

---

**79** http://images.kbr.be/multi/KBR_10066-77Viewer/imageViewer.html; Gheyn, *Catalogue des manuscrits*, 2:49–50.

**80** Silvestre, "À propos du 'Bruxellensis'," 132.

**81** Silvestre, "À propos du 'Bruxellensis'."

**82** Babcock, *The Psychomachia Codex*.

**83** Babcock, *The Psychomachia Codex*; Mareschal, "Un témoin latin illustré." Silvestre, "À propos du 'Bruxellensis'." Many grateful thanks to Baudouin Van den Abeele for enabling me to read Mareschal's thesis.

The first and third units were joined in the fifteenth century at the Abbey of Saint Laurent in Liège, as evident from the ownership mark on fol. 2v (*Liber Sancti Laurencii prope Leodium*) and the fifteenth-century list of contents on fol. 4r, which mentions texts in both parts. Eighteenth-century shelfmarks, titles, and an explicit added to fols. 4r, 65r, 80r, and 89r confirm the location as the Abbey of Saint Laurent. The similarity of the hands on fols. 4r, 4v, and 70ra suggests that part II was also added to the manuscript in the late fourteenth or fifteenth century. The remaining units were brought together prior to the eleventh century.[84] The manuscript can therefore be divided into two halves by assembly date.

The *Physiologus* in this manuscript is occasionally referred to by scholars as Brussels, Bibliothèque royale de Belgique, MS lat. 10074 after its classmark in the 1839 catalogue of Marchal, who gave each text in his catalogue its own separate class-number.[85] The classmark 10066–77 represents the two texts that are the first and last in the bound manuscript, according to the convention first used by the Brussels cataloguer van den Gheyn. Silvestre noted that Hanns Swarzenski regularly referred to it as 1066–77.[86]

The dating of some parts of this manuscript continues to be disputed. As O'Sullivan noted in her study of the *Psychomachia*, the present copy has been dated to the tenth century by Gaspar and Lyna, while Steinmeyer and Sievers dated both the poem and the *Physiologus* to the eleventh century, and considered fols. 4–65 to be twelfth century.[87] Van den Gheyn thought fols. 99–163 to be eleventh-century and the rest of the codex twelfth-century. The dating given by Silvestre, which is the most recent and is based on a thorough examination of the codex, has been followed here.

## COLLATION

I–VII$^{8}$, VIII–IX$^{6}$, X–XI$^{8}$, singleton, XII–XVI$^{8}$, XVII$^{6}$, XVIII–XXI$^{8}$.

## CONTENTS

1. 1r–3v: Breviary fragment.[88] From a larger fourteenth-century manuscript. Text partly trimmed away. Here serving as end-leaves (cf. fol. 162$^{bis}$).

### Part I

2. 4r: Twelfth-century table of contents. Original folio cut away and replaced in the fourteenth century. With notes from s. xiv, xv and xviii.
   *Inc.* "In ho volumine continentur subsequencia…"

---

**84** Mareschal, "Un témoin latin illustré," 25.

**85** Marchal, *Inventaire*.

**86** Gheyn, *Catalogue des manuscrits*, 2; Silvestre, "À propos du 'Bruxellensis'," 131.

**87** O'Sullivan, *Early Medieval Glosses*, 61; Steinmeyer and Sievers, *Die althochdeutschen Glossen* 2, 462–6.

**88** Listed by O'Sullivan, *Early Medieval Glosses*, fols. 163r–163v.

3. 4v: Preface to Solinus, *Collectanea rerum memorabilium* (s. xiv). In the same hand as table of contents.[89]
   *Inc.* "Doniam quidam impaciencius pocius..."

4. 5r–65r: Solinus, *Collectanea rerum memorabilium* (s. xii). In two columns on 5r–8r. The explicit on 65r may be an eighteenth-century imitation of twelfth-century script, in the same hand as the notes on 4r, 80r, and 89r. Silvestre identified five separate hands in this text on the following folios: A = 5r–11v; B = 12 and 19; C = 13r–18v, 20r–22r (line 12), 22v (line 19)–23v (line 22); D = 22r (line 13)–22v (line 18); E = 23v (line 22)–65r.
   *Inc.* "Indiculus capitulorum iulii solini rerum collectarium infra scriptarum sic. i. De origine urbis romę..."

### Part II

5. 65v: Quill sketch of a gothic rose window with twenty rays.
   Underneath: "Probatio Incausti / utrum bonum sit annon / Iudica Godefride." Folio otherwise blank. The scribe Godefrid may be the owner of hand E in Solinus' text (4).

6. 66r–70ra: Richard of Saint-Victor, *Liber exceptionum*. This first ternio (fols. 66–71), which is markedly whiter, finer, and smaller than the other quires bearing this text, contains book II, chapters 1–9 and 11–14 (also known as the *Allegoriae in vetus et novum Testamentum* I and II.1–15); the last chapter is unfinished. In two columns.[90] A hand of the late fourteenth or fifteenth century has continued the unfinished text on fol. 70ra. This continuation is also unfinished.
   *Inc.* "Incipiunt allegoriarum obscuritates..."

7. 70rb–71v: Blank, unruled except for margins, which have been ruled with ink.

8. 72r–79v: Richard of Saint-Victor, *Liber exceptionum* (*Allegoriae in vetus et novum Testamentum* IV.4–VII.23). In two columns. Book VII lacks the poem on Solomon and is also unfinished.
   *Inc.* "De altari quod construxit yosue. Iosue subuersis hostibus..."

### Part III

9. 80r–84ra: Hebrew onomastic glossary (A–I, L–P, R–Z). Some 880 entries in three columns. Tenth century. The title *KK Interpretatio dictionum hebraicarum bibliæ* in the upper margin of 80r was made by an eighteenth-century hand roughly imitating the eleventh-century script of the main text.[91]
   *Inc.* "aron fons [*sic* for mons] fortitudinis..."

---

**89** The authenticity of this preface is in doubt, as it exists only in a handful of manuscripts.

**90** Listed by Migne among the dubious works of Hugh of Saint-Victor (PL 175.634–828), though it has now been established to be the work of Richard. For a more detailed discussion of the authorship see the seminal article by Moore, "The Authorship of the *Allegoriae*."

**91** Mareschal, "Un témoin latin illustré," 27.

10. 84ra–85r: Greek–Latin glossary (A–C) in three columns.[92] Tenth century. Contains some Hebrew words. Silvestre noted that this text, if completed, would have extended to fol. 87v.
*Inc.* "Absida lucida..."

11. 85v, 86r, and 87v: *Glossae collectae* on beginning and end of the *Psychomachia*.[93] Eleventh-century, but, according to Silvestre, posterior in date to the subsequent texts on 86v–87r. The scribe of these texts would have preferred facing pages, and the clean fol. 86v to the imperfect fol. 85v, which is marked with ink that has seeped through from 85r. This conclusion is supported by the fact that the second part of the *glossae* (on the end of the *Psychomachia*) is compressed entirely onto fol. 87v, using narrower lines and even marginal space to fit the text. Fols. 85v–86r, on the other hand, have regular spacing. Had the scribe been able to use fols. 86v–87r, the first and second part of the *glossae* would not have been separated in the way that they are, and the second part would have been distributed over more than one folio.
*Inc.* "Glosę in prima et extrema parte sicyomachię [*sic*] prudentis. Septem sunt..."

12. 86v–87r: Canonical office or *historia* of St. Nicholas (fragment with neumes). Eleventh-century. In two columns.[94] Heading in uncial.
*Inc.* "In natale sancti nicholai pontificis ad uersus. O pastor aeterne o clemens..."

13. 87r: Matins antiphon from an office for the Assumption of the Virgin (with neumes). Eleventh-century.[95] Heading in uncial.
*Inc.* "Antiphonum de assumptione sanctae mariae ad noctem canendo. Specialis virgo inter agmina virginum..."

14. 87va: Homily fragment (9 lines).
*Inc.* "Septem columnis septem signantur dona spiritus sancti..."

## Part IV

15. 88r–v: Single leaf isolated from adjoining quaternions and probably once serving as cover. Various small tenth and eleventh-century texts, drawings, neumes, and *probationes pennae*, including:

a. A paraphrased extract from Macrobius, *In somnium Scipionis*.[96]
*Inc.* "Fructum uirtutis in conscientia ponenti..."

---

**92** Described in Silvestre, "Une copie."

**93** From line 10. O'Sullivan, *Early Medieval Glosses*, 61–2, does not record glosses on fol. 87v. Pr. Silvestre, "Aperçu."

**94** Pr. in Jones, *The Saint Nicholas Liturgy*. See also a brief discussion of the authorship of the *historia*, with references, in Silvestre, "À propos du 'Bruxellensis'," 136.

**95** As Silvestre, "Antiennes de matines."

**96** Pr. in Silvestre, "Une adaptation du Commentaire de Macrobe."

b. 2 lines of verse in hexametre (a logogriph or word puzzle for the word *Saturnus*).[97]
*Inc.* "Si caput abstuleris remanebit fortis in armis / et fit homo sanus medium si subtrahis intus."

c. Glossary to works by Vergil, Juvenal, Horace, and Lucan, with Germanic and Romance words. These derive from Wallonia, perhaps even from the region around Liège. Also on fols. 139v, 161va, and 162r. Eleventh-century.[98]
*Inc.* "Omelia sermo ad populum..."

16. 89r–111v: *Glossae collectae* on books of the Bible. Draws extensively on the *Etymologiae*. In two columns from 89v. Title in the upper margin of 89r by a fifteenth-century hand imitating the tenth-century script of the main text. Includes three Greek words on 99vb, 100rb, and 100vb. Tenth-century.
*Inc.* "Hieronimus secundum quosdam super expositionem dictionum difficilium bibliæ. Prologus id est praelocutio prohemium initium dicendi..."

## Part V

17. 112r–139r: Prudentius, *Psychomachia*. Tenth-century, with contemporary glosses. Two twelfth- and thirteenth-century glosses written in the margins of fols. 137v and 138r, on the jewels adorning the temple of the soul (an allegorising tradition that derives from Bede's *Apocalypse*). Its illustrations cover fifty of the poem's tales, but the illustrative program was not completed. The illustrations were made before the text.[99]
*Inc.* "Senex [uetulus] fidelis [bonus] prima..."

---

**97** Pr. in. Silvestre, "Un cliché peu étudié," 256n3. The solution, *Saturnus*, is provided in the other manuscript known to contain this puzzle: Paris, Bibliothèque nationale de France, MS lat. 8088, fol. 195v. The wording there is more complete: "Si caput abstuleris remanebit fortis in armis / et fit homo plenus fuerit si calce minutus / panditur et sanus modium si subtrahis intus." See also Silvestre, "Une adaptation du Commentaire de Macrobe," 95n3. Finally, a similar riddle appears on the first page of the ninth-century manuscript Vatican City, Biblioteca Apostolica Vaticana, MS Vat. lat. 1260, as noted in Arévalo, *Sancti Isidori* 2 (1850), 846:8.

**98** See De Cesare, "Su di un gruppo."

**99** Stettiner, *Die illustrierten Prudentius-Handschriften* I, 61–9. Stettiner saw the manuscript while one of its folios was detached; he consequently incorrectly stated that 46 verses (404–449) were missing between fols. 126 and 127. The missing folio was subsequently reattached as fol. 126bis so as not to interrupt the existing foliation. Silvestre further clarified that the last 23 verses (verses 427–449) on the verso of 126bis were a s. xi–xii addition. This means that 126bis was not the only missing folio, and that this lost folio was already absent by the eleventh or twelfth century. As Silvestre notes, this missing folio need not ever have formed part of the completed manuscript: it may have been lost when the codex was put together or passed over accidentally during the copying stage. He suggests three possible collations for quire XVII, which contains 126bis and the missing folio (labelled 126ter), in Silvestre, "À propos du 'Bruxellensis'," 139–40. O'Sullivan, *Early Medieval Glosses*, 62, noted that the text is missing verses 284–309 and 326–43. See also Woodruff, "The Illustrated Manuscripts," 33–79.

18. 139v: Glossary (see no. 3, 88r). Eleventh-century.
*Inc.* "Hornix gallina..."

## Part VI

19. 140r–156v: *Physiologus*, illustrated, in 37 chapters (A family). Tenth century. Title in uncial. Fol. 146, containing the chapters *De uulpe* and *De unicorni*, is a twelfth-century replacement. Its illustrations cover thirteen chapters, but as with the *Psychomachia* the illustrative program was not completed.
*Inc.* "Incipit physiologus de naturis animalium et bestiarum. De leone rege bestiarum..."

20. 157r: Gerbert of Aurillac, *Scholium ad Boethii arithmeticam institutionem* I.1–3. Eleventh-century.
*Inc.* "Locus in arithmetica hic quem quidam inuictum estimant sic resoluitur..."

21. 157v–158r: Anonymous commentary on Boethius' *De consolatione Philosophiae.*[100] Eleventh-century. In two columns and three hands. Influenced by the commentary of Remigius of Auxerre.[101]
*Inc.* "O qui perpetua mundum rationem gubernas..."

22. 158v–160v: Micon of Saint-Riquier, *Opus prosodiacum*. Tenth-century. In two columns, missing a number of verses.
*Inc.* "Hoc omnes discunt ante alfabeta puellae..."

23. 161r: Boethius, *Commentarius in Ciceronis Topica*. Extracts from beginnings of books I, II, IV, and VI. Eleventh-century.
*Inc.* "Exortatione tua patrici rethorum peritissime..."

24. 161vb: Calcidius, *Commentarius in Timaeum* (extract). Eleventh-century.
*Inc.* "Sunt igitur tam ignis quam..."

25. 161va–162r: Glossary (see no. 3, 88r). Eleventh-century. Several words written in bottom margin of 162r in a fourteenth- or fifteenth-century hand.
*Inc.* "Adumbium est male..."

26. 162v: *Probationes pennae* and drawings.

27. 162bis: Breviary fragment (cf. fols. 1–3).

28. 162ter (paper): Fragment about a possession in the village of Lith, North Brabant province, Netherlands. Fifteenth-century.[102]
*Inc.* "Etira luodien habet quas villa..."

---

**100** After comparing this commentary to one by Remigius of Auxerre, Courcelle argued that it showed Eriugenian influence (Courcelle, *La Consolation de Philosophie*, 27). Mareschal, *Un témoin latin illustré*, 35, followed Courcelle in identifying the author as Eriugena. However, the evidence for Eriugena's authorship is scant. He probably did write such a commentary, but it has not survived. For a brief outline of the arguments see Moran, *The Philosophy of John Scottus Eriugena*, 45.

**101** McCluskey, "Boethius's Astronomy and Cosmology," 70.

**102** Mareschal, *Un témoin latin illustré*, 28, noted that Lith is situated on the Meuse, some 18 km north of Bois-le-Duc (s' Hertogenbosch), on the northern edge of the medieval diocese of Liège.

### SCRIPT

Caroline minuscule; pre-Gothic script and Gothic Textualis in parts I and II.

### DECORATION

Decorated initials; marginal decoration: Animals on 91r+v, 96v, 101r; miniatures.

### PROVENANCE

The manuscript bears marks of ownership from the Abbey of Saint Laurent, Liège (2v, 4r, 80r, all from s. xv–xviii). It also contains library stamps from the Bibliothèque de Bourgogne (3r, 4r) and the Bibliothèque nationale de France (1r, 3r, 161v). In the library of the Dukes of Burgundy by the eighteenth century.[103]

## 5. Chartres, Médiathèque L'Apostrophe, 63 (125)[104]

i + 92 fols. | 31 × 22.5 cm | 2 columns.

This tenth- or eleventh-century manuscript from France, like many others, was destroyed in the bombing of Chartres on 26 June 1944. No images survive. The below list of contents derives from the nineteenth-century catalogue, which lists *explicits* for many of the entries. They are also included here.

i. "Hic est liber Sancti Petri Carnotensis cenobii, quem fratres caritative de suis caritatibus emerunt a quodam Langobardo monacho. Quem si quis furaverit vel ingeniose abstulerit, pereatcum eis qui dixerunt domino Deo: recede a nobis." Note of ownership from s. xiii[in].

1. 3r: Ambrose, *Hexameron*.
   *Inc.* "Tractatus sancti Ambrosii in Exameron. Tantumne opinionis adsumpsisse..."

2. 46r: Ambrose, *Oratio summa et incomprehensibilis natura*.
   *Inc.* "Oratio sancti Ambrosii perpulera. Summa et incomprehensibilis natura..."

3. 48v: Vision of St. Epiphanius on the Mount of Olives and the discovery of St. James' tomb in Jerusalem.
   *Inc.* "Apparitio sanctorum Jacobi apostoli et primi archiepiscoporum atque sacerdotum Symonis et Zachariae..."

4. 50r: Commentary on Genesis, making use of three levels of seeing (literal, moral, spiritual).[105]
   *Inc.* "Commemoratio Geneseos. Populus Irahel, audito decalogo ab ore Domini, rogavit Moysen..." *Expl.* "...pro nobis homo factus est."

---

**103** O'Sullivan, *Early Medieval Glosses*, 62.

**104** Omont, Molinier, Couderc, and Coyecque, *Catalogue général* 11, 30–32. Also previously described in Chasles and Rossard de Mianville, *Catalogue des manuscrits*, 26–27.

**105** Also appears in the late ninth-century manuscript Verona, Biblioteca Capitolare, MS XXVII (25), 99r–138v.

5. 67v: *Physiologus*. Unknown recension (perhaps C family).
   *Inc.* "Leo fortis significat Deus fortis, quia Deus verax est...Item de leone. Leo, cum dormit, oculi eius vigilant..." Further chapters of the *Physiologus* are listed: "De laena, de audolaps, de pelicano, de nocticorax, de aquila, de fenice, de uppupa, de vipera, de serpente, de formica, de onocentauro..."

6. 70r: Isidore of Seville, *Eymologiae*, book I on grammar.
   *Inc.* "Domino et filio Sisebuto Isidorus... Dum te prestantem ingenio... De diebus. Dies est solisorientis presentia..." *Expl.* "...transitu ita lucere. Finit liber primus Isidori."

7. 83r: Isidore of Seville, *Differentiae*.
   *Inc.* "Incipit differentiarum liber sancti Ysidori. Inter Deumet Dominum ita quidem..." *Expl.* "...peperit me mater mea. Explicit..."

8. 88v: a. St. Jerome, epistle 26 to Marcella on nouns used in the Bible. Incomplete.
   *Inc.* "Marcelle Hieronimus. Nuper, cum pariter essemus, non per epistolam..." *Expl.* "...iam constet."

   b. Unidentified text on the ten names of God.
   *Inc.* "Incipit de decem nominibus Dei. Nonagesimum psalmum legens in eodem loco..." *Expl.* "...Athenas, Thebas, Salonas. Expc. de decem nominibus."

   c. Unidentified text on the *diapsalma* (musical pause).
   *Inc.* "Incipit de diapsalmatae. Que acceperis reddendacum fœnore sunt... *Expl.* "... scienciam nesciencium. Explicit."

10. 89v: St. Jerome, epistle 29 to Marcella, explaining the Hebrew words "Ephod bad" and "Teraphim."
    *Inc.* "Incipit de Efot et Terafim. Marcelle Hieronimus. Infronte epistole tue, posueras..." *Expl.* "...fuisse monstratur. Explicit Derafim."

11. 90r: Paschal cycle attributed to Victorius of Aquitaine.
    *Inc.* "Ratio paschalis Theophili, Cesariensis Palestine episcopi, ad ceteros episcopos CCCXVIII... ex jussione Victorii, pape urbis Rome. Cum omnes apostoli..." *Expl.* "...nec major quam XI."

12. 91v: Excerpts from Isidore of Seville, *Etymologiae*, book 13 (on the world and its parts), beginning with chap. 36 on the winds.
    *Inc.* "Ex XIII Ethimologiarum Isidori. Aer commotus et agitatus et per diversis partibus nomina diversa sortitur..." *Expl.* "...utilitatis austri."

## 6. Montecassino, Archivio dell'Abbazia, MS 316 + MS 323[106]

Montecassino, Archivio dell'Abbazia, MS 323: i + 110 pp. | 25.7 × 18 cm | 34–35 lines.

Montecassino, Archivio dell'Abbazia, MS 316: i + i + 142 pp. + i + i | 25.3 × 17.5 cm | 35 lines | pp. 23–37 and 49–142 in two columns.

These two units, 323 and 316 (always given in the discussion in reverse numerical order to prioritize the original quire order) date from the second quarter of the ninth century.[107] Newton believed that they were originally two separate volumes produced as a pair. He gave their origin as France, perhaps because, as he noted, their folio collation—in which hair side faces flesh side—was unusual in southern Italy at that time.[108] The appearance of the script and decoration strongly suggests Italy rather than France, however, so Bischoff's identification has been followed here.

Post-medieval ink foliation is now replaced by post-medieval pencil pagination in both manuscripts. Ruling is clearly visible on most pages. The fore-, top-, and tail-edges of both parts are painted black. The quire numbering is sixteenth- or seventeenth-century; the only original quire number is to be found on p. 54 in Montecassino, Archivio dell'Abbazia, MS 316, where it is marked as "XXI." As this is actually quire XI in the present manuscript, at least ten quires were lost in either or both codices at some stage before the manuscript was paginated. To judge from the whiteness of the last page in both manuscripts, more quires have been lost from the end of both.

### COLLATION

Montecassino, Archivio dell'Abbazia, MS 323: I–IV$^{8}$, V$^{8}$ (wants 1, 8), VI–VII$^{8}$.

Montecassino, Archivio dell'Abbazia, MS 316[109]: VIII$^{6}$, IX (structure unclear), X–XI$^{8}$, XII$^{6}$, XIII$^{10}$, XIV$^{6}$, XV$^{10}$, XVI–XVII$^{6}$.

---

**106** Inguanez, *Codicum Casinensium manu scriptorum catalogus*, 150, 161–3 (cat. no. for MS 316: F; ext. 316 e 458; int. 316 e 333. Cat no. for MS 323: KK; ext. 323 e 452; int. 817). There is a short entry in Bischoff, *Katalog der festländischen Handschriften*, 2:195–6. The *incipits* and *explicits* have been printed in Reifferscheid, *Bibliotheca patrum latinorum italica* III, 411–13 (for Montecassino, Archivio dell'Abbazia, MS 323) and 367–9 (for Montecassino, Archivio dell'Abbazia, MS 316). The palaeography of both codices is also briefly described in Newton, *The Scriptorium and Library*, 373.

**107** Bischoff, *Katalog der festländischen Handschriften*, 2:196.

**108** Newton, *The Scriptorium and Library*, 412 and 373.

**109** The second quire appears to be made up of three singletons and a bifolium, but the binding obstructs a clear identification of the structure.

## CONTENTS

### Montecassino, Archivio dell'Abbazia, MS 323:

1. Blank.
2. 2–34. Isidore of Seville, *De ortu et obitu patrum.*
   *Inc.* "Incipit uita uel obitus sanctorum qui in domino prȩcesserunt..."
3. 34–56. Isidore of Seville, *Allegoriae quaedam sacrae scripturae.*
   *Inc.* "Domino sancto ac reuerentissimo fratri horosio esidorus. Quadam notissima nomina legis et euangeliorum..."
4. 56. Extracts from Jesse of Amiens, *Epistula de baptismo* (on baptism) and Hrabanus Maurus, *De ecclesiastica disciplina* (on the imposition of hands by a bishop).
   *Inc.* "Ideo nouissime per impositione manus a summo sacerdote..."
5. 56–7. Explanation of the mass and its purpose, with extracts from Hrabanus Maurus, *De rerum naturis* IV.1 (on humanity).
   *Inc.* "Ordo misse a sancto petro apostolum institutus est..."
6. 57–61. Explanation of Greek words on baptism and of a basic credal statement, to be recited by children in the children's question and response section of an Alemannic Holy Saturday service (*Ordo in Sabbato sancto).* On baptism.
   *Inc.* "De illa uerba greca quod uertunt in latina baptistari. Baptizo te hoc est intingo te..."
7. 62. a. On the seven winds.
   *Inc.* "Incipit dicta de septem celos. Nomina eorum primus aereus...."

   b. On the sun.
   *Inc.* "Sol pro qua dicitur. Hoc est lumen christi..."
8. 62–3. On thunder.
   *Inc.* "De t[r]onitrua. T[r]onitrua uero inter ȩreo..."
9. 64. Blank, except a *probatio pennae* of the letter *k* and a quire number.
10. 65–110. *Physiologus* in 34 chapters (B family).
    *Inc.* "Iacob benedicens filium suum ait..."

### Montecassino, Archivio dell'Abbazia, MS 316:

11. 1–22. Eucherius, *Instructiones* II.
    *Inc.* "Quoniam fili carissimȩ superiore librum propositionibus tuis..."
12. 23–30. Latin glossary C–U, Z. In two columns: One containing the lemma, the other the *interpretamentum.*
    *Inc.* "Caius. Humane fragilitas..."
13. 30–49. Pseudo-Cicero, *Synonyma.* Alphabetical A–G, K, L–U.
    *Inc.* "Incipit sinonima ciceronis..."

14. 49–118. Glossaries on Old Testament books: Genesis, Exodus, Numbers, Deuteronomy, Joshua, Judges, Ruth, Kings I–IV, Isaiah, Micah, Joel, Jonah, Nahum, Habakkuk, Zephaniah, Zachariah, Malachi, Jeremiah, Daniel, Ezekiel, and Job.
    *Inc.* "Phylo uir dissertissimus iudeorum. Origenis quoque testimonio..."
15. 118–142. Glossaries on New Testament books: Matthew, Mark, John, Acts; on the General epistles of Peter, John, and Jude; on the epistles of Paul to the Romans, Corinthians (I and II), Galatians, Ephesians, Philippians, Colossians, and Thessalonians; on the epistles, attributed to Paul, to Timothy (I and II), and Titus; on Paul's epistles to Philemon and the Hebrews; and on Revelation (Apocalypse).
    *Inc.* "De nouo testamento. De matheo incipiunt homina ebreica. Abraham pater uidens populum..."

## SCRIPT

Bischoff cautiously identified at least 2 hands ("anscheinend von wenigstens 2 Händen"), of which one is more experienced, and which appears in both parts. Both Inguanez and Newton identified a single hand, however.[110]

Bischoff characterized the script as a short, broad Caroline minuscule with a slight slant. There are thirteenth-century marginalia from Montecassino throughout.[111] Erasures have been preferred to expuncts.

## DECORATION

Both manuscripts display a clear hierarchy of script. The largest titles are rustic capitals with the inner spaces of the letter curves, ascenders and descenders filled in red, blue, and yellow. The smaller uncial titles are generally filled in red or yellow, or in two colours (e.g. blue and yellow); occasionally they are also multicoloured. Rubricated uncial is used only for *Physiologus* chapter titles. In Montecassino, Archivio dell'Abbazia, MS 316 an early modern hand has sketched a large decorated hollow initial *A* fourteen lines tall in black ink (fol. 10).

## PROVENANCE

At Montecassino since the early Middle Ages.

110 Newton, *The Scriptorium and Library*, 250.

111 Newton, *The Scriptorium and Library*, 250.

## 7. Munich, Bayerische Staatsbibliothek, MS Clm 14388[112]

Part I: I + 1–112 fols. | 26.5–27.5 × 16.5 cm | 27–31 lines.

Part II: 113–238 fols. | 26.5–27 × 16.5–17 cm | 27–31 lines | 2 or 4 columns from 184r.

This is a composite manuscript in two parts, bound together in s. $xv^{2}$. Part I (fol. 8 onwards) was copied from Munich, Bayerische Staatsbibliothek, Clm 13038 (fols. 262v–387v), which was made at the abbey of St. Emmeram, Regensburg. Part I was thought by Bischoff to have been made not at St. Emmeram but in an unknown scriptorium in southwestern Bavaria at the beginning of the ninth century. Folios 1–7, obviously inserted to replace the missing first part of Jerome's text, are in a twelfth-century hand. Part II (fols. 113–238) dates from the middle of the ninth century and was tentatively located by Bischoff to northwestern Germany.

### COLLATION

Part I: $I^{8}$ (wants 1), $II^{6}$, $III–IV^{8}$, $V^{6}$, $VI–VIII^{8}$, $IX^{2}$, $X^{8}$, $XI^{8}$ (wants 1, 3), $XII^{8}$, $XIII–XIV^{2}$, $XV^{8}$, $XVI^{2}$.

Part II: $I^{8}$ (wants 1), $II^{8}$, $III–IV^{2}$, $V–IX^{8}$, $X^{6}$ + 2 after 4 (wants 1 of the 2), $XI^{8}$, $XII^{10}$ (wants 2, 8), $XIII–XVII^{8}$.

### CONTENTS

#### Part I

1. 1r–112v: Jerome, commentary on the letters of Paul to the Ephesians and Titus.
   *Inc.* "<E>x prefatione Hieronimi in epistolam pauli ad effesios. <S>i quicquam est..."

#### Part II

2. 113r–172r: Pseudo-Jerome, *Expositio quattuor euangeliorum.*
   *Inc.* "In nomine d*omi*ni summi. Incipit expositio euangelii..."

3. 172v–183v: *Physiologus* in 48 chapters (A family). The *explicit* attributes the text to John chrysostom: "Explicit de natura bestiarum sancti iohanni constantinobolitane urbis antistinie. Liber conscriptus deo gratias amen."[113]
   *Inc.* "Incipiunt capitula. i. De leone..."

---

**112** http://daten.digitale-sammlungen.de/~db/0002/bsb00022465/images/index.html?seite=00001&l=de. Halm, von Laubmann, and Meyer, ed., *Catalogus* 4, 165; Bierbrauer, *Die vorkarolingischen und karolingischen Handschriften*, 123; Bischoff, *Katalog der festländischen Handschriften*, 2:164; Bergmann and Stricker, *Katalog* 3:1115–6; Helmer, *Katalog*, 414–19. See also Dorofeeva, "Strategies of Knowledge Organisation."

**113** "Here ends the work on the nature of beasts by St John, bishop of the city of Constantinople. The book is written by the grace of God, amen." "Antistinie" is a mis-rendering of "antistes," a Latin term for "bishop," here presumably intended to be the genitive "antistitis."

4. 183v: Description of spikenard with a recipe for unguent.
   *Inc.* "De pigmentis nardi spicatæ. Nardus est arbor habet fructum..."
5. 184r–222v: "Accipe" Latin glossary (A–P). With added Greek words.
   *Inc.* "Accipe audi cognosce..."
6. 223r–225v: Latin–Hebrew glossary based on Eucherius of Lyon, *Instructiones*, bk. 2. With added Greek words.
   *Inc.* "Incipiunt nomina interpretatio de hebraico in latino quod exposuit beatus hieronimus. Adonai dominus significat..."
7. 225v–227v: Greek/Hebrew–Latin glossary based in part on Eucherius of Lyon, *Instructiones*, bk. 2
   *Inc.* "De grecis nominibus. AΓω sanctus..."
8. 227v–228r: "De ponderibus." Contains many excerpts, some paraphrased, from book 16 of Isidore of Seville's *Etymologiae*.
   *Inc.* "Incipiunt nomina hebreorum de ponderibus. Minima pars est..."
9. 228r–230r: "Abauus" Latin glossary (A–C, F).
   *Inc.* "Incipiunt glossae de A. Abauu pater proaui id est auis aui..."
10. 230r–238v: Pseudo-Cicero, *Synonyma*.
    *Inc.* "Incipiunt synnonima ticeronis. Cycero ueterio sio [*sic*]..."

## SCRIPT

The early medieval folios are in ninth-century Caroline minuscule: by several hands in part I, by one hand in part II.

## DECORATION

Initial P in brown ink pen on 1v. Several decorated initials, including two small, striated initials on 133v and 162r.[114]

## PROVENANCE

According to an entry in the catalogue made by the abbey librarian Dionysius Menger, in 1500 or 1501 the codex was at St. Emmeram. In 1811, as a result of the secularisation of the Napoleonic period, the codex was transferred to the Bavarian State Library.[115]

114 Bergmann and Stricker, *Katalog* 3, 1116.

115 Bergmann and Stricker, *Katalog* 3, 1116.

## 8. Munich, Bayerische Staatsbibliothek, MS Clm 19417[116]

I + 103 fols. | 15 × 11 cm | 14–15 lines.

This codex was produced in the first third of the ninth century, in southern Bavaria or Switzerland according to Bischoff, in southern Bavaria according to Bierbrauer.[117] This manuscript has quire numbering, which begins anew with the change of hand on fol. 28v. This means that the two halves of the manuscript are numbered I–III and I–X. Quires I (in the first part), and V, VIII, and XIII (in the second part) lack quire numbers.

### COLLATION

$I^{10}$, $II\text{–}VII^{8}$, $VIII^{6}$, $IX\text{–}XIII^{8}$.

### CONTENTS

1. Ir–v: *De sancto mauritio*, eleventh-century liturgical hymn with neumes.
   *Inc.* "V Dulce carmen et melodum…"
2. Iv and 1r: Two prayers for blessing and protection. Eleventh century. The incipit is faded to near-illegibility. Both prayers are additions to fill the space at the bottom of each page, and do not form part of the main text.
   *Inc.* "Deus filius nos benedicere…"
3. 1r–14r: *Constitutio et fides Nicenii concilii.* Title in half-uncial. Begins with the Creed of the First Council of Nicaea (325 CE) and continues with the 20 canons of this council.
   *Inc.* "Incipit constitutio et fides niceni concilii subditis capitulis suis. Facta est autem…"
4. 14r–24v: Venantius Fortunatus, *Commentus in symbolum Athanasianum*.[118]
   *Inc.* "Quicumque uult saluus esse, ante omnia opus est ut teneat catholicam fidem…"
5. 24v–28r: Commentary on Nicene-Constantinopolitan Creed.[119] Originally composed in Spain in s. vii or France in s. viii, or perhaps southern Gaul in s. vi, against the Adoptionists.[120]
   *Inc.* "Credo in unum deum sanctam trinitatem…"

---

**116** Online at: www.digitale-sammlungen.de/en/view/bsb00036883?page=,1.

**117** Bischoff, *Die südostdeutschen* Schreibschulen 1, 164; Bierbrauer, *Die vorkarolingischen und karolingischen* Handschriften, 89. Bierbrauer saw a resemblance in Munich, Bayerische Staatsbibliothek, MS Clm 19417 to manuscripts from Benediktbeuern-Kochel in south Bavaria.

**118** Pr. in Burn, *The Athanasian Creed*, 28–39. Keefe, *A Catalogue of Works*, 155, noted that Scherrer was mistaken in thinking this was not Fortunatus' commentary. See Scherrer, *Verzeichniss*, 88.

**119** Pr. in Jacobi, "Eine noch ungedruckte Bearbeitung," 288–90; Hahn and Hahn, ed., *Bibliothek der Symbole*, 349–51; Parmentier, "Trying to Unravel Jacobi's Unknown Creed"; and PL Suppl. 4, cols. 1519–20.

**120** For further information see Keefe, *A Catalogue of Works*, 93.

6. 25v–28r: Invocations against snakes. Tenth or eleventh-century addition in lower margins. The same scribe may have contributed to item 13.[121]
*Inc.* "In nomine sancta ales in nomine sancta sol ficiunt. In nomine sancti apostoli..."

7. 28v–71r: *Physiologus* in 47 chapters (Y family). Rubricated. Chapter 32 on beaver missing due to page loss, as is part of chapter 28 on the lion and panther. Explicit written in a cipher: "EXALKCKT: DFNBTXRA BFSTK.B.RXM. SCK̄ KPHBNNF CPNSTBNTKNP. CPNSCRKASKT DP̄ GRBTKBS. AMFN." This resolves to: "Explicit de natura bestiarum sancti iohanne constantino conscripsit deo gratias amen."[122]
*Inc.* "Incipimus loqui de leone primum rege omnium bestiarum. Iacob benedicens iudam..."

8. 71r–74r: *Joca monachorum*.[123]
*Inc.* "Interrogatio. Quis est quod tangitur et non uidetur..."

9. 74r–100r: *Glossae collectae* from the councils of Nicaea, Ancyra, Neocaesarea, Gangres, Antioch, Laodicea, Chalcedon, Sardica, Carthage, and Africa; and from the decretals of Popes Syricus, Innocent, Zosimus, Boniface, Caelestinus, Leo, Hilarius, Simplicius, Gelasius, and Symmachus.[124] Not rubricated but underlined in red. On fols. 74r–98v the Latin is mixed with 180 glosses in a Bavarian dialect.[125]
*Inc.* "Glose canonum. De canone apostolorum..."

10. 100v–101r: Riddle about *lapis* and *apis* using cryptography (*notae Bonifatii*), linked by a kind of basic numerology to a following short text about the unity of the Trinity, excerpted from Augustine's first sermon on the mystery of the Trinity and of the Incarnation (*sermo* 145, par. 2 and 5).
*Inc.* "Dic mihi quinque in lege leguntur..."

11. 101r–102v: Pseudo-Augustine, sermon 245 on the virginity of the Virgin Mary.[126] Excerpted and rearranged. The order of paragraphs as printed by Migne is as follows:

Paragraph 2: *Inc.* "Tria pariter adesse uidentur...," *expl.* "...operatio tribus constat."
Paragraph 5: *Inc.* "Esaias propheta apertius mariam sanctam...," *expl.* "...intus fuit."
Paragraph 3: *Inc.* "Et detestandus Manichaeus...," *expl.* "...nec uirginis partum."

---

**121** The penultimate line on fol. 28r contains the word "saikarom," which appears to be a garbled form of the magic word "sarisarco," also found in a similar text in the eleventh-century manuscript Luxembourg, Bibliothèque nationale, MS 109 (see Steinmeyer, *Die kleineren althochdeutschen Sprachdenkmäler*, 392).

**122** "Here ends the work on the nature of beasts written by St John of Constantinople. Thanks be to God, amen."

**123** Pr. in Wilmanns, "Ein Fragebüchlein." Discussed as a text in Wright, "From Monks' Jokes to Sages' Wisdom." Munich, Bayerische Staatsbibliothek, MS Clm 19417 is listed in Suchier, *Das mittellateinische Gespräch*, 90.

**124** This list appears in Hahn, ed., *Catalogus*, 244, and in McKitterick, *The Frankish Church*, 41.

**125** Bergmann and Stricker, *Katalog* 3, 1256.

**126** Part of the sermon *Legimus sanctum Moysen*, printed by Migne as Augustine of Hippo's sermon 245 (PL 39.2196–98). Fassler noted that this sermon is actually a composite work, whose first half

12. 102v–103r: Athanasian Creed. In a different hand.
    *Inc.* "Quicumque uult saluus esse..." Ends abruptly one third of the way through at "Ita deus pater..."

13. 103v: Two prayers in two late tenth- or eleventh-century hands. The second hand is possibly identical with that of item 6.
    *Inc.* "Deo gracias semper tibi domine..."

### SCRIPT

Two hands (1: 1r–28r; 2: 28v–102v), both characterized by clubbed ascenders, with item 12 in a third hand. The first hand uses a bold uncial in brown ink as part of its script gradation; the second uses red ink for lines of minuscule or majuscule script. The letters of the second hand are not especially carefully formed (descenders of differing lengths, for instance). Words well separated, frequent use of interpuncts.

### DECORATION

Large decorated initial F in shades of red on 1r, which is typical of southern Bavaria and serves as the basis for locating the manuscript to that region.[127] Initials mostly small, in brown and occasional red ink, modestly decorated.

### PROVENANCE

Benedictine monastery at Tegernsee, according to a fifteenth-century ownership inscription on the front cover board (also fifteenth century). Appears in two catalogues compiled by the monastery's librarians: Ambrosius Schwerzenbeck's 1483 catalogue (as CXX) and Konrad Sartori's 1500 catalogue (as C55).[128]

## 9. Oxford, Bodleian Library, Auct. T.2.23 (S. C. 20618)[129]

i + 156 fols. + i | 20 × 17 cm | 20–22 lines.

According to the *Summary Catalogue*, which does not give a place of origin, the entire manuscript is tenth-century. According to Bischoff's catalogue, fols. 82–156 are s. ix$^{2/3}$ and originated in eastern France ("ostfranzösisches Zentrum"); according to Bischoff's research elsewhere, fols. 88v–90r are from Tours or "one of the places influenced by Tours" ("einer von Tours beeinflußten Stätte").[130] Eastern France has been preferred here.

---

is taken from a letter written in 437 by the African Antonius Honoratus, and whose second half comes from *Sanctus hie*, another pseudo-Augustinian sermon (Fassler, "Mary's Nativity," 414n88).

127 Bierbrauer, *Die vorkarolingischen und karolingischen Handschriften*, 169.

128 Bierbrauer, *Die vorkarolingischen und karolingischen Handschriften*, 169.

129 Hunt, Madan, and Record, *A Summary Catalogue* 4 (1897), 434–5. Also described in Pächt and Alexander, *Illuminated Manuscripts* 1, 34. Fols. 82–156 are described in Bischoff, *Katalog der festländischen Handschriften*, 2:359.

130 Bischoff, *Katalog der festländischen Handschriften*, 2:359; Bischoff, "Die lateinischen Übersetzungen," 151.

The present manuscript is made up of four separate parts listed by the *Summary Catalogue* as A, B, C, and D. These parts are reproduced below with relevant notes from the catalogue. Pächt and Alexander modify the *Summary Catalogue* entry in that they tentatively date the manuscript to s. xi², list its place of origin simply as "East" and give its ninth-century provenance as St. Médard at Soissons. The whole manuscript cannot have a ninth-century provenance if it originated in the eleventh century, but it is unclear whether the cataloguers refer to the whole or to individual sections or folios, and what these sections are. Their catalogue pre-dates Bischoff's identification of parts C and D (fols. 82–156, containing the *Physiologus*) as being ninth-century.

My list agrees with the *Summary Catalogue* and Bischoff (the *Summary Catalogue* has been followed for parts A–C and Bischoff for part D). The divisions given in the *Summary Catalogue* are supported by the appearance of the parchment: in parts A and B (until fol. 82r) it is poorly cut, with obvious hair sides and uneven thickness from folio to folio. The quality improves in parts C and D, and is of a relatively high standard by the *Physiologus*.

## COLLATION

A: $I^{2}$, $II^{8}$, $III^{10}$, $IV–V^{8}$.

B: $VI^{10}$, $VII–XI^{8}$.

C: $XII–XVI^{8}$, $XVII^{2}$, $XVIII^{8}$.

D: $XIX–XXI^{8}$, $XXII^{8}$ (structure unclear).

## CONTENTS

1. 1v: Early modern list of contents:

   Cassiodorus de anima
   Augustinus de animae quantitate
   S. augustini sermo de adventv domini. Tomo VI contra Iudaeos, Paganos et Arrianos cap. XI
   Dicta sibvllae magae
   Ordo officii de vita christi
   S ambrosii de nabvthe
   Physiologvs de natvra animalivm
   Collectae pro rotlindis tribulata

### Part A

2. 2v–33r: Cassiodorus, *De anima*. Tenth-century. Titles, chapter numbers, and historiated initials in rubricated uncial. 2r is blank except for a title in a sixteenth-century hand: "Cassiodori de anima." Several faces and *maniculae* appear in the margins of the text.

   *Inc.* "Liber magni aurelii cassiodori senatoris de anima capitula. i. Quid amici requisierint...."

## Part B

3. 33r–81r: Augustine, *De animae quantitate*. Fols. 33r–v, and a title on fol. 32v, are a twelfth-century addition, containing Augustine's *Retractatio* on the same text. A few twelfth-century interlinear and marginal glosses are scattered throughout. The upper margin has been trimmed almost completely away. Punctuation has been inserted on some folios in a darker ink and some phrases have been encircled or underlined in pencil (e.g. fol. 76r). Frequent use of the *nota* sign in the margins. *Litterae notabiliores* in rustic capitals. Copied in a variety of hands. Bottom margin of final folio contains partially erased Latin alphabet (A–H).
   *Inc.* "Incipit retractacio b. augustini super librum de quantitate anime ab eo disputatum..."

4. 81v: a. Collect for St. Pancras, with neumes.[131]
   *Inc.* "Iesu rex noster benigne qui beatum pancratium in primeuo flore..."

   b. Hrabanus, *De rerum naturis* VII.8.298 (not in the *Etymologiae*). Paraphrased.
   *Inc.* "Verba sancti hieronimi de anima. Sanguis animae origo est ut quidam uolunt de cuius natura disputandi..."

   c. First line of St. Augustine, sermon 343.i: "Quid est invidia, nisi odium felicitatis alienae?"[132] In bottom left margin.
   *Inc.* "Inuidia est odium felicitatis alienę..."

## Part C

5. 82r–88v: Quodvultdeus, *Sermo de symbolo ad catechumenos*, chap. 11–16 (the last chapter abbreviated after the lines of the prophecy and the beginning of chap. 17).[133] Includes an acrostic Sibylline prophecy in hexametres from 87v (*inc.* "Iudicii signum..."). With rubricated titles and rubricated rustic capital initials at the beginning of each line of the prophecy.
   *Inc.* "Incipit sermo sancti augustini de aduentu domini. Uos inquam conuenio iudei..."[134]

6. 88v–93r: Two Sibylline prophecies. Contain:

   a. A cento (includes acrostic CRISTUS in verses 77–83, 91v–92r). Ends on 90r.
   *Inc.* "Incipit dicta sybillae magae. Non multi..."

---

131 Tenth-century hand according to the *Summary Catalogue* but twelfth-century according to Pächt and Alexander; the correct dating is that of the *Summary Catalogue*.

132 PL 39.1561.

133 Muntané, "Del 'Iudicii signum'," 167.

134 Attributed to Augustine in the Middle Ages but now considered to have been written by the fifth-century African bishop Quodvultdeus. See Harmless, *Augustine and the Catechumenate*, 262.

b. The medieval poem *Prophetiae Sibillae magae.*[135] In a single column, with rubricated titles and rubricated rustic capital initials at the beginning of each line. Third line left blank for a missing verse.
*Inc.* "Mundus origo mea..."

7. 93r–96v: Night office for weekdays. Only the first lines of the antiphons, verses, psalms etc. have been written down. The script was judged to be "regellos" ("irregular") by Bischoff. Tenth-century addition.
*Inc.* "Regem uenturum dominum. Antiphona missus est gabriel angelus..."

8. 97r–126r: Ambrose, *De nabutha israelita.* Contains faces and *maniculae* of the same style as in item 2, as well as contemporary marginal and interlinear glosses. *Litterae notabiliores* in rustic capitals. A set of neumes have been drawn in the left-hand margin of fol. 120v.
*Inc.* "Incipit de nabuthae. Nabuthae historia tempore uetus est usu cotidiana..."

9. 126v: Magical recipes: "for menstruation" (*inc.* "ad profluium"), "for a woman in childbirth" (*inc.* "ad partum mulieris") and several small others. Twelfth century. The bottom half of the folio is cut away, apparently without text loss.

**Part D**

10. 127r–155r: *Physiologus* in 30 chapters (B family). Dated to ca. 900 by Madan.[136] A large part of the first folio lower margin has been cut off. *Litterae notabiliores* in rustic capitals.
*Inc.* "Incipit liber physiologus de natura animalium uel auium seu bestiarum. i. De natura leonis..."

11. 155r–v: Mass for Rotlindis. Addition made at the end of the ninth century (according to Bischoff; ca. 1000 according to the *Summary Catalogue*; eleventh-century according to Pächt and Alexander).
*Inc.* "Dimite deus peccata nostra..."

12. 155v: Antiphon for St. Lambert, Bishop of Tongres (near Liège), with neumes. Eleventh- or twelfth-century according to Bischoff (twelfth-century according to the older catalogues).
*Inc.* "Magna uox laude sonora..."

13. 156r–v: *Probationes pennae.*

**135** Not a translation from the Greek but rather an original creation, possibly from seventh-century Spain according to Dronke, "Medieval Sibyls," 615. Title given by Bischoff, who discovered it, in "Die lateinischen Übersetzungen." Bischoff noted that the text also exists in Valenciennes, Bibliothèque municipale, MS 404 (386), fols. 61v–62r (s. x) and Prague, Národní knihovna České republiky, MS XIII.G.18, fol. 230v (s. xv); and that the manuscripts are closely related to Oxford, Bodleian Library, MS Auct. T.2.23 and to each other (p. 164). For a history of the texts, see Lendinara, "The *Versus Sibyllae*," 93; and for an analysis of their contents, Dronke, *"Medieval Sibyls."*

**136** Hunt, Madan, and Record, *A Summary Catalogue* 4, 435.

### SCRIPT

Various hands from various centuries, principally writing Caroline minuscule. The script of section D was characterized by Bischoff as a calligraphic minuscule. Modern foliation in pencil.

### DECORATION

A sketch of a decorative line for a border appears on fol. 156r. Doodles of faces appear in the left-hand margins of fols. 40r, 97v and 105v, also acting as indications of lemmas. *Litterae notabiliores*, chapter numbers and titles are rubricated throughout parts A, C, and D. Part B is undecorated and unrubricated.

### PROVENANCE

This manuscript belonged to the Jesuit College of Clermont at Paris by the sixteenth century, as attested by a note on fol. 2v: "Collegij Parisiensis Societatis Jesu." Its classmark then was MS 489.[137] It was acquired with a large number of others by Geraard Meerman in 1763–64, as evidenced by a note on fol. 3r: "Paraphé au desir de l'arrest du 5 Juillet 1763 Mesnil."[138] Geraard bequeathed his collection to his son Jan Meerman (1753–1815). Jan's library was auctioned at the Hague in 1824. The present codex was one of 58 volumes bought by the Bodleian.

## 10. Oxford, Bodleian Library, MS Laud. Misc. 129[139]

i + 160 fols. + i | 28.5 × 22 cm| 26 lines.

This ninth-century manuscript from the Main river valley or its environs is largely unmentioned in *Physiologus* and bestiary scholarship.[140]

### COLLATION

I–XX$^{8}$. A quire is missing at the beginning, as indicated by two (?) different systems of quire numbering, one in Roman numerals ('ii', fol. 8v) and the other alphabetic, presumably starting at 'a', but appearing only intermittently at the beginning of quires (I, fol. 57; l, fol. 73; m, fol. 81; p, fol. 105; q, fol. 113; s, fol. 129; t, fol. 137).[141]

---

**137** Pächt and Alexander, *Illuminated Manuscripts*, 34.

**138** For an explanation of this note, see Franklin, *Les anciennes bibliothèques* 2, 275.

**139** Digitized as part of the Polonsky German collection at https://digital.bodleian.ox.ac.uk/objects/b96fd583-aba5-4848-affe-e067455c711c/surfaces/a1959129-797c-4d85-8f0a-a560cbd932cd/. The catalogue description for this manuscript is based on Coxe, *Laudian Manuscripts*. It was updated during the digitization and is available online at https://medieval.bodleian.ox.ac.uk/catalog/manuscript_6939. See this description for full information about the secondary literature. More recent printed catalogues that mention this manuscript include Bischoff, *Katalog der festländischen Handschriften*, 2:373, no. 3834; Pächt and Alexander, *Illuminated Manuscripts* III, no. 15, pl. I.

**140** Except Faraci, *Il bestiario medio inglese*. I am very grateful to Matthew Holford at the Bodleian Library for drawing my attention to this manuscript. I have not examined it in person.

**141** My heartfelt thanks go to Matthew Holford for providing the collation.

## CONTENTS

1. 1r–2r: Gregory I, *Homiliae in Ezechielem*, Hom. 8 (20), extract (7–10) on bodily resurrection.
   *Inc.*: "Nam semina arborum odorem uel saporum [*sic*] habent..."
2. 2r–4r: Homily on the meaning of biblical trumpets.
   *Inc.*: "Sententie prophete sicut ipse dominus per illum testatur sacerdotibus. Clama ne cesses quasi tuba..."
3. 4r–5r: Homily on the fear and love of God.
   *Inc.*: "De timore Domini. Seruite Domino in timore et exaltate ei cum tremore..."
4. 5r–8v: Caesarius of Arles, *Sermo* 10, extracts describing good Christian living as outlined in the Gospel; on the soldier; and on the fruiting vine.
   *Inc.*: "In euuangelium. Igitur si quis uult discipulus esse mandata sua custodiat, humilitatem discat..."
5. 8v–11r: Caesarius of Arles, *Sermo* XVII, extracts describing the Passion and how it relates to human salvation and sin.
   *Inc.*: "Omelia sancti Augustini. Fratres karissimi, ad memoriam uestram reducimus..."
6. 11r–16r: Life of the Seven Sleepers of Ephesus.
   *Inc.*: "Maximianus, Malcus, Martianus, Dionisius, Iohannes, Serapion, Constantinus. In illo tempore..."
7. 16r–v: Various questions and answers, including a long discussion of six sins.
   *Inc.*: "Incipit de diuersis interpretationum. Interrogo te pro quid dicitur homo?"
8. 16v–22v: Life of St. Ciriacus (also known as the *Acta Cyriaci* or *Inventio sanctae crucis*); paraphrased and abbreviated.
   *Inc.*: "Incipit inuentio sanctae crucis. Anno ducentissimo trigissimo tertio regnante uenerabili dei cultore magno uiro Constantino..."
9. 22v–23r: Sermon contrasting peace and deceit.
   *Inc.*: "Discretio tenenda. Pax uera tenenda est quae de filiis diabuli filios dei efficit, et de inimicis amicos, et de adversariis fratres..."
10. 23r–24r: Homily on heaven.
    *Inc.*: "De uocat(ione). Tribus uocationibus uocauit Deus hunc mundum ad uitam. Prima uocatio in ueteri testamento per prophetam in spiritu dicentem, Ueni te filii..."
11. 24r–25r: Homily on the temptation of evil, using the story of the Tree in the Garden of Eden, and how it may be resisted, with Gospel proverbs.
    *Inc.*: "De suggestione diabuli. Dixit Deus ad Adam de ligno scienti boni et mali: Ex eo morte morieris et mortuus est, et factus est per suggestionem diabuli..."
12. 25r–27v: Homily on Matt. 10:16: "Behold, I send you forth, as sheep among wolves; be you therefore wise as serpents, and harmless as doves."
    *Inc.*: "In euuangelio. Ecce ego mitto uos sicut oues in medio luporum..."

13. 27v–31r: Eusebius Gallicanus, *Homilia ad monachos* V (serm. 40), on the ways in which the temptation and pleasure of sin can lead the soul astray.
    *Inc.*: "Omelia sancti Caesarii episcopi. Scimus quidem spritale [*sic*] militiae cui nos mancipauimus..."

14. 31v–32v: Eusebius Gallicanus, *Homilia ad monachos* IV (serm. 41).
    *Inc.*: "Omelia item uerbis. Fratres karissimi ad hoc istum locum conuenimus ut domino nostro uacare possimus..."

15. 32v–34v: Eusebius Gallicanus, *Homilia ad monachos* IX (serm. 44).
    *Inc.*: "Sancti Caesaris. Uidete uocationem uestram fratres karissimi..."

16. 34v–36v: Caesarius of Arles, *Sermo* 233.
    *Inc.*: "Omelia sancti Cesarii. Sanctus et uenerabilis pater uester relegiosam fidem..."

17. 38v–41r: Eusebius Gallicanus, *Homilia ad monachos* (*Sermo extravagans* 6).
    *Inc.*: "Incipit omelia sancti Cesarii. Ad hunc locum, fratres karissimi, non ad quietem, non ad securitatem, sed ad pugnam..."

18. 41r–42r: Ps.-Augustine, sermon about ten denarii.
    *Inc.*: "Incipit interpretatio sancti Augustini de x talentis. Auditor hoc tempus per mulierem illam sollicitam quae habuit decem denarios..."

19. 42r–44r: Extracts from an anonymous Old Testament (Vulgate) commentary.

    a. 42r–43r: Commentary on God's promise to Abraham that his offspring would be as numerous as the stars (Gen. 15:5).
    *Inc.*: "Erit semen tuum sicut arena maris, Et sicut stelle caeli. Haec loquitur pater omnium creaturarum..."

    b. 43r–44r: Commentary on the Dream of Jacob's Ladder (Gen. 28:10–19).
    *Inc.*: "De scala illa quam uidit Iacob. Uidit Iacob scalam..."

20. 44r–86r: Isidore of Seville, *Quaestiones super uetus testamentum.*
    *Inc.*: "De vietatibus [*sic* for aetatibus] mundi..."

21. 86r–87r: Text praising the singing of Psalms, beginning with a citation from the Life of St. Theodosia.
    *Inc.*: "Psalmorum canticis laus ad animam refocilare. Quid enim in psalmis inuenitur quod non proficiat ad aedificationem..."

22. 87r–88r: Riddling text presenting facts about Christ's birth as opposing pairs.
    *Inc.*: "De natiuitate Domini et interpretatio. Sciendum est quare natus est iesus in octava kal. Ianuarii..."

23. 88r–89v: Ps.-Maximus of Turin, On the Assumption of the Virgin Mary I (serm. 11).
    *Inc.*: "In natiuitate sancte Marie semper uirginis. Illum sacratissimum uterum ex quo homine [*sic*] deus unigenitus apparuit..."

24. 89v–91v: Sermon on Matt. 8:1, the large crowd following Christ when he descended from the mountain.
    *Inc.*: "Descendit Dominus de monte ubi erant turbe..."

25. 91v–93v: Petrus Chrysologus, serm. 143 on the Annunciation of the Lord.
    *Inc.*: "Incipit sermo de natiuitate Domini. Fratres karissimi ineffabile natiuitatis dominice sacramentum iam credi magis conuenit quam referri..."

26. 94r–96r: Allegorical interpretations of the branch emerging from Jesse's roots (Isaiah 11:1).[142]
    *Inc.*: "De Iesse. Exiet uirga de radice Iesse..."

27. 96r–97r: Pseudo-Augustinian sermon on Easter.[143]
    *Inc.*: "Omelia sancti Augustini de pascha. Gaudemus, fratres karissimi, quia dominus noster Ihesus Christus uenit..."

28. 97v–99r: Explanation of the significance of the number and names of the 12 Apostles.
    *Inc.*: "De nominibus apostolorum interpretatio. Sciendum est tamen..."

22. 99r–100v: Homily on the rewards of just death, based on Ps. 116:15.
    *Inc.*: "De pretiosa morte iustorum. Pretiosa est in conspectu Domini mors sanctorum eius..."

23. 100v–101v: Anonymous sermon on the Wise and Foolish Builders (Matt. 7:24–27).[144]
    *Inc.*: "Explanatio sancti Gregorii pape. Omnis ergo qui audit..."

24. 101v–103r: Maximus of Turin, serm. 12 on the holy martyrs.[145]
    *Inc.*: "Omelia de sanctorum martyrum. Cum omnium sanctorum martyrum, fratres karissimi, natalem deuotissime celebrare debemus..."

25. 103r–104r: Caesarius of Arles, serm. 50. On the importance of the health of the soul over the body, and the avoidance of sorcery.
    *Inc.*: "Omelia de sanitate animae. Nostis, fratres karissimi, omnes homines sanitatem corporis querere..."

26. 104r–105v: Augustine, serm. 326. On the birthday of the twenty martyrs.[146]
    *Inc.*: "Sermo sancti Augustini de sanctorum martyrum. Sollemnitas beatissimorum martyrum..."

27. 105v–107r: Caesarius of Arles, serm. 227 on the feast of a church.[147]
    *Inc.*: "Omelia in natiuitate et dedicatio ecclesie. Quotienscumque, fratres karissimi, altaris uel templi festiuitatem colimus..."

---

**142** A similar commentary exists in Orléans, Médiathèque, MS 313, pp. 188–90.

**143** See Ps.-Augustine, *sermo* 156 (*De passione domini VII*), PL 39.2053–55.

**144** This sermon is also found in Munich, Bayerische Staatsbibliothek, MS Clm 27152, fols. 66r–67r (s. ix½, www.digitale-sammlungen.de/view/bsb00031770?page=1); Linz, Bibliothek der katholisch-theologischen Hochschule, MS A 1/6 (MS 1), 95v–96v (s. ix); and in Zurich, Zentralbibliothek, MS 463 [Rh. 951], fols. 69r–71v (s. x–xi). See Quain and Plante, "Catalogue," 117; Étaix, "Un manuel," 126–28.

**145** CCSL 23, from p. 41.

**146** Augustine, *The Works of Saint Augustine*, trans. Hill and Rotelle, 170–2.

**147** CCSL 104.897. Caesarius of Arles, *Sermons 3: 187 – 238*, trans. Mueller, 164–8.

28. 107r–108r: Meditation on vices and virtues, based on Galat. 5:16.
*Inc.*: "Predicatio sancti Pauli apostoli. Spiritu ambulate et desideria carnis non perficietis..."

29. 108r–109r: Homily on the fear of God.
*Inc.*: "De Dei timore. Timor Domini sicut dictum est: Initium sapientiae timor Domini..."

30. 109v–119r: Gregory I, *Dialogi*, extracts (II.4, I.12, I.11, IV.52–4, IV.51, IV.35–7, II.23, II.24, IV.57–60).
*Inc.*: "De monacho uage mentis ad salutem reducto..."

31. 119r–127r: Rufinus of Aquileia (trans.), *Historia monachorum in Aegypto* (extracts).
*Inc.*: "Incipit de sancto Eulogio..."

32. 127r–131v: Rufinus of Aquileia (trans.), *Vitas patrum* III (or *Verba seniorum*) chapters 14, 37–38, 40, 61, 68, 157, 169–171 (extracts), 172, 173–4, 197.
*Inc.*: "Erat quidam monachus et habitabat in herimo [*sic*]..."

33. 131v–133r: Ps.-Jerome, *Verba seniorum* or *Liber adhortationum sanctorum patrum*, V.39, VI.21, IX.11, X.1 (?), V.30.
*Inc.*: "Sancti patres aliquando congregati prophetauerunt de ultima generatione..."

34. 133r–137r: Rufinus of Aquileia (trans.), *Historia monachorum in Aegypto*, chap. 16.
*Inc.*: "Incipit de sancto pafnutio..."

35. 137r–v: Rufinus of Aquileia (trans.), *Vitas patrum* III (or *Verba seniorum*), chap. 29.
*Inc.*: "De sancto eulalio..."

36. 138r–149v: Isidore of Seville, *Sententiae*, selection.
*Inc.*: "De martirio" (I.23.1–3), "De antichristo" (I.25), "De die iudici" (I.27.1–10), "De gehenna" (I.28), "De penis impiorum" (I.29), "De resurrectione" (I.26.2–5), "De sapientia" (II.1, varies at end), "De fide" (II.2.1–8), "De caritate" (II.3), "De spe" (II.4, abbrev.), "De gratia" (II.5), "De conuersis" (II.7.4–11.7, abbrev.), "De conpunctione" (II.12.1–5, 8, varied), "De confessione" (II.13.1–18), "De manifestis ocultisque peccatis" (II.20.1–4), "De desperatio" (II.14), "De peccandi consuetudine" (II.23), "De cogitatione" (II.25.1–8), "De conscientia" (II.26.2–3), "De mendatio" (II.30.1–10), "De iuramento" (II.31.1–8, abbrev.), "De continentia" (II.40, abbrev.), "De tribulatione iustorum" (III.58), "De iracundis" (III.40), "De amicitia munera orta" (III.30), "De odio" (III.27), "De tollerantia diuine correptionis" (III.4), "De temptationibus diaboli" (III.5.5–30), "De inuidia" (III.25.4–8).

37. 149v–158v: *Physiologus* (Y version) in 22 chapters.
*Inc.*: "Incipit de natura animalium de leone et animalium [*sic*]..."

38. 158v–159v: St. Jerome, commentary on the Song of Songs (excerpt).
*Inc.*: "Super cantica canticorum. Rex Dominus Ihesus aeclesiam id est dei sermo animam sub coniugis titulo appellauit..."

### SCRIPT

Insular minuscule. Insular-influenced early Caroline minuscule on fols. 120v, 148v, 149r. Rubrics in uncial.

### DECORATION

Decorated initials.

### PROVENANCE

Part of William Laud's (1573–1645) third donation to the Bodleian in 1639. Medieval provenance unknown.

## 11. Paris, Bibliothèque nationale de France, MS Nouv. acq. lat. 1616[148] + Orléans, Médiathèque, MS 18 (olim 15)149 + Paris, Bibliothèque nationale de France, MS Nouv. acq. lat. 455[150]

Paris, Bibliothèque nationale de France, MS Nouv. acq. lat. 1616: i + i + 14 fols. + i + i | 25.3 × 17.2 cm | 27–32 lines.

Orléans, Médiathèque, MS 18 (*olim* 15): 188 pp. | 25 × 17 cm | 19–24 lines

Paris, Bibliothèque nationale de France, MS Nouv. acq. lat. 455: I + I + 8 fols. + I + I | 23 × 13 cm | 27 lines.

The eight-leaf quire containing the *Physiologus* in MS Nouv. acq. lat. 455 was originally bound with Orléans, Médiathèque, MS 18 (*olim* 15); and with Paris, Bibliothèque nationale de France, MS Nouv. acq. lat. 1616.[151] I was not able to examine any of these manuscripts personally. Pages 1–28 originally in the Orléans codex are now Paris, Bibliothèque nationale de France, MS Nouv. acq. lat. 1616. Paris, Bibliothèque nationale de France, MS Nouv. acq. lat. 455, which formed the end of the collection, is incomplete (ending partway through the *Physiologus* chapter on the ercine (*herinacius*)), meaning that further manuscript parts may exist. The whole Paris-Orléans collection was one of the books stolen by Guglielmo Libri (no. 45 in the original list of Libri's books, no. 74 in the account made of the thefts by Delisle).[152]

---

**148** https://archivesetmanuscrits.bnf.fr/ark:/12148/cc13524w; Bischoff, *Katalog* 3:242. The identifications of the texts are from the IrCaBriTT project: https://ircabritt.nuigalway.ie/handlist/catalogue/170.

**149** https://mediatheques.orleans-metropole.fr/ark:/77916/FRCGMBPF-452346101-01A/D18010449.locale=fr; Bischoff, *Katalog* 2:328–9; Pellegrin and Bouhot, *Catalogue des manuscrits*, 19–20; Delisle, *Catalogue des manuscrits*, 79–80.

**150** https://archivesetmanuscrits.bnf.fr/ark:/12148/cc71179s.

**151** Delisle, *Catalogue des manuscrits*, 79–80; see also Delisle, "Notice," 360. The binding of MS 18 is digitized at https://bvmm.irht.cnrs.fr/resultRecherche/resultRecherche.php?COMPOSITION_ID=19734.

**152** Cahier and Martin, *Mélanges*, 1:203; *Eighth Report of the Royal Commission on Historical Manuscripts*.

Pellegrin and Bouhot identified Paris, MS Nouv. acq. 1616 as part of the original collection and gave the place of origin for the Orléans codex as western France. Bischoff did not link any of these manuscripts to each other (Paris, MS Nouv. acq. 455 and Orléans, MS 18 were linked by Delisle).[153] He thought that Paris, MS Nouv. acq. 1616 was from Brittany due to interlinear glosses in Breton on p. 1, l. 6–10.[154] Jacopo Bisagni noted that the text containing Breton glosses may have been copied from a Breton exemplar, perhaps by a Breton scribe.[155] This makes it possible that the Paris-Orléans collection originated at Fleury, for example, where it has a relatively early provenance, although a Breton origin cannot be ruled out. Bisagni also noted the idea that Paris, MS Nouv. acq. 1616, or its copy, could have been one of the computistical manuscripts brought to England by Abbo of Fleury.[156] I have therefore given the place of origin as western France, following Pellegrin and Bouhot.

It should be noted that despite Delisle's linking of Paris, MS Nouv. acq. lat. 455 and the Orléans codex, this may not be accurate. The Paris manuscript is significantly narrower than the other two, a feature which does not seem attributable to page trimming. I have cautiously treated the Paris manuscript as part of this collection, but further careful examination of all parts is required.

Bischoff dated Paris, MS Nouv. acq. 1616 to the ninth or tenth century, or to the first half of the tenth century. Pellegrin and Bouhot dated the Orléans codex to the first half of the ninth century. My own dating for Paris, MS Nouv. acq. 455 is the last third or last quarter of the tenth century (see discussion in Chapter 5). Item 3 in Paris, MS Nouv. acq. 1616 may provide a *terminus post quem* of 903 CE (the year which may be indicated in the text as the *annus praesens*).[157]

The quires in Orléans, Médiathèque, MS 18 were rearranged before the twelfth century, when notes were added about their correct order. This order is as follows: 29–68, 81–96, 69–80, 97–227. Six quires from p. 131 have original quire signatures (A–F).

---

**153** Delisle, *Catalogue des manuscrits*, 80.

**154** These glosses were given by Delisle, *Catalogue des manuscrits*, 278 and plate 6.2.

**155** IrCaBriTT project: https://ircabritt.nuigalway.ie/handlist/catalogue/170.

**156** Juste, "Comput et divination," 111–12; IrCaBriTT project: https://ircabritt.nuigalway.ie/handlist/catalogue/170.

**157** IrCaBriTT project: https://ircabritt.nuigalway.ie/handlist/catalogue/170.

## COLLATION

Paris, Bibliothèque nationale de France, MS Nouv. acq. lat. 1616: The structure of this manuscript is unclear. The description accompanying the digital facsimile specifies only that the first quire has 3 leaves and the second quire has 11 leaves. The original pastedowns were removed and are now in Orléans, Médiathèque, MS 192 (*olim* 169).[158]

Orléans, Médiathèque, MS 18 (*olim* 15): I$^{8}$ (wants 1, 8), II$^{8}$ (wants 4, 5), III$^{8}$–VI$^{8}$, VII$^{8}$ (wants 6, 7, 8), VIII$^{4}$, IX$^{8}$–XIV$^{8}$.

Paris, Bibliothèque nationale de France, MS Nouv. acq. lat. 455: I$^{8}$.

## CONTENTS

### Paris, Bibliothèque nationale de France, MS Nouv. acq. lat. 1616:

1. 1r: Table on the lunar phases.
   *Inc.*: "Cursus lunae per XII signa..."

2. 2v: a. Computistical *argumentum*.
   *Inc.*: "Statutum invenimus in concilio Romanorum ut nec ante XI kalendas aprilis nec post VII kalendas maii pascha non debeat fieri..."

   b. Computistical *argumentum*. Contemporary addition.
   *Inc.*: "Si nosse feriam hoc est die septimane..."

3. 3r: Computistical *argumentum* for finding computistical data of the *dies praesens*.
   *Inc.*: "Quod datarum est hodie per dies anni..."

4. 3v–5r: Computistical tables and diagrams.
5. 5v: Computistical excerpt about the seasons.
   *Inc.*: "Ver a vertendo dicitur eo quods terram in flores vertit..."

6. 6r: Bede, *De temporum ratione*, 1:25–107, on finger-reckoning. With interlinear Breton glosses.
   *Inc.*: "Cum ergo dicis unum minimum in leva digitum inflectens in medium palme artum infiges..."

7. 6v: Computistical excerpt.
   *Inc.*: "De alterius modi compoto. Est et alterius modi compotus articulatim decurrens..."

8. 7r: Computistical excerpt about the lunar age in relation to the solar calendar.
   *Inc.*: Si vis scire per singuulas [sic] lune aetates..."

158 https://mediatheques.orleans-metropole.fr/ark:/77916/FRCGMBPF-452346101-01A/D18011385.locale=fr.

9. 7v: a. Brief dialogue.
   *Inc.*: "Dicito mihi unum diem qui habet XII incentiones..."

   b. Apuleian sphere.
   *Inc.*: "Ratio spere Phtagor [sic] philosophi quam Epulegus descripsit..."

10. 8r–9r: Computistical tables.
11. 9r: Computistical poem on the names of weekdays.
    *Inc.*: "Prima dies Phoebi sacrato nomine fulget..."

12. 9v–10r: Text on the invention of the clock.
    *Inc.*: "Hieronimus invenit horologium, in duobus invenit, I. est in mensura pedum..."

13. 10r:

    a. On the Nicene synod.
    *Inc.*: "Sancta senodus ubi CCC et XVIII pontifices apud Nicenam civitatem Bithinae converunt [sic]..."

    b. On calculating intercalary ratios. A summarised version of Bede, *De temporum ratione*, chap. 45, *De embolismis et communibus annis*.
    *Inc.*: "Communium et embolismorum ratio ista est..."

    c. On the date of Easter.
    *Inc.*: "Legimus in epistolis Grecorum quod post passionem apostolorum sanctus Pachomius abba in Aegipto cum monachiis [sic] suis..."

14. 10v–12r: Lunar prognostics.
    *Inc.*: "Luna prima. Qui incenditur in ipsa sanabitur..."

15. 12r: a. Egyptian days.
    *Inc.*: "Incipiunt dies egyptiali [*sic*] qui in anno observandi sunt non itinere, non ambulare, non in vineam plantare..."

    b. Blood-letting tract.
    *Inc.*: "Tres dies maximi observandi sunt in anno per omnia..."

    c. On the stages of human gestation. Contemporary addition.
    *Inc.*: "De suggesione formationis hominis in utero: VI dies in eodem calore fit, VIIII in sanguine, XII in cruore..."

16. 12v: Wind prognostic.
    *Inc.*: "Si in nocte fuerit ventus in nocte natalis Domini nostri Jhesu Christi, in hoc anno reges et pontifices peribunt..."

17. 13r–v: Computistical argumenta (acephalous compilation).
    *Inc.*: "De feria in terminis. Hoc ergo modo invenitur feria in terminis..."

18. 13v: Computistical table accompanied by Hebrew and Greek words associated with the names of God (*saddai, hel, eloim, eloe, sabaoth, elion, asiriee, adonaia, tetra gramaton*).

19. 14r: Prognostic text *De ragono subiecto* (*sic*, for *De <tet>ragono subiecto*).[159]
    *Inc.*: "De ragono subjecto. Si nosse vis de quolibet infirmo..."

20. 14v: a. Theological note. Contemporary addition.
    *Inc.*: "Trinitatis in amore alme individue consors quidem sacerdotum reminisceni [*sic*] Dominum, criminum preteritorum, futurorum operum..."

    b. Computistical note. Contemporary addition.
    *Inc.*: "III observationes sunt que nobis requirende sunt in exploratione initii, I est si annus bissexti fuerit..."

**Orléans, Médiathèque, MS 18 (*olim* 15):**

21. 29–127: Excerpts from the book of Isaiah (Is. 14:28–25:10, 28:2–63:1, 65:12-66:24) (*incipit* illegible).
22. 127–227: Excerpt from the book of Ezekiel (Ez. 1:1–30:25).
    *Inc.* "Incipit praefatio Hieronymi presbiteri in Jezechiel prophetae..."

**Paris, Bibliothèque nationale de France, MS Nouv. acq. lat. 455:**

23. 1r–3r: Isidore of Seville, *De natura rerum* 26: *De nominibus astrorum*. A title has been written at the top of fol. 1r by a nineteenth-century hand: "De quibusdam stellis et animalibus fragmentum."[160]
    *Inc.* "Legitur in iob dicente domino..."

24. 3r–8v: *Physiologus* in 16 chapters (Y family).[161] Untitled.
    *Inc.* "Incipimus loqui de leone rege bestiarum..."

## SCRIPT

Paris, Bibliothèque nationale de France, MS Nouv. acq. lat. 1616: Copied by several hands writing Caroline minuscule, of very different character, most with an uncertain ductus.

Orléans, Médiathèque, MS 18 (*olim* 15): Copied by several well-practiced hands writing Caroline minuscule.

Paris, Bibliothèque nationale de France, MS Nouv. acq. lat. 455: Copied by three hands writing Caroline minuscule with Visigothic features. See Chapter 5 for a palaeographical analysis.

159 Also found in Laon, Bibliothèque municipale Suzanne Martinet, MS 40A, fols. 136v–137r).

160 "Fragment on certain stars and animals."

161 The chapter on the *ercine* seems to be from the A recension. Further textual study is required.

## DECORATION

Paris, Bibliothèque nationale de France, MS Nouv. acq. lat. 1616: Multiple computistical tables and drawings. On 3r a face has been drawn inside the first letter, *Q*, a *littera notabilior*.

Orléans, Médiathèque, MS 18 (*olim* 15): Rubricated *litterae notabiliores* and chapter numbers throughout the text.

Paris, Bibliothèque nationale de France, MS Nouv. acq. lat. 455: Somewhat clumsy blue and yellow decorated initials on 1v and 2r; brown ink initial L on 1r filled with geometric designs. All other initials in brown ink, generally plain or lightly decorated with plant motifs.

## PROVENANCE

This collection was one of the books stolen by Guglielmo Libri from French libraries between ca. 1824 and 1848 (see introduction to this handlist entry above). It was bought from him in 1847 by Lord Ashburnham and in 1888 acquired by the Bibliothèque nationale de France. Paris, MS 1616 has twelfth-century provenance in Fleury (medieval classmark P 3), which suggests that the whole collection was once there. Similarly, the Orléans manuscript had a Fleury exlibris on p. 1 which was erased by Libri.

### 12. St. Gallen, Stiftsbibliothek, MS 230[162]

i + i + 280 pp. | 21 × 13 cm | 28–31 lines.

The original codicological whole is complete and remains in its original quire order. Another manuscript—St. Gallen, Stiftsbibliothek, Cod. Sang. 125—partially depends on St. Gallen, Stiftsbibliothek, Cod. Sang. 230 (pp. 137–248 are copies of pp. 348–437).[163] Since this latter codex is dated to the last third of the eighth century, Lowe's earlier date of s. viii$^{2}$ for St. Gallen, Stiftsbibliothek, Cod. Sang. 230 is preferable to Scherrer's date of s. ix$^{in}$. In addition, St. Gallen, Stiftsbibliothek, Cod. Sang. 40 is written in a hand very similar to the hand of St. Gallen, Stiftsbibliothek, Cod. Sang. 230, indicating a possible relationship.

The codex is tentatively given by Scherrer and others as originating at St. Denis.[164] While it contains the story of the passion of St. Dionysius and its decoration "recalls French ornamentation," according to Lowe, there is no further indication that it originated at that centre. Both Lowe and Bischoff locate the manuscript's origin to St. Gallen.

---

**162** http://www.e-codices.unifr.ch/en/list/one/csg/0230; CLA VII.933. St. Gallen, Stiftsbibliothek, Cod. Sang. 230 is only briefly mentioned in Bischoff, *Katalog der festländischen Handschriften*, 3:316. The principal catalogue for this manuscript is Scherrer, *Verzeichniss*, 83–4..

**163** For example, in Dorfbauer, "Der Codex St. Gallen, Stiftsbibl. 125," 5–7. CLA VII.933.

**164** For example, in Dorfbauer, ed., *Pseudo-Augustinus*, 232.

Nebbiai-Dalla Guarda, in her extensive study of the books of St. Denis, noted that the passion of Dionysius in St. Gallen, Stiftsbibliothek, Cod. Sang. 230 was copied from a St. Denis codex, which is now lost, at St. Gallen. The French and Swiss abbeys enjoyed a good relationship from the eighth century onwards, perhaps begun by Waldo of St. Gallen when he became head of the Abbey of St. Denis in 782. The present manuscript is one of at least two (with St. Gallen, Stiftsbibliothek, Cod. Sang. 446) that are witnesses of this relationship.[165] That the entire manuscript, including marginal notes, was copied in Alemannic minuscule also suggests that it was copied by a St. Gallen-trained scribe at St. Gallen. Finally, the different colours, layout, and form of uncial headings throughout the manuscript resemble other books from St. Gallen (for example, St. Gallen, Stiftsbibliothek, Cod. Sang. 229).

The codex has modern pagination from 1 to 571. This includes the end pastedown: although Lowe believed it to be unnumbered, it is now clearly marked "571" in pencil. The pagination is incorrect after 349, when it jumps ten to 360. Each quire is signed with hollow capitals or with minuscule or majuscule letters in alphabetical order followed by Roman numerals, some in colour and set off by dots or dashes. Most of the manuscript is written by a single hand, except page 564 onwards.

## COLLATION

I–V$^{8}$, VI$^{6}$, VII–VIII$^{8}$, IX$^{10}$, X–XXXV$^{8}$.

## CONTENTS

1. Ir: Paper note by Ildefons von Arx, monastery librarian of St. Gallen 1823–33, recording the hand of Winithar in the second half of the codex: "Posterior pars hujus codicis manifeste scripta est abs Winithario presbytero, qui sub S. Othmaro vixit, et exaravit codices n. 907, et n. 70 ubi de eo plura leges." Signed "v. Arx. Bibliothecarius et Regens Seminarii 1825." The same hand has inserted "n. 238" after "n. 70" in a darker ink. The identification of the hand which copied items 47–49 as Winithar's is doubtful.
2. Iv: Latin list of contents, unidentified nineteenth-century hand.
   *Inc.* "Yn hoc codice continentur..."
3. iir: Small paper note in a modern hand recording the former presence of 4 strips of text from the *Edictus Rothari*, a seventh-century compilation of Lombard law, in the spine of the codex: "Aus MS 230 wurden im Rücken 4 Streifen mit Text des Edictus Rothari herausgenommen." Verso blank.
4. 1–2: Original list of contents.
   *Inc.* "In nomine sancte trinitatis incipit concollectio diuinorum librorum..."

165 Nebbiai-Dalla Guarda, *La Bibliothèque de l'abbaye de Saint*-Denis, 59.

5. 2–144: Excerpts from works by Isidore of Seville: *Liber sententiarum* I (p. 2); *Liber differentiarum* II (p. 49); *Etymologiae* 3.42–71 (*On Astronomy*, p. 81) and VII (*On God, the Saints and Apostles*, p. 93); and *De ecclesiasticis officiis* I (p. 118).
   *Inc.* "In nomine sancte trinitatis incipit liber sancti ysidori yspanensis patri quod deus summus et incommutabilis sit. Summum bonum deus est..."

6. 144–83: Eucherius, *Formulae spiritalis intellegentiae.*
   *Inc.* "In nomine domini nostri iesu christi incipit liber sancti eucherii de grecis nominibus uel hebracis. Eucherius uerano filio in christo salutem dicit. Formulas spiritalis..."

7. 183–203: Eucherius, *Instructiones*, book II.
   *Inc.* "Incipit interpretatio nominum liber secundus. Adam homo...."

8. 203–69: Eucherius, *Instructiones*, book I.
   *Inc.* "Incipit de questionibus ueteris et noui testamenti. Apocrifa recondita..."

9. 269–316: Pseudo-Augustine, *Dialogus quaestionum*. Dialogue on 65 theological questions between Augustine and Orosius.[166]
   *Inc.* "Incipit prologus de questiunculis sancti augustini. Licet multi et probatis..."

10. 316–25: Questions and responses based on the works of Isidore, especially books 6 and 11 of the *Etymologiae*, encompassing the books of the Bible, man, and portents; and on the *Differentiae* II.5.13, encompassing the Trinity.
    *Inc.* "Incipiunt sententias de floratibus diuersis. Interrogatio..."

11. 325–31: "Octo pondera de quibus factus est Adam," from the Anglo-Saxon prose dialogue *Solomon and Saturn*.
    *Inc.* "Incipiunt de octo pondera. Octo pondera de quibus factus est adam..."

12. 331–41: The two letters of Jerome and Damasus from the *Liber pontificalis.*
    *Inc.* "Damasi ad hieronymum. Quid sibi uult quae in genesi..."

13. 341–48: Anonymous commentary on selected passages from the Gospels, falsely (according to Migne) attributed to Arnobius of Sicca.[167]
    *Inc.* "Incipit træ euangelii excarpsum hieronimi presbiteri. In principio erat uerbum..."

14. 348–61: Jerome, *Commentarii in euangelium Matthaei* III.21 and 16 (from p. 349).
    *Inc.* "Sermo de tractatu sancti hieronimi presbiteri ex euangelio mathei. Euangelica lectio quae nobis hodie recitata est..."

15. 361–64: Augustine, *Sermo II de Scripturis*, attributed sermon on the calling of Abraham.
    *Inc.* "Humilia sancti agustini de uocatione abraham. Modo cum diuina lectio legitur..."

---

**166** Also known as *Liber quaestionum Augustini*, *Quaestiunculae Augustini* or *Quaestiones Orosii et responsiones Augustini*. For a discussion of the history of this text, see Dorfbauer, "Eine Untersuchung"; Dorfbauer, ed., *Pseudo-Augustinus*, 232–3; and Keskiaho, *Dreams and Visions*, 169–73.

**167** See PL 53.569.

16. 364–67: Augustine, *Sermo VI de Scripturis*, attributed sermon on the immolation of Isaac.
    *Inc.* "Incipit sermo sancti agustini episcopi de abraham et isac filium suum. Lectio ille fratres karissimi in qua beatus abraham isaac filium suum in holocausto..."

17. 367–70: Augustine, *Sermo XIII de Scripturis*, attributed sermon on Joseph.
    *Inc.* "Incipit humilia de sancto ioseph. Quotiens uobis fratres karissimi lectiones de testamentis ueteri recitantur..."

18. 370–74: Augustine, *Sermo XVI de Scripturis*, attributed sermon on Joseph's death and the sons of Israel.
    *Inc.* "Incipit sermo de eo quod scriptum est mortuus ioseph et filii israel creuerunt. Audiuimus in lectione..."

19. 374–408: Gregory I, extracts from the *Homiliaria in Evangelio*.
    *Inc.* "Incipiunt sententias excarapsas de humilias sancti gregori. Ecce mitto angelum meum..."

20. 408: Anonymous commentary on the Lord's Prayer.[168]
    *Inc.* "Incipit expositio de oratione domini. Pater noster qui es in caelis. Patrem inuocemus deum..."

21. 408–12: John Cassian, *Collationes patrum in scetica eremo* IX.18–23. Exposition of the Lord's Prayer.
    *Inc.* "Item expositio de oracione domini. Pater noster. Cum ergo uniuersitatis deum..."

22. 412–17: Anonymous commentary on the Lord's Prayer.
    *Inc.* "Item expositio de oracione domini. Pater noster. Hic confessio intellegendum..."

23. 417–19: Anonymous commentary on the Lord's Prayer.[169]
    *Inc.* "Item expositio de oracione domini. Pater noster qui es in caelis. Ut filius merearis..."

24. 419: a. Anonymous passage on the development and purpose of the Mass.
    *Inc.* "De missa pro quid cantat sacerdos. Missam caelebrare primum a sancto petro..."

    b. Ordinal of Christ in the Hibernian Chronological version.[170]
    *Inc.* "De septem gradibus ecclesie. Primus gradus lector..."

25. 419–20: List of the significance and link with Christ's life of each hour's sung *cursus* in the Divine Office.
    *Inc.* "De sancto matheo pro quales uirtutes cantatur omnis cursus. Primus cursus nocturnus propter hoc cantatur pro illa uirtute..."

---

**168** Listed, though without reference to this manuscript, in Bloomfield et al., eds., *Incipits of Latin Works*, 647.8861. This is the same commentary as Orléans, Médiathèque, MS 313 (*olim* 266), text 33.

**169** Listed, though without reference to this manuscript, in Bloomfield et al., eds., *Incipits of Latin Works*, 643.8807.

**170** As Reynolds, *The Ordinals of Christ*, 58.

26. 420–38: Isidore of Seville, *Allegoriae quaedam sacrae scripturae.*
*Inc.* "In christi nomine incipit prologus sancti hysidori de floratibus. Domino sancto hac reuerentissimo..."

27. 438–41: Anonymous Irish text introducing fourfold exegesis of the Gospels by discussing the four sacred rivers, the tetramorph etc., with a short extract from the book of the genealogy of Christ (*Liber generationis*) and a passage on Jacob.[171]
*Inc.* "Incipit de quatuor euangeliis seu de aliis questionibus. Primum quidem..."

28. 441–98: Defensor, *Liber scintillarum.* Attributed to St. Eligius in the text. Ends in the middle of chapter 76 (of 80).
*Inc.* "Incipit capitulorum libri huius. i. De caritate..."

29. 498–510: Gennadius of Marseille, *De ecclesiasticis dogmatibus.*
*Inc.* "Incipit doctrine fides ecclesie. i. De fide trinitatis..."

30. 510–518: *Physiologus* in 14 chapters (Y family).
*Inc.* "Incipit interpretatio spiritalis de libro bestiarum. De leone rege..."

31. 518–19: Anonymous commentary on the Song of Songs, including a comparison with Is. 63:1.
*Inc.* "Interpretatio de canticis canticorum. Quis est iste qui uenit ex edom..."

32. 520–21: Anonymous exegetical passage on law, introducing David the law-maker; interpreting ten coins from Luke 15:8–10 as ten prophets and ten virtues, with the woman as holy law; and listing seven Christian churches and their virtues.
*Inc.* "Gloriosus et inlustris propheta dauid..."

33. 521–26: Isidore of Seville, *Sententiae* II:1, *On Wisdom*, II:7, *On Conversion* and II:29, *On the Sermon.*
*Inc.* "Hisidorus dixit de sapientia. Omnis qui secundum deum sapiens est beatus est..."

34. 526–29: Anonymous text on virginity, patience, sobriety, and drunkenness, the four virtues, the virtues of the saints, and the house of the Lord.
*Inc.* "De uirginitate agustinus dixit. Si uis regnare cum christo..."

35. 529–34: Proverbs and wisdom sentences from the Bible.
*Inc.* "Incipiunt sententias salamonis. Initium sapientiae timor domini..."

36. 534: List of the Ten Commandments.
*Inc.* "De decem uerba legis. Audi Israhel. Dominus deus tuus deus unus est..."

37. 534–35: List of the symbols of the four Evangelists.
*Inc.* "De quattuor euangelistis. Multo ante predicta hie..."

38. 535–36: Allegorical interpretation of the four Evangelists.
*Inc.* "Interpretatio quatuor euangelistis. Olim ostensum est denique ex genesi..."

---

171 Identified by Bischoff, "Wendepunkte," 244. The text is partially dependent on Ps-Jerome, *Expositio quattuor evangeliorum.*

39. 536–37: Allegorical interpretation of the contents of Christ's tomb.
*Inc.* "De camara christi. Cum hominem dei hic est..."

40. 537–38: Jerome, *Commentarii in euangelium Matthaei* IV.25.1. On the parable of the ten virgins (Matt. 25).
*Inc.* "Incipit lectio de tractatu sancti hieronimi presbyteri de euangelii sancti matthaei. Paulo ante cum euangelica..."

41. 539–44: *Decretum Gelasianum de libris recipiendis et non recipiendis.*
*Inc.* "Incipit decretale sancti gelasii pape urbis rome. Post propheticas..."

42. 544–46: Anonymous question and answer text on the use of the psalms by King David.[172]
*Inc.* "Incipit inquisitio qualis psalmus primus a dauid fuisset cantatus. Psalterium inquirendum est..."

43. 546–47: Numerical list of the kinds of psalms.[173]
*Inc.* "Dauid filius iesse..."

44. 547–48: Text praising the singing of psalms.[174]
*Inc.* "Agustinus dixit. Canticum psalmorum..."

45. 548–49: Isidore of Seville, *Mysticorum expositiones sacramentorum seu quaestiones in vetus Testamentum. In Leviticum* XIII.
*Inc.* "Item sententias hisidori super leuiticum. Inter haec etiam iubetur..."

46. 549–63: Explanatory text on the letters of the alphabet. Alphabetical A–O with frequent reference to Jerome and Isidore. The beginning is taken from Donatus, *Ars maior.*
*Inc.* "In nomine sancte trinitatis incipit uolumen de litteris abcnariis [*sic*] XVII. De littera. Littera est pars minima..."

47. 564–68: On the passion of St. Dionysius (incomplete). Later eighth- to ninth-century addition (by Winithar? See item 1).
*Inc.* "Passio sancti diunisi cum sociis suis quod est VII ides octubris. Post domini nostri..."

48. 569–70: Jerome, *Commentarii in euangelium Matthaei* III.21. The same text as on pp. 348–9. Later eighth- to ninth-century addition (by Winithar? See item 1).
*Inc.* "Sermo de tractatu sancti hieronimi presbyteri ex euangelio mathei. Euangelica lectio..."

49. 571 (paste-down): Isidore of Seville, *De officiis* I:27. On Palm Sunday. Later eighth- to ninth-century addition (by Winithar? See item 1).
*Inc.* "Dies palmarum ideo celebratur quia in eo dominus..."

172 As Ayuso Marazuela, *La Vetus Latina Hispana*, 342–3.

173 As Ayuso Marazuela, *La Vetus Latina Hispana*, 295–6.

174 As Ayuso Marazuela, *La Vetus Latina Hispana*, 324–5.

### SCRIPT

The script is Alemannic minuscule. The headings and *litterae notabiliores* are in uncial. The single scribe separates words clearly, and frequently uses the *nt* ligature mid-word, as well as abbreviations such as the insular symbols for *enim* and *est*. Minor corrections have been made throughout. Other hands from the second half of the eighth century and the early ninth century have written the text from fol. 564 onwards.

### DECORATION

Most titles and initials are multi-coloured or rubricated uncial hollow letters. Beautifully decorated large initials are often used at the beginning of texts (e.g. p. 332). These contain motifs of birds, fishes, leaves, and sinuous lines.

### PROVENANCE

No information is available on the provenance of the codex. It is listed in the mid-ninth century catalogue of the library of St. Gallen abbey (St. Gallen, Stiftsbibliothek, Cod. Sang. 728, p. 10), indicating that it was in St. Gallen within approximately fifty years of its creation, if not from the beginning.[175] It is probable that it has remained at the abbey library ever since.

## 13. Vatican City, Biblioteca Apostolica Vaticana, Pal. lat. 1074[176]

> I + 30 fols. i + i + 25 fols. + I | 17 × 10.5–11 cm (part I), 13.5 × 9.5 cm (part 2) | 30–31 lines (part I), 18 lines (part II).

Part I (1r–30av) of this manuscript was written in an area where West Romance was spoken, possibly in Catalonia, in s. x–xi.[177] It is foliated in a post-medieval hand; a modern hand has re-traced the numbers in pencil. Early medieval quire numbers were added to the end of each quire. The text contains accents (see the discussion in Chapter 5).

Part II (30br–54v) was copied on paper in France in s. xvi². It has seventeenth- or eighteenth-century foliation, with additional sixteenth-century foliation from fol. 30b. Folios 30a and 30b are twentieth-century paper inserts separating the two manuscript parts. The Roman binding is from between 1869 and 1878, when the two parts were first put together.

Walz lists the contents of fols. 23v–30r under the heading "Dogmatica in processionem sancti spiritus." This is an umbrella heading for a group of texts that are given as separate items below.

---

**175** http://www.e-codices.unifr.ch/en/csg/0728/10.

**176** http://digi.ub.uni-heidelberg.de/diglit/bav_pal_lat_1074?ui_lang=ger. The primary catalogue is Walz, *Die historischen und philosophischen Handschriften*, 255–7.

**177** Located to Catalonia by Bischoff in a private letter to Walz.

## COLLATION

Part I: I–III$^{8}$, IV$^{6}$.

Part II: I$^{10}$ + 1, II$^{10}$, III$^{4}$.

## CONTENTS

### Part I

1. 1r–21r: *Physiologus* in 37 chapters (B family). Sixteenth-century title ("De naturis animalium").
   *Inc.* "Incipit liber phisiologi seu expositio de natura animalium et bestiarum. De tres naturis leonis..."

2. 21v–22r: Excerpts from Isidore of Seville, *Etymologiae* 16.10.1 on *margarita*, 12.7.24 on *psitacus* (with additions from various texts, including Persius, *Saturae*, prol. 1–13), 12.7.31 on *ercine*, 12.7.64 on *coturnix*.
   *Inc.* "De ethymologiis. Margarita prima candidarum gemmarum..."

3. 22r–23v: Anonymous, *Commentum in visiones Danielis* (Dn. 10.4–5; 10.7).
   *Inc.* "Cum essem inquid iuxta fluuium magnum tigris..."

4. 23v–24r: Profession of faith sent to Spanish bishops in 794 after the Council of Frankfurt in response to the Adoptionist controversy.[178]
   *Inc.* "Credimus in sanctam trinitatem id est patrem..."

5. 24v–27r: Venantius Fortunatus, *Commentum in symbolum Athanasianum*.[179]
   *Inc.* "Incipit expositio in fide catolica. Quicumque uult saluus esse..."

6. 27r–30r: Smaragdus of Saint-Mihiel, *Epistola de processione Spiritus Sancti*.[180]
   *Inc.* "Questio que de spiritum sancti processione est..."

7. 30r–30v: Charlemagne, *Epistola ad Alcuinum anno 798 scripta*.[181] The text breaks off at "ut aestimamus" and the remainder is lost.
   *Inc.* "Incipit responsio de septuagesima, sexagesima, quinquagesima et quadragesima. Carolus gratia dei rex..."

30a (paper): Blank.

### Part II

30b (paper): Blank.

8. 30br–54r: Pierre Gilles, *Descriptio noua elephanti*. Printed in Hamburg in 1614.
   *Inc.* "Elephanti tam terrestris quam marini atque equi fluuiatilis..."

---

**178** As MGH Conc. 2, 163–4.

**179** Pr. Burn, *The Athanasian Creed*, 28–39.

**180** Pr. Mai, *Scriptorum veterum nova collectio* 7, 252–5. As MGH Conc. 2.1, 236–9.

**181** As MGH Epp., 228–29.

## SCRIPT

Part I is written in a variety of hands using Caroline minuscule. There is frequent use of the *punctus uersus* and *punctus eleuatus*, and very frequent use of the acute accent throughout. The common mark of abbreviation is 2-shaped. Occasional text titles in uncial. *Litterae notabiliores* are in rustic capitals or uncial. Abbreviations are frequent and ascenders generally long and straight, with narrow, almost cramped letter forms. Many contemporary glosses and marginal notes. Part II is printed using a humanistic cursive type.

## DECORATION

Plain, except for occasional modest pen drawings on *litterae notabiliores*. Rubricated chapter titles in the *Physiologus*.

## PROVENANCE

Belonged to the castle library of Otto-Henry (Ottheinrich), Elector Palatine (d. 1559), then to the Heidelberg Heiliggeistkirche from ca. 1556.[182] At Heidelberg in 1622 when the city was captured as part of the 30 Years' War by Maximilian of Bavaria; in recompense for the support given him by the papacy, Pope Gregory XV asked for the famous manuscripts of the Bibliotheca Palatina. Some 11,000 volumes, of which 3,600 manuscripts, were taken to Rome in 1623. This manuscript is one of them.

# 14. Wolfenbüttel, Herzog August Bibliothek, MS Gud. lat. 148[183]

I + 124 fols. + I | 24 × 18.5 cm | 27–8 lines.

Identified by Bischoff as deriving from the same French scriptorium as Paris, Bibliothèque de l'Arsenal, MS 852, Paris, Bibliothèque nationale de France, MS lat. 2731A and Munich, Bayerische Staatsbibliothek, MS Clm 5250.[184] Butzmann's identification of the place of origin as eastern France is used here.

This manuscript has been dated very differently by Bischoff (s. ix$^{4/4}$, France), Butzmann (s. ix$^{med}$), and Michsack (s. x) Bischoff's dating of the manuscript is the most recent and seems the most accurate, so it has been adopted here.

Quire signatures range from Q I A to Q XV P (where the *Q* presumably stands for *quaterni/quaternum*), indicating that the original codicological whole is unaltered and almost complete. The last pages of the last quires are missing and so lack a quire sig-

182 Walz, *Die historischen und philosophischen Handschriften*, xxiv.

183 http://diglib.hab.de/?db=mss&list=ms&id=148-gud-lat. The two principal catalogues for this manuscript are Butzmann, "Die Weissenburger Handschriften," 300–302 and Milchsack, *Die Handschriften*, 164–65. It also has a short entry in Bischoff, *Katalog der festländischen Handschriften*, 3:502. Also pr. in Cahier and Martin, eds., *Mélanges* 2–4: In part 2 on pp. 107–223 (to chap. 14); in part 3 on pp. 238–85 (to chap. 19); and in part 4 on pp. 57–70 (to chap. 24). Translated and described in Steiger and Homburger, *Physiologus Bernensis*.

184 Bischoff, *Katalog der festländischen Handschriften*, 3:502.

nature. 1v has *.f.* as a signature letter in the top margin. This was a s. xv$^{2}$ addition to all Weissenburg monastery books to divide them by subject. *F* was the group for religious literature not included in earlier categories (biblical texts and biblical commentaries).

## COLLATION

I$^{8}$ (wants 1), II–XV$^{8}$, XVI$^{8}$ (wants 6, 7, 8).

## CONTENTS

1. End-leaf from an eleventh-century missal (one other leaf of which is the last end-leaf of Weissenburg 73).
   *Inc.* "Gratiam sancti spiritus..."
2. Ar: Sixteenth-century Weissenburg ownership inscription. "Folios A et 1–123" written just underneath in a more modern hand. Otherwise blank, ruled.
   *Inc.* "Liber monasterii sanctorum petri pauli apostolorum in wisszenburg In claistro."
3. Av: Fifteenth-century marginal list of contents.
   *Inc.* "Juliani Episcopi Toletani prognosticorum..."
4. Av–59v: Julian of Toledo, *Prognosticum futuri saeculi.*[185] The preface is put together from the beginning of a letter of Julian to Idalius, Bishop of Barcelona, and the end of the letter the latter sent in reply.[186] Lists of chapters have been added after the preface and before each new book.
   *Inc.* "In nomine domini et saluatoris nostri iesu christi incipit liber pronostocorum..."
5. 60r: Isidore of Seville, *Etymologiae* 1.37.22–34. Short Latin glossary that explains the words *allegoria, enigma, tropologia, parabola, paradigma, prosa, dialogus, apologeticum.*
   *Inc.* "Allegoria est alieni [alienum] eloquium..."
6. 60r–60v: Isidore of Seville, *Mysticorum expositiones sacramentorum seu quaestiones in vetus Testamentum. In Deuteronomium* XVI.3. On the eight vices.
   *Inc.* "De octo uiciis. De gastromargia namque nascuntur commessationes aebrietates..."
7. 60v–82r: Phaedrian fables in the Romulus tradition. 63 prose fables in 5 books, beginning with the letter of Aesop to Rufus. Also known as the "Weissenburg Aesop," and manuscript *W* in the family. Spaces for miniatures left between fables.
   *Inc.* "Incipit liber Ysopi [Esopi] magistro rufo aesopus salutem..."

---

**185** Sometimes also called the *Prognosticon futuri saeculi.* I follow the editor of Julian's works in calling it *Prognosticum* (CCSL 115).

**186** PL 96.455 and PL 96.460.

8. 82r–98r: Alcuin of York, *Compendium in Cantica canticorum Salomonis*.[187]
*Inc.* "Hunc cecinit salomon mira dulcedine librum. Qui tenet egregias..."

9. 84r: Note in upper margin recording the appearance of a new star and of a rainbow. The same hand has corrected and annotated the main text. Contemporary addition.
*Inc.* "Mense septembri .die VIII. [sabb. written above] .luna xxv. in die apparuit stella..."

10. 98v–108v: *Physiologus* in 25 chapters (C family). Butzmann noted that the text is the same version as that in Bern, Burgerbibliothek, MS 318. Spaces left for miniatures in and between chapters.
*Inc.* "De natura leonis. Est leo regalis [rex] omnium animalium et bestiarum..."

11. 108v–123v: *Liber monstrorum*. Text includes a preface as well as books II and III of the *Liber*, entitled *De beluis* and *De serpentibus*.
*Inc.* "Incipit liber monstruorum [monstrorum] de diuersis generibus. De oculto orbis..."

## SCRIPT

Well-formed Caroline minuscule in light brown ink, one hand. Frequent uncial *d*. Very many corrections by a Weissenburg hand of s. ix$^2$ (that is, a contemporary hand), and many words erased.[188] All headings and *litterae notabiliores* in red or black rustic capitals. *Incipit* and *explicit* lines in large rustic capitals.

## DECORATION

On Ar black and red lines interchange; 4 lines are in majuscules and 6 lines in rustic capitals.

## PROVENANCE

In Weissenburg abbey library in the fifteenth century (judging from the inscription on Ar). This library became a collegiate foundation in 1524. The manuscript was acquired by scholar and archaeologist Marquard Gude (d. 1689). He left his library to his son Peter Marquard, who auctioned it off gradually. In 1710 most of the manuscripts from this library, including Wolfenbüttel, Herzog August Bibliothek, MS Gud. lat. 148, were bought by Gottfried Wilhelm Leibniz, the librarian of Duke Anton Ulrich von Braunschweig, and housed in the Herzog August Bibliothek Wolfenbüttel, where they have remained.

---

**187** PL 100.639–64.

**188** Butzmann, "Die Weissenburger Handschriften," 300.

Appendix II

# *PHYSIOLOGUS* FAMILIES

**A:** Brussels, Bibliothèque royale de Belgique, MS lat. 10066–77; Munich, Bayerische Staatsbibliothek, MS Clm 14388.

**B:** Bern, Burgerbibliothek, MS 225 + Bern, Burgerbibliothek, MS 233 + Orléans, Médiathèque, MS 313; Montecassino, Archivio dell'Abbazia, MS 316 + MS 323; Oxford, Bodleian Library, MS Auct. T.2.23; Vatican City, Biblioteca Apostolica Vaticana, MS Pal. lat. 1074.

**C:** Bern, Burgerbibliothek, MS 318; Wolfenbüttel, Herzog August Bibliothek, MS Gud. lat. 148.

**Y:** Bern, Burgerbibliothek, MS 611 + Paris, Bibliothèque nationale de France, MS lat. 10756; Oxford, Bodleian Library, MS Laud Misc. 129; Munich, Bayerische Staatsbibliothek, MS Clm 19417; Paris, Bibliothèque nationale de France, MS Nouv. acq. lat. 1616 + Orléans, Médiathèque, MS 18 (*olim* 15) + Paris, Bibliothèque nationale de France, MS Nouv. acq. lat. 455; St. Gallen, Stiftsbibliothek, Cod. Sang. 230.

**Unknown:** Chartres, Médiathèque L'Apostrophe, MS 63 (125) (perhaps C family): "De laena, de audolaps, de pelicano, de nocticorax, de aquila, defenice, de uppupa, de vipera, de serpente, de formica, de ono centauro" (partial list from catalogue).

## *Physiologus* Chapters by Family and Manuscript

The chapters are listed in the order they appear in the main text of each *Physiologus*. If a manuscript copy has a list of contents, any chapters missing from the list are indicated. Where chapters do not have rubrics in the main text, they are taken either from the list of contents, or directly from the chapter if no contents list is available. Square brackets indicate chapters in the main text that are not listed in the list of contents. Angled brackets indicate missing or added text. Spelling is preserved as it appears in the manuscripts. The number of each chapter refers only to its numerical order in the main text of the manuscript (not to the numbers assigned to it by the scribe, or in the list of contents, as these are often inaccurate).

## A FAMILY

### Brussels, Bibliothèque royale de Belgique, MS lat. 10066–77

| No. | Fol. | Chapter |
|---|---|---|
| 1. | 140r | leo |
| 2. | 140v | autolops |
| 3. | 141v | lapides igniferi |
| 4. | 142r | serra |
| 5. | 142r | caladrius |
| 6. | 143r | pelicanus |
| 7. | 143v | nycticorax |
| 8. | 144r | aquila |
| 9. | 144v | phoenix |
| 10. | 145v | formica |
| 11. | 146r | syrenae et onocentauri |
| 12. | 146bisr | uulpis |
| 13. | 146bisr | unicornis |
| 14. | 147r | castor |
| 15. | 147v | hyaena |
| 16. | 148r | dorca |
| 17. | 148v | onager |
| 18. | 149r | hydrus |
| 19. | 149r | simia |
| 20. | 149v | perdix |
| 21. | 150r | asida / isida |
| 22. | 150v | salamandra |
| 23. | 150v | turtur |
| 24. | 151r | columbae |
| 25. | 152r | epops |
| 26. | 152v | onager |
| 27. | 152v | uipera |
| 28. | 153r | serpens |
| 29. | 153v | herinacius |
| 30. | 154r | arbor peridexion |
| 31. | 154v | elephans |

| No. | Fol. | Chapter |
|---|---|---|
| 32. | 155v | lapis achatis |
| 33. | 155v | margarita |
| 34. | 155v | lapis adamantinus |
| 35. | 156r | lapis sindicus |
| 36. | 156r | herodius |
| 37. | 156v | leo et panthera |

### Munich, Bayerische Staatsbibliothek, MS Clm 14388

| No. | Fol. | Chapter |
|---|---|---|
| 1. | 173r | De leone |
| 2. | 173r | De autolope |
| 3. | 173v | De lapidibus pirobolis |
| [4. | 173v | De serra mecrina] |
| 4. | 174r | De charadra |
| 5. | 174v | De pelicano |
| 6. | 174v | De nycticorace |
| 7. | 175r | De aquila |
| 8. | 175r | De penice |
| 9. | 175v | De epopo |
| 10. | 175v | De onagro |
| 11. | 175v | De uipera |
| 12. | 176r | De serpente |
| 13. | 176r | De formica |
| 14. | 176v | De syrena onocentauro |
| 15. | 177r | De erinace |
| 16. | 177r | De hibice |
| 17. | 177v | De uulpe |
| <missing> | | De arbore peridexion et columbis[1] |
| 18. | 178r | De elephanto |
| 19. | 178r | De dorchon |
| 20. | 178r | De achate lapide |

1 Since "De uulpe" has text missing from the end and "De elephanto" from the start, the scribe has evidently jumped a page in their exemplar without noticing.

| No. | Fol. | Chapter |
|---|---|---|
| 21. | 178r | De lapide sostoros et de margarita |
| 22. | 178v | De lapide adamantino |
| 23. | 178v | De alia natura onagri et simis |
| 24. | 178v | De lapide senditichos |
| 25. | 179r | De herudion id est fulicet |
| 26. | 179r | De psicomora |
| 27. | 179v | De leone et pantera |
| 28. | 179v | De ceto id est aspidoceleon |
| 29. | 180r | De perdice |
| 30. | 180r | De uulturæ |
| 31. | 180v | De myrmicoleon |
| 32. | 181r | De mustela |
| 33. | 181r | De monoceraton |
| 34. | 181r | De castore |
| 35. | 181v | De uena hoc est belua |
| 36. | 181v | De niluo |
| 37. | 181v | De echinemon |
| 38. | 182r | De cornicola |
| 39. | 182r | De detture |
| 40. | 182r | De hyrundine |
| 41. | 182r | De ceruo |
| 42. | 182v | De rana |
| 44. | 182v | De alia idem salamandra |
| 45. | 182v | De lapide magniteri |
| 46. | 182v | De lapide adamantino |
| 47. | 183r | De columbis |
| 48. | 183r | De saura alia hoc est anguilla solis |

## B FAMILY

### Bern, Burgerbibliothek, MS 225 + Bern, Burgerbibliothek, MS 233 + Orléans, Médiathèque, MS 313

| No. | Fol. | Chapter |
|---|---|---|
| 1. | 1v | De natura leonis |
| 2. | 2r | De auta lobs |
| 3. | 2r | De caerobolim lapides igniferi |
| 4. | 2v | De serra in mare |
| 5. | 2v | De caladrius |
| 6. | 3r | De pellicano |
| 7. | 3v | De necticorce |
| 8. | 3v | De aquila |
| 9. | 4r | De uolatilae fenix |
| 10. | 4v | De uppuba |
| 11. | 4v | De formice natura |
| 12. | 5v | De serene et uno centauris |
| 13. | 6r | De herenaciis |
| 14. | 6v | De hibes |
| 15. | 7r | De uulpe |
| 16. | 7v | De moneceron |
| 17. | 8r | De animal casto |
| 18. | 8v | De hyena que belua |
| 19. | 8v | De hildris |
| 20. | 9r | De corcon |
| 21. | 9v | De onagro |
| 22. | 9v | De folica |
| 23. | 9v | De patera |
| 24. | 11r | De aspedocalone |
| 25. | 11v | De perdicae |
| 26. | 11v | De mustela |
| 27. | 11v | De aspide |
| 28. | 12v | De turtura |
| 29. | 12v | De ceruo item in psalmo |
| 30. | 13r | De salamandra |
| 31. | 13r | De simia |
| <missing> | | De carnium esu uel piscium |

**Montecassino, Archivio dell'Abbazia, MS 316 + MS 323**

| No. | Page | Chapter |
|---|---|---|
| 1. | 65 | Leo |
| 2. | 66 | Achiles |
| 3. | 66 | Lapides igniferi |
| 4. | 67 | Serra |
| 5. | 68 | Caladrius |
| 6. | 69 | De aue pelicano |
| 7. | 70 | De aue necticorace |
| 8. | 71 | De aquila |
| 9. | 71 | De aue fenice |
| 10. | 73 | De aue upupa |
| 11. | 73 | Formica |
| 12. | 76 | De serenis et honocentaurus |
| 13. | 77 | De erinacus |
| 14. | 77 | De aues hybis |
| 15. | 79 | De uulpe |
| 16. | 80 | De monoceron quod latine dicitur unicornis |
| 17. | 81 | Item de castora |
| 18. | 83 | De uelba |
| 19. | 84 | De yllo |
| 20. | 84 | De caprea |
| 21. | 86 | De onagro |
| 22. | 86 | De panthera |
| 23. | 90 | De aspidatestugine |
| 24. | 92 | De perdice |
| 25. | 93 | De mirmicoleon |
| 26. | 95 | De mustela |
| 27. | 96 | De asida quod est strution |
| 28. | 97 | De turtura |
| 29. | 98 | De cerbo |
| 30. | 99 | De stilione |
| 31. | 100 | De natura columbarum expositio |
| 32. | 104 | De elefantis |
| 33. | 107 | De lapide adamantine |
| 34. | 110 | De lapide concus |

**Oxford, Bodleian Library, MS Auct. T.2.23**

| No. | Fol. | Chapter |
|---|---|---|
| 1. | 128r–v | De natura leonis |
| 2. | 129r | De autalops |
| 3. | 129v | De terrebolim lapidis igniferi |
| 4. | 130r | De serra in mare |
| 5. | 130v | De calidrio |
| 6. | 131v | De pellicano |
| 7. | 132r | De necticorace |
| 8. | 132v | De aquila |
| 9. | 133r | De fenice |
| 10. | 134r | De upupa |
| 11. | 136r | De natura formicae |
| 12. | 136v | De serenis et onocentauris |
| 13. | 137r | De hirenaciis |
| 14. | 138r | De ibisohc |
| 15. | 139r | De uulpe |
| 16. | 140r | De monoceras |
| 17. | 141r | De castore |
| 18. | 142r | De belua |
| 19. | 143r | De hydris |
| 20. | 143v | De dorcon |
| 21. | 144v | De onager<o> |
| 22. | 145r | De simia |
| 23. | 145v | De fulicæ |
| 24. | 146r | De pantera |
| 25. | 149r | De aspedocelone quia aspedo testago naturus ii |
| 26. | 150r | De perdice |
| 27. | 151r | De mustela |
| 28. | 152r | De asidia |
| <missing> | | De turture |
| 29. | 152v | De coruo |
| 30. | 154v | De salamandra |
| <missing> | | De columbarum natura |
| <missing> | | De arborę perdixion |

| No. | Fol. | Chapter |
|---|---|---|
| <missing> | | qui circa dextra (*sic*) |
| <missing> | | De heliphanto |
| <missing> | | De amos propheta |
| <missing> | | De mirmicole |

## Vatican City, Biblioteca Apostolica Vaticana, MS Pal. lat. 1074

| No. | Fol. | Chapter |
|---|---|---|
| 1. | 1r | De natura leonis |
| <missing> | | De austolops |
| 3. | 2r | De teroboli lapidem igniferum |
| 4. | 2r | De serra |
| 5. | 2v | De caladrio |
| 6. | 3r | De pellicano |
| 7. | 3v | De nectcoraie |
| 8. | 3v | De aqila |
| 9. | 4r | De phenice |
| 10. | 4v | De upupo |
| 11. | 4v | De tribus naturis formice |
| 12. | 5v | De serena et onocentauro |
| 13. | 6r | De renacho |
| 14. | 6v | De ibice |
| 15. | 7r | De uulpis |
| 16. | 7v | De monacero |
| 17. | 8r | De castore |
| 18. | 8v | De henna |
| 19. | 9r | De ridris |
| 20. | 9v | De drochon |
| 21. | 10r | De honagro |
| 22. | 10v | De simia |
| 23. | 10v | De fulica |
| 24. | 11r | De pantera |
| 25. | 13r | De duabus naturis aspidiscelonis |
| 26. | 13v | De perdice |
| 27. | 14r | De mustela |

| No. | Fol. | Chapter |
|---|---|---|
| 28. | 14v | De assida et strucione |
| 29. | 15r | De turture |
| 30. | 15r | De ceruo |
| 31. | 15v | De salamandra |
| 32. | 16r | De columbarum naturis |
| 33. | 16r | De arboe prexioti |
| 34. | 17v | De helefante |
| 35. | 18v | De amos propheta |
| 36. | 19r | De adamate lapide |
| 37. | 20r | De mercholion |

## C FAMILY

### Bern, Burgerbibliothek, MS 318

| No. | Fol. | Chapter |
|---|---|---|
| 1. | 7r | Est leo regalis |
| 2. | 8v | De natura animalium aesaure |
| 3. | 9r | De natura uolatile quae dicitur calatrius |
| 4. | 9v | Pelicanus |
| 5. | 10r | De nocticoracis |
| 6. | 10v | De natura uolatile aquile |
| 7. | 11r | De natura uolatile quæ dicitur yppopus |
| 8. | 11v | De natura uiperae |
| 9. | 11v | De natura serpentis secunda |
| 10. | 12v | De natura formicae |
| 11. | 13v | De natura serenae et honocentauris |
| 12. | 14r | De natura yricii |
| 13. | 14v | De natura uulpis |
| 14. | 15r | De animale qui dicitur pantherius |
| 15. | 15v | De cęto magno aspidohelunes |
| 16. | 16v | De animale unicornium |
| 17. | 17r | De ceruo |
| 18. | 17v | De natura animalis qui dicitur salamandra |
| 19. | 18r | De arbore qui dicitur peredexion |
| 20. | 18r | De animale qui dicitur antelups |

| No. | Fol. | Chapter |
|---|---|---|
| 21. | 18v | De natura piscis maximo qui dicitur serra |
| 22. | 19r | De elifanto et mandragora |
| 23. | 20v | De lapide agato |
| 24. | 21r | De lapide indico |
| 25. | 21v | De galli cantu |
| 26. | 22v | Caballus |

## Wolfenbüttel, Herzog August Bibliothek, MS Gud. lat. 148

| No. | Fol. | Chapter |
|---|---|---|
| 1. | 98v | De natura leonis |
| 2. | 99v | De natura esauriliace |
| 3. | 99v | De natura calidrii |
| 4. | 100r | De animal qui dicitur pellicanus |
| 5. | 100v | De nocti<c>orace |
| 6. | 101r | De natura aquile |
| 7. | 101r | De natura uolatile quod dicitur ipopus |
| 8. | 101v | De natura uipere |
| 9. | 102v | De quarta natura serpentis |
| 10. | 103r | De formica exigua |
| 11. | 103v | De natura serene maris |
| 12. | 103v | De natura ericii |
| 13. | 104r | Uulpis |
| 14. | 104v | De panterio |
| 15. | 104v | Cetus |
| 16. | 105v | Unicornis |
| 17. | 105v | Ceruus |
| 18. | 106r | Salamandra |
| 19. | 106r | Columba |
| 20. | 106v | Antelups |
| 21. | 106v | Serra |
| 22. | 107r | Elifans |
| 23. | 107v | Agates |
| 24. | 108r | De lapide indico |
| 25. | 108r | Galli cantus |

## Y FAMILY

### Bern, Burgerbibliothek, MS 611 + Paris, Bibliothèque nationale de France, MS lat. 10756

| No. | Fol. | Chapter |
|---|---|---|
| 1. | 117v | De leone bestiarum |
| 2. | 119r | De autolaps |
| 3. | 119r | De lapide pyropoli |
| 4. | 119v | De serra |
| 5. | 120r | De caradrione |
| 6. | 120v | De pellegano |
| 7. | 121r | De necticora |
| 8. | 121v | De aquila |
| 9. | 122r | De faenice |
| 10. | 122v | De oppope |
| 11. | 123r | De onagro |
| 12. | 123r | De uipera |
| 13. | 124r | De serpente |
| 14. | 124v | De formica |
| 15. | 125v | De serine et onocentauri |
| 16. | 126r | De herenacio |
| 17. | 127r | De uulpe |
| 18. | 128r | De unicorni |
| 19. | 129r | De castur |
| 20. | 129v | De simea |
| 21. | 130r | De panthera |
| 22. | 130v | De arbore perindex |
| 23. | 131r | De elephante |
| 24. | 132v | De dorcon siui cabriola |
| 25. | 133r | De adamans |
| 26. | 133v | De honager |
| 27. | 133v | De senditico |
| 28. | 134r | De herudio siui folicae |
| 29. | 134v | De achatae natura |
| 30. | 135r | De coeto magno |
| 31. | 135v | De perdice |

| No. | Fol. | Chapter |
|---|---|---|
| 32. | 135v | De uultori et ambrone |
| 33. | 136v | De mustella |
| 34. | 136v | De sullo |
| 35. | 137r | De chineomone |
| 36. | 137r | De turture |
| 37. | 137r | De herundene |
| 38. | 137v | De ceruo |
| 39. | 138r | De lapide magnite |
| 40. | 138r | De lapide adamantino |

## Munich, Bayerische Staatsbibliothek, MS Clm 19417

| No. | Fol. | Chapter |
|---|---|---|
| 1. | 29v | De leone |
| 2. | 31r | De autolope |
| 3. | 31v | De lapidibus piropolis |
| 4. | 32r | De serra marina |
| 5. | 33r | De charadrio |
| 6. | 34v | De pelicano |
| 7. | 35v | De nocticarace |
| 8. | 36r | De aquila |
| 9. | 37r | De phenice |
| 10. | 38v | De epopo |
| 11. | 39r | De onagro |
| 12. | 39v | De uipera |
| 13. | 40v | De serpente |
| 14. | 42r | De formica |
| 15. | 44r | De syrena onocentauro |
| 16. | 45v | De hibice |
| 17. | 47r | De uulpe |
| 18. | 48r | De arbore peridexion et columbis |
| 19. | 49v | De elephanto |
| 20. | 52v | De dorchon |

| No. | Fol. | Chapter |
|---|---|---|
| 21. | 53v | De agathe lapide |
| 22. | 53v | De lapide sosturos et de margarita |
| 23. | 54v | De lapide adamantino |
| 24. | 55r | De alia natura onagri et simi |
| 25. | 55v | De lapide senditichos |
| 26. | 56v | De herodion id est fulice |
| 27. | 57r | De psycomora |
| 28. | 58v | De leone et pantera |
| 29. | 59r | De ceto id *est* aspido celeon |
| 30. | 60r | De perdice |
| 31. | 60v | De uulture |
| 32. | 62r | De mirmice |
| 33. | 62v | De mustela |
| 34. | 63v | De monoceraton |
| <missing> | | De castore |
| 35. | 64r | De yena hoc est belua |
| [36. | 64v | De niluo] |
| 37. | 65r | De echinemon |
| 38. | 65v | De cornicola |
| 39. | 66r | De turture |
| 40. | 66v | De hyrundine |
| 41. | 67r | De ceruo |
| [42. | 67v | De rana] |
| 43. | 68r | De salamandra |
| 44. | 68v | De lapide magniteri |
| 45. | 69r | De lapide adamantino |
| 46. | 69v | De columbis |
| 47. | 70r | De saura eliace hoc est anguilla solis |

## Oxford, Bodleian Library, MS Laud. Misc. 129

| No. | Fol. | Chapter |
|---|---|---|
| 1. | 149v | De leone[2] |
| 2. | 150r | De autolape |
| 3. | 150v | De pirabolo |
| 4. | 150v | De serra |
| 5. | 150v | De caratrio |
| 6. | 151r | De pellicano |
| 7. | 151v | De noctecoracae |
| 8. | 151v | De foenicae |
| 9. | 152r | De upupo |
| 10. | 152v | De onagro |
| 11. | 152v | De serpente |
| 12. | 153v | De aquila |
| 13. | 154r | De formica |
| 14. | 154v | De uulpe |
| 15. | 155r | De arbore perennicae |
| 16. | 155v | De elifante |
| 17. | 156v | De agate |
| 18. | 157r | De lapide adamantino |
| 19. | 157r | De pardice |
| 20. | 157v | De ultore |
| 21. | 158r | De renocae |
| 22. | 158r | Lagates lapis[3] |

**2** This is part of the rubric, which reads: "Incipit de natura animalium de leone et animalium et enim iacob bene dicens ait catulus leonis iuda." The scribe has clearly conflated the original rubric, the heading for the chapter on the lion, and the first line of that chapter, which reads: "Iacob, benedicens Iudam filium suum, ait: Catulus leonis Iuda." See Carmody, *Physiologus latinus versio Y*, 95–134.

**3** No rubric; title taken from chapter.

**Paris, Bibliothèque nationale de France, MS Nouv. acq. lat. 1616 + Orléans, Médiathèque, MS 18 (*olim* 15) + Paris, Bibliothèque nationale de France, MS Nouv. acq. lat. 455**

| No. | Page | Chapter |
|---|---|---|
| 1. | 3r | de leone |
| 2. | 3v | de autulpe |
| 3. | 4r | de lapide piropolo |
| 4. | 4r | de serra |
| 5. | 4r | de caratrio |
| 6. | 4v | de pellicano |
| 7. | 5r | de nocticorace |
| 8. | 5r | de aquila |
| 9. | 5v | de fehice |
| 10. | 6r | de opupa |
| 11. | 6r | de onagro |
| 12. | 6r | de vipera |
| 13. | 6v | de serpente |
| 14. | 7v | de formica |
| 15. | 8r | de syrene[4] |
| 16. | 8v | de erenacio |

4 Includes onocentaur.

### St. Gallen, Stiftsbibliothek, MS 230

| No. | Page | Chapter |
|---|---|---|
| 1. | 510 | De leone rege bestiarum |
| 2. | 511 | De elifanto |
| 3. | 512 | De talapo |
| 4. | 513 | De serra |
| 5. | 513 | De erenatio |
| 6. | 514 | De formica |
| 7. | 515 | De bestis quodam in monte orientis |
| 8. | 515 | De uipera |
| 9. | 515 | De serpente |
| 10. | 516 | De auibus de caradrio |
| 11. | 517 | De pellicano |
| 12. | 517 | De nictycorax |
| 13. | 517 | De aquila |
| 14. | 517 | De quodam arbore uel columba |

# BIBLIOGRAPHY

## Manuscripts


Albi, Médiathèque Pierre-Amalric, RES MS 39, https://cecilia.mediatheques.grand-albigeois.fr/idurl/1/114

Bamberg, Staatsbibliothek, MS Msc. Patr. 61, https://bavarikon.de/object/bav:SBB-KHB-00000SBB00000157

Bern, Burgerbibliothek, MS 3, www.e-codices.unifr.ch/en/list/one/bbb/0003

Bern, Burgerbibliothek, MS 4, www.e-codices.unifr.ch/en/list/one/bbb/0004

Bern, Burgerbibliothek, MS 109

Bern, Burgerbibliothek, MS 225

Bern, Burgerbibliothek, MS 233

Bern, Burgerbibliothek, MS 318, www.e-codices.unifr.ch/en/list/one/bbb/0318

Bern, Burgerbibliothek, MS 611, www.e-codices.unifr.ch/en/list/one/bbb/0611

Bern, Burgerbibliothek, MS 847.9, www.e-codices.unifr.ch/en/list/one/bbb/0847-09

Brussels, Bibliothèque Royale, MS lat. 10066–77, https://uurl.kbr.be/1332800

Cambridge, Corpus Christi College, Parker Library, MS 448, https://parker.stanford.edu/parker/catalog/kk233ff6415

Cambridge, Trinity College, Wren Library, MS R.5.33, https://mss-cat.trin.cam.ac.uk/Manuscript/R.5.33

Chartres, Médiathèque L'Apostrophe, MS 63 (125)

Colmar, Bibliothèque municipale, MS 33 (olim 41), https://bvmm.irht.cnrs.fr/sommaire/sommaire.php?reproductionId=1946

Cologne, Erzbischöfliche Diözesan- und Dombibliothek, MS 85, https://digital.dombibliothek-koeln.de/hs/content/titleinfo/168968

Exeter, Cathedral Library, MS 3501, https://theexeterbook.exeter.ac.uk/

Fano, Archivio del Capitolo della Cattedrale, MS 5

Florence, Biblioteca Medicea Laurenziana, MS Plut. XVII 20, http://mss.bmlonline.it/s.aspx?Id=AWOHzoytI1A4r7GxMCQv&c=Evangelia%20quae%20toto%20anno%20leguntur%20in%20ecclesia#/book

Florence, Biblioteca Medicea Laurenziana, MS San Marco 650

Gotha, Forschungsbibliothek, memb. I.85 or I.75

Hannover, Museum August Kestner, MS W. M. XXIa 36

Hannover, Niedersächsische Landesbibliothek, MS I 189, http://digitale-sammlungen.gwlb.de/index.php?id=6&tx_dlf%5Bid%5D=9889&tx_dlf%5Bpage%5D=1

Laon, Bibliothèque municipale Suzanne Martinet, MS 40A

Leiden, Universiteitsbibliotheek, MS BPL 114

Leiden, Universiteitsbibliotheek, MS Voss. lat. Q.29

Leiden, Universiteitsbibliotheek, MS Voss. lat. Q.122

Leipzig, Deutsche Nationalbibliothek, Deutsches Buch- und Schriftmuseum, Klemm-Sammlung, MS fragm. 1966/Bl/95

Linz, Bibliothek der katholisch-theologischen Hochschule, MS A 1/6 (MS 1)

London, British Library, MS Add. 34890, www.bl.uk/manuscripts/FullDisplay.aspx?ref=Add_MS_34890

London, British Library, MS Arundel 155, www.bl.uk/manuscripts/FullDisplay.aspx?ref=Arundel_MS_155
London, British Library, MS Cotton Vespasian A.I, www.bl.uk/manuscripts/FullDisplay.aspx?ref=Cotton_MS_Vespasian_A_I
London, British Library, MS Harley 603, www.bl.uk/manuscripts/FullDisplay.aspx?ref=Harley_MS_603
London, British Library, MS Harley 2637, www.bl.uk/manuscripts/FullDisplay.aspx?ref=Harley_MS_2637
London, British Library, MS Harley 3017, www.bl.uk/manuscripts/FullDisplay.aspx?ref=Harley_MS_3017
London, British Library, MS Royal 13.C.V, www.bl.uk/manuscripts/FullDisplay.aspx?ref=Royal_MS_13_C_V
London, British Library, MS Stowe 944, www.bl.uk/manuscripts/FullDisplay.aspx?ref=Stowe_MS_944
London, Society of Antiquaries, MS 60
Luxembourg, Bibliothèque nationale, MS 109
Merseburg, Domstiftsbibliothek, MS 136, https://archive.thulb.uni-jena.de/korax/receive/Korax_cbu_00000880
Milan, Biblioteca Ambrosiana, MS E 16 sup., https://ambrosiana.comperio.it/opac/detail/view/ambro:catalog:70286
Mons, Archives de l'État, Cartulaire 12
Montecassino, Archivio dell'Abbazia, MS 316
Montecassino, Archivio dell'Abbazia, MS 323
Montpellier, Bibliothèque de l'École de Médecine, MS H 412, https://bvmm.irht.cnrs.fr/iiif/8142/canvas/canvas-1264182/view (partial digitisation)
Munich, Bayerisches Hauptstaatsarchiv, Hochstift Passau, Inneres Archiv 5, https://bavarikon.de/object/bav:GDA-OBJ-00000BAV80016806
Munich, Bayerische Staatsbibliothek, MS Clm 5250, www.digitale-sammlungen.de/view/bsb00001969?page=1
Munich, Bayerische Staatsbibliothek, MS Clm 13038, https://www.digitale-sammlungen.de/view/bsb00042776?page=1
Munich, Bayerische Staatsbibliothek, MS Clm 14222, www.digitale-sammlungen.de/view/bsb00036222?page=1
Munich, Bayerische Staatsbibliothek, MS Clm 14388, www.digitale-sammlungen.de/view/bsb00022465?page=1
Munich, Bayerische Staatsbibliothek, MS Clm 19417, www.digitale-sammlungen.de/view/bsb00036883?page=1
Munich, Bayerische Staatsbibliothek, MS Clm 27152, www.digitale-sammlungen.de/view/bsb00031770?page=1
New York, Columbia University Library, MS Plimpton 58
New York, Morgan Library and Museum, MS M.397, https://ica.themorgan.org/manuscript/thumbs/112333 (partial digitisation)
Orléans, Médiathèque, MS 18 (*olim* 15), https://mediatheques.orleans-metropole.fr/ark:/77916/FRCGMBPF-452346101-01A/D18010449.locale=fr
Orléans, Médiathèque, MS 192 (*olim* 169), https://mediatheques.orleans-metropole.fr/ark:/77916/FRCGMBPF-452346101-01A/D18011385.locale=fr
Orléans, Médiathèque, MS 313, https://mediatheques.orleans-metropole.fr/ark:/77916/FRCGMBPF-452346101-01A/D18012396.locale=fr

Oxford, Bodleian Library, MS Auct. D.2.19, https://digital.bodleian.ox.ac.uk/objects/b708f563-b804-42b5-bd0f-2826dfaeb5cc/
Oxford, Bodleian Library, MS Auct. T.2.23, https://digital.bodleian.ox.ac.uk/objects/ee32da36-f00d-441a-b87a-3625af92a757/ (partial digitisation)
Oxford, Bodleian Library, MS Junius 25, https://digital.bodleian.ox.ac.uk/objects/4df4cdd2-14d8-4256-ae07-a09917b3408e/
Oxford, Bodleian Library, MS Laud misc. 126, https://digital.bodleian.ox.ac.uk/objects/9fed4442-f4b6-439d-b359-41df38d9c9d4/
Oxford, Bodleian Library, MS Laud Misc. 129, https://digital.bodleian.ox.ac.uk/objects/b96fd583-aba5-4848-affe-e067455c711c/
Paris, Bibliothèque de l'Arsenal, MS 852, https://archivesetmanuscrits.bnf.fr/ark:/12148/cc80081v
Paris, Bibliothèque nationale de France, MS gr. 835, https://archivesetmanuscrits.bnf.fr/ark:/12148/cc239354
Paris, Bibliothèque nationale de France, MS lat. 614A, https://archivesetmanuscrits.bnf.fr/ark:/12148/cc62069fParis, Bibliothèque nationale de France, MS lat. 2175, https://archivesetmanuscrits.bnf.fr/ark:/12148/cc60068x
Paris, Bibliothèque nationale de France, MS lat. 4629, https://archivesetmanuscrits.bnf.fr/ark:/12148/cc34028b
Paris, Bibliothèque nationale de France, MS lat. 7581, gallica.bnf.fr/ark:/12148/btv1b90769340
Paris, Bibliothèque nationale de France, MS lat. 2731A
Paris, Bibliothèque nationale de France, MS lat. 7900A, https://archivesetmanuscrits.bnf.fr/ark:/12148/cc94410h
Paris, Bibliothèque nationale de France, MS lat. 8088
Paris, Bibliothèque nationale de France, MS lat. 10612
Paris, Bibliothèque nationale de France, MS lat. 10756, https://archivesetmanuscrits.bnf.fr/ark:/12148/cc724569
Paris, Bibliothèque nationale de France, MS lat. 10837, https://archivesetmanuscrits.bnf.fr/ark:/12148/cc72512p
Paris, Bibliothèque nationale de France, MS lat. 11561, https://archivesetmanuscrits.bnf.fr/ark:/12148/cc730943
Paris, Bibliothèque nationale de France, MS lat. 12048, https://archivesetmanuscrits.bnf.fr/ark:/12148/cc130152
Paris, Bibliothèque nationale de France, MS lat. 13159, https://archivesetmanuscrits.bnf.fr/ark:/12148/cc91798
Paris, Bibliothèque nationale de France, MS lat. 13348, https://archivesetmanuscrits.bnf.fr/ark:/12148/cc74374b
Paris, Bibliothèque nationale de France, MS lat. 14935, https://archivesetmanuscrits.bnf.fr/ark:/12148/cc75664s
Paris, Bibliothèque nationale de France, MS Nouv. acq. lat. 455, https://archivesetmanuscrits.bnf.fr/ark:/12148/cc71179s
Paris, Bibliothèque nationale de France, MS Nouv. acq. lat. 1616, https://archivesetmanuscrits.bnf.fr/ark:/12148/cc13524w
Prague, Národní knihovna České republiky, MS XIII.G.18
Prague, Knihovna Národního muzea, MS XIV.A.12
St. Gallen, Stiftsbibliothek, Cod. Sang. 18, www.e-codices.unifr.ch/en/list/one/csg/0018
St. Gallen, Stiftsbibliothek, Cod. Sang. 40, www.e-codices.unifr.ch/en/list/one/csg/0040

St. Gallen, Stiftsbibliothek, Cod. Sang. 125, www.e-codices.unifr.ch/en/list/one/csg/0125
St. Gallen, Stiftsbibliothek, Cod. Sang. 225, www.e-codices.unifr.ch/en/list/one/csg/0225
St. Gallen, Stiftsbibliothek, Cod. Sang. 229, www.e-codices.unifr.ch/en/list/one/csg/0229
St. Gallen, Stiftsbibliothek, Cod. Sang. 230, www.e-codices.unifr.ch/en/list/one/csg/0230
St. Gallen, Stiftsbibliothek, Cod. Sang. 236, www.e-codices.unifr.ch/en/list/one/csg/0236
St. Gallen, Stiftsbibliothek, Cod. Sang. 237, www.e-codices.unifr.ch/en/list/one/csg/0237
St. Gallen, Stiftsbibliothek, Cod. Sang. 446, www.e-codices.unifr.ch/en/list/one/csg/0446
St. Gallen, Stiftsbibliothek, Cod. Sang. 728, www.e-codices.unifr.ch/en/list/one/csg/0728
St. Gallen, Stiftsbibliothek, Cod. Sang. 878, www.e-codices.unifr.ch/en/list/one/csg/0878
Stuttgart, Württembergische Landesbibliothek, MS Bibl. fol. 23, https://digital.wlb-stuttgart.de/index.php?id=6&tx_dlf%5Bid%5D=8680&tx_dlf%5Bpage%5D=1
Trier, Domschatz, MS 61
Trier, Stadtbibliothek, MS 40/1018 8°
Valenciennes, Bibliothèque municipale, MS 404
Vatican City, Biblioteca Apostolica Vaticana, MS Barb. lat. 671, https://digi.vatlib.it/view/MSS_Barb.lat.671
Vatican City, Biblioteca Apostolica Vaticana, MS Ottob. lat., 339
Vatican City, Biblioteca Apostolica Vaticana, MS Pal. lat. 577, https://doi.org/10.11588/diglit.9726
Vatican City, Biblioteca Apostolica Vaticana, MS Pal. lat. 1074, https://doi.org/10.11588/diglit.10935
Vatican City, Biblioteca Apostolica Vaticana, MS Pal. lat 1357, https://doi.org/10.11588/diglit.10051
Vatican City, Biblioteca Apostolica Vaticana, MS Reg. lat. 190
Vatican City, Biblioteca Apostolica Vaticana, MS Vat. lat. 1260
Verona, Biblioteca Capitolare, MS XXVII (25)
Vienna, Österreichische Nationalbibliothek, MS 652, http://data.onb.ac.at/dtl/7223619
Wolfenbüttel, Herzog August Bibliothek, MS Gud. lat. 148, https://diglib.hab.de/?db=mss&list=ms&id=148-gud-lat&lang=en
Wolfenbüttel, Herzog August Bibliothek, MS Guelf. 102 Weiss, https://diglib.hab.de/?db=mss&list=ms&id=102-weiss
Würzburg, Universitätsbibliothek, MS M.p.th.f.40, http://vb.uni-wuerzburg.de/ub/mpthf40/ueber.html
York, Minster Library, MS Add. 1
Zurich, Zentralbibliothek, MS 463 [Rh. 951]

## Digital Resources

IrCaBriTT project: https://ircabritt.nuigalway.ie/handlist/catalogue/170.
*Liber glossarum*: http://liber-glossarum.huma-num.fr/exist/apps/libgloss/index.html.

## Primary sources

### *Physiologus*

Carmody, Francis J., ed. *Physiologus latinus: éditions préliminaires, versio B.* Paris: Droz, 1939.

Carmody, Francis J., ed. *Physiologus latinus versio Y.* University of California Publications in Classical Philology 12 (Berkeley: University of California Press, 1941).

Curley, Michael J., trans. *Physiologus: A Medieval Book of Nature Lore.* Chicago: University of Chicago Press, 2009.

Eden, Paul T. ed. *Theobaldi "Physiologus." Edited with Introduction, Critical Apparatus, Translation, and Commentary.* Leiden: Brill, 1972.

Heider, Gustav, trans. *Der Physiologus, nach einer Handschrift des XI. Jahrhunderts herausgegeben und erläutert.* Archiv für Kunde österreichischer Geschichtsquellen 5. Vienna: Aus der kaiserlich-königlichen Hof- und Staatsdruckerei, 1851.

Kaimakis, Dimitris, ed. *Der Physiologus nach der ersten Redaktion.* Beiträge zur Klassischen Philosophie. Meisenheim am Glan: Hain, 1974.

Morris, Richard, ed. *An Old English Miscellany: Containing a Bestiary, Kentish Sermons, Proverbs of Alfred, Religious Poems of the Thirteenth Century, etc.* London: Trübner, 1872.

Offermanns, Dieter, ed. *Der Physiologus nach den Handschriften G und M.* Beiträge zur klassischen Philologie. Meisenheim am Glan: Hain, 1966.

### Other Primary Sources

Aristotle. *Politics.* Edited by C.D.C. Reeve. Indianapolis: Hackett, 1998.

Augustine. *De Civitate Dei. Book V.* Translated by P. G. Walsh. Aris & Phillips Classical Texts. Oxford: Aris and Phillips, 2009.

——. *The Works of Saint Augustine: A Translation for the 21st Century. Sermons Part III (306–340A). On the Saints.* Translated by Edmund Hill and John E. Rotelle. Brooklyn: New City Press, 1990.

Augustine, pseudo. *Pseudo-Augustinus. De oratione et elemosina, De sobrietate et castitate, De incarnatione et deitate Christi ad Ianuarium, Dialogus quaestionum.* Edited by Lukas J. Dorfbauer. Corpus Scriptorum Ecclesiasticorum Latinorum 99. Vienna: Verlag der Österreichischen Akademie der Wissenschaften, 2011.

Bede. *De Arte Metrica et De Schematibus et Tropis.* Translated by Calvin B. Kendall. Bibliotheca Germanica, Series Nova 2. Saarbrücken: AQ, 1991.

——. *On the Nature of Things and On Times.* Translated by Calvin B. Kendall and Faith Wallis. Translated Texts for Historians 56. Liverpool: Liverpool University Press, 2010.

Bible. *La Vetus Latina Hispana V. El Salterio. Introducción General y Edición Crítica.* Edited by Teófilo Ayuso Marazuela. 3 vols. Textos y Estudios del Seminario Filológico Cardenal Cisneros. Madrid: Instituto Francisco Suarez del CSIC, 1962.

Boethius. *The Consolation of Philosophy.* Translated by David R Slavitt. Cambridge, MA: Harvard University Press, 2010.

——. *Theological Tractates. The Consolation of Philosophy.* Translated by Hugh F. Stewart, Edward K. Rand, and Stanley J. Tester. The Loeb Classical Library 74. Cambridge, MA: Harvard University Press (Loeb), 1973.

Caesarius of Arles. *Sermons. 3: 187 – 238.* Translated by Mary Magdeleine Mueller. The Fathers of the Church 66. Washington, DC: Catholic University of America Press, 2004.

Cassiodorus. *Cassiodorus, Explanation of the Psalms*. Translated by P.G. Walsh. 2 vols., Vol. 1. Ancient Christian Writers 51. New York: Paulist Press, 1990.

——. *Institutions of Divine and Secular Learning*. Edited by James W. Halporn and Mark Vessey. Liverpool: Liverpool University Press, 2004.

Eriugena, Johannes Scottus. *Periphyseon: The Division of Nature*. Translated by John Joseph O'Meara and I. P Sheldon-Williams. Montréal: Bellarmin, 1987.

Fredegar. *Fredegarii Chronicorum liber quartus cum continuationibus. The Fourth Book of the Chronicle of Fredegar: With its Continuations*. Translated and edited by John Michael Wallace-Hadrill. Nelson's Medieval Texts. London: Thomas Nelson, 1960; repr. Greenwood Press, 1981.

Horace. *Satires. Epistles. The Art of Poetry.* Translated by H. Rushton Fairclough. Loeb Classical Library 194. Cambridge, MA: Harvard University Press, 1926. DOI: 10.4159/DLCL.horace-satires.1926.

Isidore of Seville. *The Etymologies of Isidore of Seville*. Translated by Stephen A. Barney et al. Cambridge: Cambridge University Press, 2006.

——. *On the Nature of Things*. Translated by Calvin B. Kendall and Faith Wallis. Translated Texts for Historians 66. Liverpool: Liverpool University Press, 2016.

*Les 'Ordines romani' du Haut Moyen-Âge*. Edited by Michel Andrieu. 5 vols. Spicilegium Sacrum Lovaniense. Études et Documents 11. Louvain: Spicilegium Sacrum Lovaniense Bureaux, 1931–61.

Priscian. *Institutiones: Grammatici latini 2. Prisciani institutionum grammaticarum libri I–XII*. Edited by Heinrich Keil. Leipzig: Teubner, 1855.

——. "Praeexercitamina Prisciani grammatici ex Hermogene versa." In *Rhetores Latini minores: ex codicibus maximam partem primum adhibitis*. Edited by Karl Halm, 551–60. Leipzig: Teubner, 1863.

St. Augustine. *On Christian Teaching*. Translated by Roger P. H. Green. Oxford World's Classics. Oxford: Oxford University Press, 2008.

Servius. *Servii grammatici qui feruntur in Vergilii carmina commentarii*. Edited by Georg Thilo and Hermann Hagen. 3 vols. in 5 parts. Leipzig: Teubner, 1878–1902.

## Secondary sources

Aerts, Willem J., and Georg A. A. Kortekaas, eds. *Die Apokalypse des Pseudo-Methodius: Die ältesten griechischen und lateinischen Übersetzungen*. 2 vols. Corpus Scriptorum Christianorum Orientalium 569–570. Subsidia 97–98. Leuven: Peeters, 1998.

Agazzi, Evandro. *Scientific Objectivity and its Contexts*. New York: Springer, 2014.

Ahern, Eoghan. *Bede and the Cosmos: History, Theology and the Natural World*. Routledge Studies in Early Medieval Britain and Ireland. Abingdon: Routledge, 2020.

Ahrens, Karl. *Zur Geschichte des sogenannten Physiologus*. Plön: S. W. Hirt, 1885.

Alberto, Paulo F. "The Textual Tradition of the *Carmen de uentis* (AL 484): Some Preliminary Conclusions with a New Edition." *Aevum: Rassegna di scienze storiche, linguistiche e filologiche* 83 (2009): 341–75.

Allport, Jeffrey L. "Three Early Christian Interpretations of Nature and Scripture: The Physiologus, Origen, and Basil." M. Div. thesis, Princeton University, 1984.

Alpers, Klaus. "Physiologus." In *Theologische Realenzyklopädie: Paris – Polen*, vol. 26, edited by Gerhard Müller et al., 596–602. Berlin: De Gruyter, 1996.

——. *Untersuchungen zum Griechischen Physiologus und den Kyraniden*. Hamburg: Friedrich Wittig, 1984.

Amsler, Mark. *Etymology and Grammatical Discourse in Late Antiquity and the Early Middle Ages*. Amsterdam Studies in the Theory and History of Linguistic Science 44. Amsterdam, PA: J. Benjamins, 1989.

Anderson, Benjamin. *Cosmos and Community in Early Medieval Art*. New Haven: Yale University Press, 2017.

Andrist, Patrick, Paul Canart, and Marilena Maniaci. *La syntaxe du codex: essai de codicologie structurale*. Bibliologia 34. Turnhout: Brepols, 2013.

Arnaldez, Roger, and Jean Beaujeau, eds. *La Science antique et médiévale (des origines à 1450). Histoire générale des sciences* 1, edited by Réné Taton. Presses Universitaires de France 173. Paris: Presses universitaires de France, 1957.

Arnold, Ellen Fenzel. *Negotiating the Landscape: Environment and Monastic Identity in the Medieval Ardennes*. Middle Ages Series. Philadelphia: University of Pennsylvania Press, 2013.

Ashworth, William B. "Natural History and the Emblematic World View." In *Reappraisals of the Scientific Revolution*, edited by David C. Lindberg and Robert S. Westman, 303–32. Cambridge: Cambridge University Press, 1990.

Atkinson, Charles M. *The Critical Nexus: Tone-System, Mode, and Notation in Early Medieval Music*. AMS Studies in Music. Oxford: Oxford University Press, 2009.

Babcock, Robert Gary. *The Psychomachia Codex from St Lawrence (Bruxellensis 10066–77) and the Schools of Liège in the Tenth and Eleventh Centuries*. Bibliologia 42. Turnhout: Brepols, 2017.

Baldacchini, Lorenzo. "*In uno corpore continentur*: Le miscellanee. Per un approccio unitario a un problema della biblioteconomia del libro antico." *Bollettino AIB. Rivista Italiana di biblioteconomia e scienze dell'informazione* 45 (2005): 203–10.

Baldwin, Spurgeon W. *The Medieval Castilian Bestiary from Brunetto Latini's Tesoro. Study and Edition*. Exeter Hispanic Texts 31. Exeter: University of Exeter, 1982.

Barker, Andrew. "Words for Sounds." In *Science and Mathematics in Ancient Greek Culture*, edited by Christopher Tuplin and Tracy E. Rihll, 22–35. Oxford: Oxford University Press, 2002. DOI: 10.1093/acprof:oso/9780198152484.003.0002.

Batiffol, Pierre. *History of the Roman Breviary*. Translated by Atwell Baylay. London: Longmans, Green and Co., 1898.

Baxter, Richard. "Learning from Nature: Lessons in Virtue and Vice in the Physiologus and Bestiaries." In *Virtue & Vice: The Personifications in the Index of Christian Art*, edited by Colum Hourihane, 29–41. Publications of the Department of Art and Archaeology, Princeton University. Index of Christian Art Resources. Princeton: Princeton University Press, 2000.

Baxter, Ron. *Bestiaries and Their Users in the Middle Ages*. London: Sutton Publishing, 1998.

Becker, Gustav. *Catalogi bibliothecarum antiqui*. Bonn: M. & F. Cohen, 1885.

Behringer, Wolfgang. *Kulturgeschichte des Klimas: Von der Eiszeit bis zur globalen Erwärmung*. Munich: Beck, 2007.

Belli, Giovanna. *Il Physiologus: L'ermetismo attraverso i simboli degli animali*. Milan: Edizione Kemi, 1991.

Benjamin, Francis S. "Review: Florence McCulloch. *Mediaeval Latin and French Bestiaries*." *Speculum* 39 (1964): 726–28. DOI: 10.2307/2854774.

Bergmann, Rolf, and Stephanie Stricker, with Yvonne Goldammer and Claudia Wich-Reif. *Katalog der althochdeutschen und altsächsischen Glossenhandschriften*. 6 vols. Berlin: De Gruyter, 2005. https://glossen.germ-ling.uni-bamberg.de/pages/1. DOI: 10.1515/9783110918250.

Berschin, Walter. "Vier karolingische Exlibris." In *Mittellateinische Studien I*, edited by Walter Berschin, 169-178. Heidelberg: Mattes Verlag, 2005.

Bibolet, Françoise. "Bibliotheca Pithoeana. Les manuscrits des Pithou: Une histoire de fraternité et d'amitié." In *Du copiste au collectionneur: Mélanges d'histoire des textes et des bibliothèques en l'honneur d'André Vernet*, edited by Donatella Nebbiai-Dalla Guarda and Jean-François Genest, 497–521. Bibliologia 18. Turnhout: Brepols, 1998.

Bierbrauer, Katharina. *Die vorkarolingischen und karolingischen Handschriften der Bayerischen Staatsbibliothek*. Katalog der illuminierten Handschriften der Bayerischen Staatsbibliothek in München 1. Wiesbaden: L. Reichert, 1990.

Biggs, Frederick M., "The Eschatological Conclusion of the Old English Physiologus." *Medium Aevum* 58 (1989): 286–97. DOI: 10.2307/43629257.

Bintley, Michael D.J. *Settlements and Strongholds in Early Medieval England: Texts, Landscapes, and Material Culture.* Studies in the Early Middle Ages 45. Turnhout: Brepols, 2020. DOI: 10.1484/M.SEM-EB.5.116729.

Bischoff, Bernhard. "Die lateinischen Übersetzungen und Bearbeitungen aus den Oracula Sibyllina." In *Mittelalterliche Studien: Ausgewählte Aufsätze zur Schriftkunde und Literaturgeschichte* 1, 150–70. Stuttgart: Hiersemann, 1966.

——. "Eine Sammelhandschrift Walahfrid Strabos (Cod. Sangall. 878)." In *Mittelalterliche Studien: Ausgewählte Aufsätze zur Schriftkunde und Literaturgeschichte* 2, 34–51. Stuttgart: Hiersemann, 1967.

——. "Epitaphienformeln für Äbtissinnen (achtes Jahrhundert)." *Anecdota Novissima: Texte des 4. bis 16. Jahrhunderts* 11 (1984): 150–53.

——. *Katalog der festländischen Handschriften des neunten Jahrhunderts (mit Ausnahme der wisigotischen). Vol. 1, Aachen – Lambach*. Edited byby Birgit Ebersperger. Veröffentlichungen der Kommission für die Herausgabe der Mittelalterlichen Bibliothekskataloge Deutschlands und der Schweiz. Wiesbaden: Harrassowitz, 1998.

——. *Katalog der festländischen Handschriften des neunten Jahrhunderts (mit Ausnahme der wisigotischen, Vol. 2, Laon – Paderborne*. Edited by Birgit Ebersperger. Veröffentlichungen der Kommission für die Herausgabe der Mittelalterlichen Bibliothekskataloge Deutschlands und der Schweiz. Wiesbaden: Harrassowitz, 2004.

——. *Katalog der festländischen Handschriften des neunten Jahrhunderts (mit Ausnahme der wisigotischen), Vol. 3, Padua – Zwickau*. Edited by Birgit Ebersperger. Veröffentlichungen der Kommission für die Herausgabe der Mittelalterlichen Bibliothekskataloge Deutschlands und der Schweiz. Wiesbaden: Harrassowitz, 2014.

——. *Paläographie des romischen Altertums und des abendländischen Mittelalters. Edited by* Hugo Moser and Hartmut Steinecke. Grundlagen der Germanistik 24. Berlin: Erich Schmidt, 1979. Trans. as *Latin Palaeography: Antiquity and the Middle Ages*, by Dáibhí Ó Cróinin and David Ganz. Cambridge: Cambridge University Press, 1990.

——. "Wendepunkte in der Geschichte der lateinischen Exegese im Frühmittelalter'. In *Mittelalterliche Studien: Ausgewählte Aufsätze zur Schriftkunde und Literaturgeschichte*, 1:205–73; repr. from *Sacris Erudiri* 6 (1954): 189–279. Stuttgart: Hiersemann, 1966.

Bischoff, Bernhard, and Michael Lapidge. *Biblical Commentaries from the Canterbury School of Theodore and Hadrian*. Cambridge: Cambridge University Press, 2006.

Bishop, Terence A. M. *English Caroline Minuscule*. Oxford Palaeographical Handbooks. Oxford: Clarendon Press, 1971.

Bitterling, Klaus. "Physiologus und Bestiarien im englischen Mittelalter." *Mittellateinisches Jahrbuch: internationale Zeitschrift für Mediävistik* 40 (2005): 153–70.

Blair, John. *Building Anglo-Saxon England*. Princeton: Princeton University Press, 2018.

Blan, Noah. "Charlemagne's Peaches: A Case of Early Medieval European Ecological Adaptation." *Early Medieval Europe* 27 (2019): 521–45.

Bloch, Hermann. "Ein karolingischer Bibliotheks-Katalog aus Kloster Murbach." In *Strassburger Festschrift zur XLVI. Versammlung deutscher Philologen und Schulmänner*, 257–85. Strassburg: K. J. Trübner, 1901.

Bloomfield, Morton W., et al., eds. *Incipits of Latin Works on the Virtues and Vices, 1100–1500 A.D.: Including a Section of Incipits of Works on the Pater Noster.* Mediaeval Academy of America 88. Cambridge, MA: Mediaeval Academy of America, 1979.

Bologna, Corrado. "La tradizione manoscritta del Liber monstrorum de diversis generibus." *Cultura Neolatina* 34 (1974): 337–46.

Bonura, Christopher. "A Forgotten Translation of Pseudo-Methodius in Eighth-Century Constantinople: New Evidence for the Dispersal of the Greek Apocalypse of Pseudo-Methodius during the Dark Age Crisis." In *From Constantinople to the Frontier: The City and the Cities.* Edited by Nicholas S. M. Matheou, Theofili Kampianaki, and Lorenzo M. Bondioli, 260–76. The Medieval Mediterranean 106. Leiden: Brill, 2016.

Borst, Arno. *Computus: Zeit und Zahl in der Geschichte Europas.* Berlin: Wagenbach, 2004.

——. *Die karolingische Kalenderreform.* Monumenta Germaniae Historica Schriften 46. Hannover: Hahnsche, 1998.

——. ed. Schriften zur Komputistik im Frankenreich von 721 bis 818. Monumenta Germaniae Historica 21. Hannover: Hahnsche Buchhandlung, 2006.

——. *Der Streit um den karolingischen Kalender.* Hannover: Hahnsche, 2004.

Bremmer, Rolf H., and Kees Dekker, eds. *Foundations of Learning: The Transfer of Encyclopaedic Knowledge in the Early Middle Ages.* Mediaevalia Groningana New Series 9. Leuven: Peeters, 2007.

——. *Practice in Learning: The Transfer of Encyclopaedic Knowledge in the Early Middle Ages.* Mediaevalia Groningana New Series 16. Leuven: Peeters, 2010.

——. *Fruits of Learning: The Transfer of Encyclopaedic Knowledge in the Early Middle Ages.* Mediaevalia Groningana New Series 21. Leuven: Peeters, 2016.

Brooke, John L. *Climate Change and the Course of Global History: A Rough Journey.* Studies in Environment and History. Cambridge: Cambridge University Press, 2014.

Brown, Giles. "Introduction: The Carolingian Renaissance." In *Carolingian Culture: Emulation and Innovation*, edited by Rosamond McKitterick, 1–51. Cambridge: Cambridge University Press, 1993.

Brown, Warren C. "The *gesta municipalia* and the Public Validation of Documents in Frankish Europe." In *Documentary Culture and the Laity in the Early Middle Ages*, edited by Warren C. Brown, Marios Costambeys, Matthew Innes and Adam J. Kosto, 95–124. Cambridge: Cambridge University Press.

Bruce, Scott G. "Introduction: Hoffmann in The Historiography of Environmental History." In *Ecologics and Economies in Medieval and Early Modern Europe: Studies in Environmental History for Richard C. Hoffmann*, edited by Scott G. Bruce, 1–21. Brill's Series in the History of the Environment 1. Leiden: Brill, 2010.

Bruce, Scott G., ed. *Ecologies and Economies in Medieval and Early Modern Europe: Studies in Environmental History for Richard C. Hoffmann.* Brill's Series in the History of the Environment. Boston: Brill, 2010.

Buc, Philippe. *The Dangers of Ritual: Between Early Medieval Texts and Social Scientific Theory.* Princeton: Princeton University Press, 2009.

Budny, Mildred. *Insular, Anglo-Saxon, and Early Anglo-Norman Manuscript Art at Corpus Christi College, Cambridge: An Illustrated Catalogue*, 2 vols. Kalamazoo: Medieval Institute Publications, Western Michigan University, 1997.

Bullough, Donald A. *Carolingian Renewal: Sources and Heritage*. Manchester: Manchester University Press, 1991.

Buringh, Eltjo. *Medieval Manuscript Production in the Latin West: Explorations with a Global Database*. Global Economic History Series 6. Leiden: Brill, 2011.

Burn, Andrew E. *The Athanasian Creed and its Early Commentaries*. Texts and Studies: Contributions to Biblical and Patristic Literature 4.1. Cambridge: Cambridge University Press, 1896.

Burridge, Claire. *Carolingian Medical Knowledge and Practice, c. 775–900: New Approaches to Recipe Literature*. Leiden: Brill, 2022.

Butterfield, David J. *The Early Textual History of Lucretius' De rerum natura*. Cambridge: Cambridge University Press, 2013.

Butzmann, Hans. "Die Weissenburger Handschriften." Kataloge der Herzog-August Bibliothek Wolfenbüttel: Neue Reihe 10. Frankfurt am Main: Klostermann, 1964.

Cahier, Charles, and Arthur Martin, eds. *Mélanges d'archéologie, d'histoire et de littérature, rédigés ou recueillis: collection de mémoires sur l'orfévrerie et les émaux des trésors d'Aix-la-Chapelle, de Cologne, etc.* 4 parts in 2 vols. Paris: Vve Poussielgue-Rusand, 1847–1856.

Carey, Frederick M. "De Scriptura Floriacensi." PhD thesis, Harvard University, 1923.

——. "The Vatican Fragment of Phaedrus." *Transactions and Proceedings of the American Philological Association* 57 (1926): 96–106, plates XLIV–XLV.

Carmody, Francis J. "De bestiis et aliis rebus and the Latin Physiologus." *Speculum* 13 (1938): 153–59.

——. *Quotations in the Latin Physiologus from Latin Bibles Earlier than the Vulgate*. University of California Publications in Classical Philology 13. Berkeley: University of California Press, 1944)

Carruthers, Mary J. *The Book of Memory: A Study of Memory in Medieval Culture*. Cambridge Studies in Medieval Literature 10. Cambridge: Cambridge University Press, 1990.

Centro italiano di studi sull'alto medioevo, ed. *I problemi della civiltà carolingia: atti della settimana di studio 26 marzo–1 aprile 1953*. Settimane di studio del centro italiano di studi sull'alto medioevo 2. Spoleto: CISAM – Centro italiano di studi sull'alto medioevo, 1954.

Chadwick, Henry. "Origen." In *The Cambridge History of Later Greek and Early Medieval Philosophy*, edited by Arthur Hilary Armstrong, 182–92. Cambridge: Cambridge University Press, 1967.

Chambert-Protat, Pierre. "Le manuscrit Montpellier 157 de Mannon de Saint-Oyen et la collection *De pascha* de Florus de Lyon." *Revue Bénédictine* 128, no. 1 (June 2018): 95–141.

Chandler, Cullen. *Carolingian Catalonia: Politics, Culture, and Identity in an Imperial Province, 778–987*. Cambridge: Cambridge University Press, 2018.

Chatelain, Émile. *Introduction à la lecture des notes tironiennes*. Paris: Privately printed, 1900.

Chazelle, Celia Martin, and Burton van Name Edwards, eds. *The Study of the Bible in the Carolingian Era*. Medieval Church Studies 3. Turnhout: Brepols, 2003.

Chroust, Anton-Hermann. "The Fundamental Ideas in St Augustine's Philosophy of Law" *The American Journal of Jurisprudence* 18 (1973): 57–79 (repr. in Richard O Brooks and James B Murphy, eds. *Augustine and Modern Law*. Philosophers and Law. Farnham: Ashgate, 2011).

Clark, Willene B. *A Medieval Book of Beasts: The Second-Family Bestiary: Commentary, Art, Text and Translation*. Woodbridge: Boydell, 2006.

——. *The Medieval Book of Birds: Hugh of Fouilloy's Aviarium*. Medieval and Renaissance Texts and Studies 80. Binghamton: Center for Medieval and Early Renaissance Studies, 1992.

Clark, Willene B., and Meradith T. McMunn. *Beasts and Birds of the Middle Ages: The Bestiary and its Legacy*. Middle Ages Series. Philadelphia: University of Pennsylvania Press, 1990.

Clason, Chrisopher A. "Animals, Birds and Fish in the Middle Ages." In *Handbook of Medieval Culture: Fundamental Aspects and Conditions of the European Middle Ages*, 3 vols, edited by Albrecht Classen, 1:18–54. Berlin: De Gruyter, 2015.

Coates, Peter A. *Nature: Western Attitudes Since Ancient Times*. Berkeley: University of California Press, 2005.

Collins, Roger. *Die Fredegar-Chroniken*. Monumenta Germaniae Historica Studien und Texte 44. Hannover: Harrassowitz, 2007.

Collins, Samuel W. *The Carolingian Debate Over Sacred Space*. New York: Palgrave Macmillan, 2012.

Conner, Patrick W. "On Dating Cynewulf." In *Cynewulf: Basic Readings*, edited by Robert E. Bjork, 23–56. Basic Readings in Anglo-Saxon England 4. New York: Garland, 1996.

Conrad-O'Briain, Helen, Anne Marie D'Arcy, and V. John Scattergood, ed. *Text and Gloss: Studies in Insular Learning and Literature Presented to Joseph Donovan Pheifer*. Dublin: Four Courts Press, 1999.

Contreni, John J. "The Pursuit of Knowledge in Carolingian Europe." In *Learning and Culture in Carolingian Europe: Letters, Numbers, Exegesis, and Manuscripts*, 106–41. Farnham: Ashgate, 2011.

——. "John Scottus, Martin Hiberniensis, the Liberal Arts, and Teaching." In *Insular Latin Studies: Papers on Latin Texts and Manuscripts of the British Isles, 550–1066*, edited by Michael W. Herren, 23–44. Papers in Mediaeval Studies 1. Toronto: Pontifical Institute of Mediaeval Studies, 1981 (repr. In John J. Contreni, *Carolingian Learning, Masters, and Manuscripts*, 1–22. Hampshire, MA: Variorum, 1992).

——. "The Carolingian Renaissance." In *Renaissances Before the Renaissance: Cultural Revivals of Late Antiquity and the Middle Ages*, edited by Warren T. Treadgold. Stanford: Stanford University Press, 1984.

——. "The Carolingian Renaissance: Education and Literary Culture." In *The New Cambridge Medieval History*. Vol. 2, *c. 700–c. 900*, edited by Rosamond McKitterick. Cambridge: Cambridge University Press, 1995.

——. "What was Emperor Augustus Doing at a Carolingian Banquet (Anth. Lat.[2] 719f)?" *Rheinisches Museum für Philologie* 146 (2003): 372–94.

Corrigan, Kathleen. "The Smyrna Physiologos and Eleventh-Century Monasticism." In *Work and Worship at the Theotokos Evergetis, 1050–1200: Papers of the Fourth Belfast Byzantine International Colloquium, Portaferry, Co. Down, 14–17 September 1995*, edited by Margaret Mullett, 201–12. Belfast Byzantine Texts and Translations 6.2. Belfast: Belfast Byzantine Enterprises, 1997.

Costambeys, Marios, Matthew Innes, and Simon MacLean. *The Carolingian World*. Cambridge Medieval Textbooks. Cambridge: Cambridge University Press, 2011.

Côté, André. "Un manuscrit oublié du Physiologus (New York, P. Morgan M. 397)." *Scriptorium* 28 (1974): 276–77.

Courcelle, Pierre. *La Consolation de Philosophie dans la tradition littéraire: antécédents et postérité de Boèce*. Paris: Études augustiniennes, 1967.

Coussemaker, Ignace de. *Cartulaire de l'Abbaye de Cysoing et de ses dépendances*. Lille: Société de Saint-Augustin, 1880.

Cox, Patricia. "The Physiologus: A Poiēsis of Nature." *Church History* 52 (1983): 433–43.

Coxe, Henry Octavius. *Laudian Manuscripts*. Bodleian Library Quarto Catalogues 2. Oxford: Bodleian Library, 1973 (reprinted from the edition of 1858–85).

Crane, Susan. *Animal Encounters: Contacts and Concepts in Medieval Britain*. Middle Ages Series. Philadelphia: University of Pennsylvania Press, 2013.

Crouse, Robert D., Willemien Otten, Walter Hannam, and Michael Treschow, eds. *Divine Creation in Ancient, Medieval, and Early Modern Thought: Essays Presented to the Rev'd Dr. Robert D. Crouse*. Brill's Studies in Intellectual History 151. Leiden: Brill, 2007.

Cuissard, Charles. *Catalogue général des manuscrits des bibliothèques publiques de France. Départements: Orléans*. Vol. 12. Paris: Plon, 1889.

——. *Inventaire des manuscrits de la Bibliothèque d'Orléans: Fonds de Fleury*. Orléans: H. Herluison, 1885.

Cuppo, Luciana. "Felix of Squillace and the Dionysiac Computus I: Bobbio and Northern Italy (MS Ambrosiana H 150 Inf.)." In *The Easter Controversy of Late Antiquity and the Early Middle Ages: Its Manuscripts, Texts, and Tables. Proceedings of the 2nd International Conference on the Science of Computus, Galway, 18–20 July, 2008*, edited by Immo Warntjes and Dáibhí Ó Cróinín, 110–36. Studia Traditionis Theologiae 10. Turnhout: Brepols, 2011.

Curley, Michael J. "Physiologus, Fisiologia and the Rise of Christian Nature Symbolism." *Viator* 11 (1980): 1–10.

Dale, Corinne. *The Natural World in the Exeter Book Riddles*. Nature and Environment in the Middle Ages. Cambridge: D. S. Brewer, 2017.

Da Rold, Orietta. "Making the Book: Cambridge, University Library, MS Ii. 1. 33." *New Medieval Literatures* 13 (2011): 237–88.

Daston, Lorraine, and Peter Galison. *Objectivity*. New York: Zone Books, 2010.

Davis, Jennifer R., and Michael McCormick, "The Early Middle Ages: Europe's Long Morning," in *The Long Morning of Medieval Europe: New Directions in Early Medieval Studies*, edited by Jennifer R. Davis and Michael McCormick, 1–10. Aldershot: Ashgate, 2008.

Davis, Raymond, trans. *The Book of Pontiffs (Liber Pontificalis): The Ancient Biographies of the First Ninety Roman Bishops to AD 715*. Rev. ed. Translated Texts for Historians 6. Liverpool: Liverpool University Press, 2000.

D'Avray, David. "Symbolism and Medieval Religious Thought." In *The Medieval World*, edited by Peter Linehan and Janet L. Nelson, 267–78. London: Routledge, 2007.

De Cesare, Raffaele. "Su di un gruppo di glosse antico-francesi in un lessico latino dell'xi secolo." *Aevum: Rassegna di scienze storiche, linguistiche e filologiche* 27 (1953): 430–48.

Delhaye, Pierre. "'Grammatica' et 'Ethica' au XIIe siècle." *Recherches de théologie ancienne et médiévale* 25 (1958): 59–110.

Delisle, Léopold. *Catalogue des manuscrits des fonds Libri et Barrois*. Paris: H. Champion, 1888.

——. *Inventaire des manuscrits latins conservés à la bibliothèque nationale sous les numéros 8823–11503 et faisant suite à la série dont le catalogue a été publié en 1744*. Paris: Auguste Durand, 1863.

——. "Notice sur plusieurs manuscrits de la bibliothèque d'Orléans." *Notices et extraits des manuscrits* 31:1 (1884), 357–439.

Derolez, Albert. *The Palaeography of Gothic Manuscript Books from the Twelfth to the Early Sixteenth Century*. Cambridge Studies in Palaeography and Codicology 9. Cambridge: Cambridge University Press, 2006.

Devroey, Jean-Pierre. *La nature et le roi: environnement, pouvoir et société à l'âge de Charlemagne (740–820)*. Collection "L'évolution de l'humanité." Paris: Albin Michel, 2019.

Diekstra, Frans N. M. "The Physiologus, the Bestiaries, and Medieval Animal Lore." *Neophilologus* 69 (1985): 142–55.

Dines, Ilya. "A Critical Edition of the Bestiaries of the Third Family." PhD thesis, The Hebrew University, 2008.

——. "The Problem of the Transitional Family of Bestiaries." *Reinardus: Yearbook of the International Reynard Society* 24 (2013): 29–52.

Dolbeau, François. "Du nouveau sur un sermonnaire de Cambridge." *Scriptorium* 42 (1998): 255–57.

Dorfbauer, Lukas J. "Der Codex St. Gallen, Stiftsbibl. 125 und sein Verhältnis zu St. Gallen, Stiftsbibl. 230. Mit Edition eines karolingischen Florilegs zur Diskussion um die Trinität." *Codices Manuscripti: Zeitschrift für Handschriftenkunde* 85/86 (2012): 1–15.

Dorofeeva, Anna. "Miscellanies, Christian Reform and Early Medieval Encyclopaedism: A Reconsideration of the Pre-Bestiary Physiologus Manuscripts." *Historical Research* 90, no. 250 (October 2017): 665–82.

——. "Reading Early Medieval Miscellanies." In *Scribes and the Presentation of Texts (from Antiquity to c. 1550) Proceedings of the 20th Colloquium of the Comité International de Paléographie Latine. Beinecke Rare Book & Manuscript Library, Yale University (New Haven, September 6-8, 2017)*, edited by Consuelo Dutschke and Barbara A. Shailor, 495–514. Bibliologia 65. Turnhout: Brepols, 2021.

——. "Strategies of Knowledge Organisation in Early Medieval Latin Glossary Miscellanies: The Example of Munich, Bayerische Staatsbibliothek, Clm 14388." In *Writing the Early Medieval West*, edited by Elina Screen and Charles West, 146–68. Cambridge: Cambridge University Press, 2018.

——. "Visualizing Codicologically and Textually Complex Manuscripts." *Journal for Manuscript Studies* 4 (2019): 334–60.

——. "What is a Vademecum? The Social Logic of Early Medieval Compilation." In *The Art of Compilation: Manuscripts and Networks in the Early Medieval Latin West*, edited by Anna Dorofeeva and Michael Kelly. Binghamton: Gracchi Books, forthcoming in 2024.

Dronke, Peter. "Medieval Sibyls: Their Character and Their 'auctoritas'." *Studi medievali* 36 (1995): 581–615 (repr. in *Forms and Imaginings: From Antiquity to the Fifteenth Century*, edited by Peter Dronke, 13–46. Storia e letteratura. Raccolta di studi e testi 243. Rome: Edizioni di storia e letteratura, 2007).

Drout, Michael D. C. "'The Partridge" is a Phoenix: Revising the Exeter Book Physiologus." *Neophilologus* 91 (2007): 487–503.

Dumville, David N. *English Caroline Script and Monastic History: Studies in Benedictinism, A.D. 950–1030*. Studies in Anglo-Saxon History 6. Woodbridge: Boydell, 1993.

——. "English Square Minuscule Script: The Background and Earliest Phases." *Anglo-Saxon England* 16 (1987): 147–79.

Eastwood, Bruce S. *Ordering the Heavens: Roman Astronomy and Cosmology in the Carolingian Renaissance*. History of Science and Medicine Library 4. Leiden: Brill, 2007.

——. *The Revival of Planetary Astronomy in Carolingian and Post-Carolingian Europe*. London: Routledge, 2018.

Effros, Bonnie. *Caring for Body and Soul: Burial and the Afterlife in the Merovingian World*. University Park: Penn State University Press, 2010.

*Eighth Report of the Royal Commission on Historical Manuscripts. Appendix. Part III: The Manuscripts of the Earl of Ashburnham*. London: Eyre and Spottiswoode, 1881

Elfassi, Jacques. "Presence of Augustine of Hippo in Isidore of Seville: Some Provisional Remarks." In *Framing Power in Visigothic Society*, edited by Eleonora Dell'Elicine and Céline Martin, 23–50. Amsterdam: Amsterdam University Press, 2020.

Embach, Michael. "Trierer Zauber- und Segenssprüche des Mittelalters." *Kurtrierisches Jahrbuch* 44 (2004): 29–76.

Erikson, Alvar. "The Problem of Authorship in the Chronicle of Fredegar." *Eranos* 63 (1965): 47–76.

Esders, Stefan. "The Merovingians and Byzantium: Diplomatic, Military, and Religious Issues, 500–700." In *The Oxford Handbook of the Merovingian World*, edited by Bonnie Effros and Isabel Moreira, 345–69. Oxford: Oxford University Press, 2020.

Estes, Heide. *Anglo-Saxon Literary Landscapes: Ecotheory and the Environmental Imagination*. Environmental Humanities in Pre-Modern Cultures. Amsterdam: Amsterdam University Press, 2017.

Étaix, Raymond. "Un manuel de pastorale de l'époque carolingienne (Clm 27152)." *Révue Bénédictine* 1 (1981): 105–30.

Euw, Anton von. *Rhin-Meuse: Art et civilisation 800–1400. Catalogue de l'exposition organisée du 14 mai au 23 juillet 1972 à Cologne et du 19 septembre au 31 octobre à Bruxelles*, edited by Jacques Stiennon and Rita Lejeune. Brussels: Ministère de l'Education Nationale et de la Culture, 1972.

Evdokimova, Ludmila. "Deux traductions du 'Physiologus': le sens allégorique de la nature et le sens allégorique de la Bible." *Reinardus: Yearbook of the International Reynard Society* 11 (1998): 53–66.

Everett, Nicholas. "'The Interrogationes de littera et de singulis causis': An Early Medieval School Text." *The Journal of Medieval Latin* 16 (2006): 227–75.

Faraci, Dora, ed. *Il bestiario medio inglese (ms Arundel 292 della British Library)*. L'Aquila: Japadre, 1990.

——. "The Bestiary and Its Sources. Some Examples." *Reinardus: Yearbook of the International Reynard Society* 7 (1994): 31–43.

Fassler, Margot. 'Mary's Nativity, Fulbert of Chartres, and the Stirps Jesse: Liturgical Innovation circa 1000 and its Afterlife'. *Speculum* 75 (2000): 389–434.

Figueira, Dorothy Matilda. *The Hermeneutics of Suspicion: Cross-Cultural Encounters with India*. Bloomsbury Research in Continental Philosophy. London: Bloomsbury Academic, 2015.

Filotas, Bernadette. *Pagan Survivals, Superstitions and Popular Cultures in Early Medieval Pastoral Literature*. Studies and Texts 151. Toronto: Pontifical Institute of Mediaeval Studies, 2005.

Flammini, Giuseppe. "L''Epistula' pseudogalenica de febribus." In *Prefazioni, Prologhi, Proemi di opere tecnico-scientifiche latine*, edited by Carlo Santini, Nino Scivolette, and Loriano Zurli, 239–57. Rome: Herder, 1998.

Flechner, Roy, and Sven Meeder, eds. *The Irish in Early Medieval Europe: Identity, Culture and Religion*. New York London: Palgrave Macmillan, 2016.

Fontaine, Jacques. "Aux sources de la lexicographie médiévale: Isidore de Séville, médiateur de l'étymologie antique." In *La lexicographie du latin médiéval et ses rapports avec les recherches actuelles sur la civilisation du Moyen-Âge, Paris, 18–21 Octobre 1978*, 97–103. Paris: CNRS, 1981.

——. *Isidore de Séville et la culture classique dans l'Espagne wisigothique*. 2nd ed. 3 vols. Paris: Études augustiniennes, 1983.

——. *Isidore de Séville: genèse et originalité de la culture hispanique au temps des Wisigoths*. Turnhout: Brepols, 2000.

——. *Tradition et actualité chez Isidore de Séville*. London: Variorum, 1988.

Foot, Sarah. "Plenty, Portents and Plague: Ecclesiastical Readings of the Natural World in Early Medieval Europe." *Studies in Church History* 46 (2010): 15–41.

Fraeys de Veubeke, Anne-Catherine. "Un catalogue de bibliothèque scolaire inédit du XIIe siècle dans le MS Bruxelles, B. R. 9384–89." *Scriptorium* 35 (1981): 23–38.

Frank, Lothar. "Die Physiologus-Literatur des englischen Mittelalters und die Tradition." PhD thesis, University of Tübingen, 1971.

Franklin, Alfred. *Les anciennes bibliothèques de Paris: églises, monastères, colléges, etc.* 3 vols. Paris: Imprimerie impériale, 1870.

Fried, Johannes. *Die Formierung Europas 840–1046.* Oldenbourg Grundriss der Geschichte 6. Munich: Oldenbourg, 1991.

Gameson, Richard. "The Colophon of the Eadwig Gospels." *Anglo-Saxon England* 31 (2002): 201–22.

——. "The Earliest Books of Christian Kent." In *St. Augustine and the Conversion of England*, edited by Richard Gameson, 313–73. Phoenix Mill: Sutton, 1999.

——. "The Origin of the Exeter Book of Old English Poetry." *Anglo-Saxon England* 25 (1996): 135–85.

——. *The Scribe Speaks? Colophons in Early English Manuscripts.* H. M. Chadwick Memorial Lectures 12. Cambridge: Department of Anglo-Saxon, Norse and Celtic, University of Cambridge, 2002.

Gannon, Anna. "Coins, Images and Tales from the Holy Land: Questions of Theology and Orthodoxy." In *New Perspectives*, edited by Tony Abramson, 88–103. Studies in Early Medieval Coinage 2. Woodbridge: Boydell Press, 2011.

——. *The Iconography of Early Anglo-Saxon Coinage: Sixth to Eighth Centuries.* Oxford: Oxford University Press, 2010.

Ganz, David. "Bureaucratic Shorthand and Merovingian Learning." In *Ideal and Reality in Frankish and Anglo-Saxon Society. Studies presented to J. M. Wallace-Hadrill*, edited by Patrick C. Wormald, Donald A. Bullough and Roger Collins, 58-75. Oxford: Oxford University Press, 1983.

——. "Book Production in the Carolingian Empire and the Spread of Carolingian Minuscule." In *The New Cambridge Medieval History*, Vol. 2, *c. 700–c. 900*, edited by Rosamond McKitterick, 786–808. Cambridge: Cambridge University Press, 1995.

——. "In the Circle of the Bishop of Bourges: Bern 611 and Late Merovingian Culture." In *East and West in the Early Middle Ages. The Merovingian Kingdoms in Mediterranean Perspective*, edited by Stefan Esders, Yaniv Fox, Yitzhak Hen and Laury Sarti, 265–80. Cambridge: Cambridge University Press, 2019.

——. "The Preconditions for Caroline Minuscule." *Viator* 18 (1987): 23–44.

Ganz, David, and Walter Goffart. "Charters Earlier than 800 from French Collections." *Speculum* 65 (1990): 906–32.

Gaspar, Camille, and Frédéric Lyna, ed. *Les principaux manuscrits à peintures de la Bibliothèque royale de Belgique.* 3 vols. Paris: Société française de reproduction de manuscrits à peintures, 1937–1945 (re-edited by Christiane van den Bergen-Pantens, Brussels: Bibliothèque Royale Albert Ier, 1984–1989).

Gastle, Brian. "The Beast Fable." In *A Companion to Old and Middle English Literature*, edited by Laura C. Lambdin and Robert T. Lambdin, 69–85. Westport: Greenwood Press, 2002.

Gebert, Bent. "Der Satyr im Bad: Textsinn und Bildsinn in der Physiologus-Handschrift Cod. Bongarsianus 318 der Burgerbibliothek Bern; mit einer Edition der Versio C des 'Physiologus Latinus'." *Mittellateinisches Jahrbuch* 45 (2010): 353–403.

Geiselmann, Josef R. *Die Abendmahlslehre an der Wende der Christlichen Spätantike zum Frühmittelalter. Isidor von Sevilla und das Sakrament der Eucharistie*. Munich: Max Hueber, 1933.

Geith, Karl-Ernst, and Walter Berschin. "Die Bibliothekskataloge des Klosters Murbach aus dem IX. Jahrhundert." *Zeitschrift für Kirchengeschichte* (1972): 61–87.

Genest, Jean-François, Anne Chalandon, and Anne-Marie Genevois. *Bibliothèques de manuscrits médiévaux en France. Relevé des inventaires du VIIIe au XVIIIe siècle*. Paris: Editions du CNRS, 1987.

Gerlach, Peter. "Physiologus." In *Lexikon der christlichen Ikonographie*, Vol. 3, ed. Engelbert Kirschbaum and Günter Bandmann, cols. 432–36. Rome: Herder, 1971.

Gersh, Stephen. *Concord in Discourse: Harmonics and Semiotics in Late Classical and Early Medieval Platonism*. Approaches to Semiotics 125. Berlin: De Gruyter, 1996.

Gheyn, Joseph van den. *Catalogue des manuscrits de la Bibliothèque royale de Belgique*. 13 vols. Brussels: Lambertin, 1901–1948.

Giliberto, Concetta, and Loredana Teresi. *Limits to Learning: The Transfer of Encyclopaedic Knowledge in the Early Middle Ages*. Storehouses of Wholesome Learning 3. Leuven: Peeters, 2013.

Gillespie, Alexandra. "Medieval Books, Their Booklets, and Booklet Theory." *English Manuscript Studies 1100–1700* 16 (2011): 1–29.

Glacken, Clarence J. *Traces on the Rhodian Shore: Nature and Culture in Western Thought from Ancient Times to the End of the Eighteenth Century*. 7th ed. Berkeley: University of California Press, 1996.

Glauche, Günter. *Schullektüre im Mittelalter. Entstehung und Wandlungen des Lektürekanons bis 1200 nach den Quellen dargestellt*. Münchener Beiträge zur Mediävistik und Renaissance-Forschung 5. Munich: Arbeo-Gesellschaft, 1970.

Glendinning, Robert J. *A Critical Study of the Old High German Physiologus and its Influence*. Winnepeg: University of Manitoba, 1959.

Gneuss, Helmut, and Michael Lapidge. *Handlist of Anglo-Saxon Manuscripts: A Bibliographical Handlist of Manuscripts and Manuscript Fragments Written or Owned in England up to 1100*. Medieval and Renaissance Texts and Studies 241. Toronto: University of Toronto Press, 2014.

Goetz, Hans-Werner. *Gott und die Welt: Religiöse Vorstellungen des frühen und hohen Mittelalters*. Ser. 1, Vol. 1, *Das Gottesbild*. Orbis Mediaevalis: Vorstellungswelten des Mittelalters 13.1. Göttingen: Vandenhoeck & Ruprecht, 2012.

Goffart, Walter. "The Fredegar Problem Reconsidered." *Speculum* 38 (1963): 206–41.

Goldberg, Eric Joseph. *In the Manner of the Franks: Hunting, Kingship, and Masculinity in Early Medieval Europe*. Philadelphia: University of Pennsylvania Press, 2020.

Goldstaub, Maximilian. "Der Physiologus und seine Weiterbildung, besonders in der lateinischen und in der Byzantinischen Literatur." *Philologus Supplementband* 8 (1899): 339–404.

Goldstaub, Maximilian, and Richard Wendriner. *Ein tosco-venezianischer Bestiarius*. Halle: Max Niemeyer, 1892.

Gorla, Silvia. "Some Remarks About the Latin Physiologus Extracts Transmitted in the Liber Glossarum." *Mnemosyne* 71 (2018): 145–67.

Gorman, Michael M. "The Carolingian Miscellany of Exegetical Texts in Albi 39 and Paris Lat. 2175." *Scriptorium* 51 (1997): 336–54.

——. "A Critique of Bischoff's Theory of Irish Exegesis. The Commentary on Genesis in Munich Clm 6302 (Wendepunkte 2)." *Journal of Medieval Latin* 7 (1996): 178–233.

Gracia, Jorge, and Lloyd Newton. "Medieval Theories of the Categories." In *The Stanford Encyclopedia of Philosophy*, edited by Edward N. Zalta, n.d. https://plato.stanford.edu/archives/win2016/entries/medieval-categories/.

Graham, Daniel W. *Science Before Socrates: Parmenides, Anaxagoras, and the New Astronomy*. Oxford: Oxford University Press, 2013.

Greenfield, Concetta C. *Humanist and Scholastic Poetics, 1250–1500*. Lewisburg: Bucknell University Press, 1981.

Greenfield, Stanley B., Daniel G. Calder, and Michael Lapidge. *A New Critical History of Old English Literature*. New York: New York University Press, 1986.

Greer, Sarah, Alice Hicklin, and Stefan Esders, eds. *Using and Not Using the Past After the Carolingian Empire, c. 900–c. 1050*. London: Routledge, 2020.

Grotans, Anna A. *Reading in Medieval St. Gall*. Cambridge Studies in Palaeography and Codicology 13. Cambridge: Cambridge University Press, 2006.

Guglielmi, Nilda, ed.. *El Fisiólogo: Bestiario medieval. Introducción, traducción y notas*. Madrid: Ediciones Eneida, 2002.

——. "Review: Nikolaus Henkel, Studien zum Physiologus im Mittelalter." *Cahiers de civilisation médiévale* 84 (1978): 402–7.

Guillerme, André. *The Age of Water: The Urban Environment in the North of France, A.D. 300-1800*. Environmental History Series 9. College Station: Texas A&M University Press, 1988.

Gumbert, Johann Peter. "Codicological Units: Towards a Terminology for the Stratigraphy of the Non-Homogeneous Codex." In *Il codice miscellaneo: tipologie e funzioni. Atti del convegno internazionale, Cassino 14–17 maggio 2003*, edited by Edoardo Crisci and Oronzo Pecere, 17–42. Segno e testo 2. Turnhout: Brepols, 2004.

——. "Skins, Sheets and Quires." In *New Directions in Later Medieval Manuscript Studies: Essays from the 1998 Harvard Conference*, edited by Derek Pearsall, 81–90. Woodbridge: Boydell & Brewer, 2014.

Haeseli, Christa M. *Magische Performativität: Althochdeutsche Zaubersprüche in ihrem Überlieferungskontext*. Philologie der Kultur 4. Würzburg: Königshausen & Neumann, 2011.

Hagen, Hermann. *Anecdota helvetica quae ad grammaticam latinam spectant ex bibliothecis turicensi einsidlensi bernensi collecta*. Leipzig: Teubner, 1870.

——. *Carmina medii aevi maximam partem inedita*. Bern: G. Frobenius, 1877.

——. *Catalogus codicum Bernensium (Bibliotheca Bongarsiana)*. Bern: G. Frobenius, 1875.

——. *Zur Geschichte der Philologie und zur römischen Litteratur; vier Abhandlungen*. Berlin: Calvary, 1879.

Hahn, Georg L., and August Hahn, ed. *Bibliothek der Symbole und Glaubensregeln der alten Kirche*. 3rd ed. Breslau: E. Morgenstern, 1897.

Halm, Karl, ed. *Catalogus codicum manu scriptorum Bibliothecae Monacensis 4.3*. Munich: Bibliotheca regia, 1878 (reprinted in 1969).

Handley, Mark. "Epitaphs, Models and Texts: A Carolingian Collection of Late Antique Inscriptions from Burgundy." *Bulletin of the Institute of Classical Studies* 44 (2000): 47–56.

Harmless, William. *Augustine and the Catechumenate*. Collegeville: Liturgical Press, 1995.

Harris, Stephen J. "Anglo-Saxon Ciphers." In *A Material History of Medieval and Early Modern Ciphers: Cryptography and the History of Literacy*, edited by Katherine Ellison and Susan Kim, 65–79. New York: Routledge, 2018.

Harrison, Peter. "Hermeneutics and Natural Knowledge in the Reformers up to 1700." In *Nature and Scripture in the Abrahamic Religions*, edited by Jitse M. van der Meer and Scott Mandelbrote, 341–62. Brill's Series in Church History 36–7. Leiden: Brill, 2008.

——. *The Bible, Protestantism, and the Rise of Natural Science*. Cambridge: Cambridge University Press, 2001.

Haugh, Richard S. *Photius and the Carolingians: The Trinitarian Controversy*. Belmont: Nordland, 1975.

Head, Thomas, and Richard Landes, eds. *The Peace of God: Social Violence and Religious Response in France Around the Year 1000*. Ithaca: Cornell University Press, 1992.

Helmer, Friedrich. *Katalog der lateinischen Handschriften der bayerischen Staatsbibliothek München. Die Handschriften aus St. Emmeram in Regensburg. Clm 14261–14400*, edited by Hermann Hauke and Elisabeth Wunderle. Vol. 3. Catalogus Codicum Manu Scriptorum Bibliotheca Monacensis. Series Nova. Wiesbaden: Harrassowitz, 2011.

Henderson, Isabel. *Pictish Monsters: Symbol, Text and Image*. H. M. Chadwick Memorial Lectures 7. Cambridge: Department of Anglo-Saxon, Norse, and Celtic, University of Cambridge, 1996.

Henderson, John. "The Creation of Isidore's *Etymologies* or *Origins*." In *Ordering Knowledge in the Roman Empire*, edited by Jason König and Tim Whitmarsh, 150–74. Cambridge: Cambridge University Press, 2007.

——. *The Medieval World of Isidore of Seville: Truth From Words*. Cambridge: Cambridge University Press, 2007.

Henkel, Nikolaus. *Studien zum Physiologus im Mittelalter*. Tübingen: Niemeyer, 1976.

Hennings, Ralph. "Disputatio de Origine Animae (CPL 623,37) – or the Victory of Creatianism in the Fifth Century." In *Historica, Theologica et Philosophica, Critica et Philologica. Papers Presented at the Twelfth International Conference on Patristic Studies Held in Oxford 1995*, edited by Elizabeth A. Livingstone, 260–68. Studia Patristica 29. Leuven: Peeters, 1997.

Herlihy, David. "Attitudes Toward the Environment in Medieval Society." In *Historical Ecology: Essays on Environment and Social Change*, edited by Lester J. Bilsky, 100–116. National University Publications. Port Washington: Kennikat Press, 1980.

Hicks, Carola. *Animals in Early Medieval Art*. Edinburgh: Edinburgh University Press, 1993.

Hildebrandt, Madge M. *The External School in Carolingian Society*. Education and Society in the Middle Ages and Renaissance 1. Leiden: Brill, 1992.

Hillgarth, Jocelyn N. "The Position of Isidorian Studies: A Critical Review of the Literature 1936–1975." *Studi Medievali* 24 (1983): 817–905.

Hoek, Michelle C. "Anglo-Saxon Innovation and the Use of the Senses in the Old English Physiologus Poems." *Studia Neophilologica* 69 (1997): 1–10.

Hoffmann, Richard C. *An Environmental History of Medieval Europe*. Cambridge: Cambridge University Press, 2014.

——. "Homo et Natura, Homo in Natura: Ecological Perspectives on the European Middle Ages." In *Engaging with Nature: Essays on the Natural World in Medieval and Early Modern Europe*, edited by Barbara A. Hanawalt and Lisa J. Kiser, 11–38. Notre Dame: University of Notre Dame Press, 2008.

——. "The Protohistory of Pike in Western Culture." In *The Medieval World of Nature: A Book of Essays*, edited by Joyce E. Salisbury, 61–76. Garland Medieval Casebooks 5. New York: Garland Publishing, 1993.

Holford-Strevens, Leofranc. "Sirens in Antiquity and the Middle Ages." In *Music of the Sirens*, edited by Linda P. Austern and Inna Naroditskaya, 16–51. Bloomington: Indiana University Press, 2006.

Homburger, Otto. *Die Illustrierten Handschriften der Burgerbibliothek Bern: Die vorkarolingischen und karolingischen Handschriften*. Bern: Burgerbibliothek, 1962.

Horst, Koert van der. "The Utrecht Psalter: Picturing the Psalms of David." In *The Utrecht Psalter in Medieval Art: Picturing the Psalms of David*, edited by Koert van der Horst, William Noel, and Wilhelmina C. M. Wüstefeld, 22–84. Utrecht: HES, 1996.

Horst, Koert van der, William Noel, and Wilhelmina C. M. Wüstefeld, eds. *The Utrecht Psalter in Medieval Art: Picturing the Psalms of David*. Utrecht: HES, 1996.

Hunt, Richard W., Falconer Madan and P. D. Record. *A Summary Catalogue of Western Manuscripts in the Bodleian Library at Oxford Which Have not Hitherto Been Catalogued in the Quarto Series, with References to the Oriental and Other Manuscripts*. 7 vols. in 8 [vol. II in 2 parts]. Oxford: Clarendon Press, 1895–1953.

Hunyadi, Zsolt. "Signs of Conversion in Early Medieval Charters." In *Christianizing Peoples and Converting Individuals*, edited by Guyda Armstrong and Ian N. Wood, 105–13. Turnhout: Brepols, 2000.

Inguanez, Maurus. *Codicum Casinensium manu scriptorum catalogus: cura e studio monachorum S. Benedicti, archicoenobii Montis Casini*. Vol. 2.2. Montecassino, 1934.

Irvine, Martin. *The Making of Textual Culture: "Grammatica" and Literary Theory, 350–1100*. Cambridge Studies in Medieval Literature 19. Cambridge: Cambridge University Press, 1994.

James, Montague R. *A Descriptive Catalogue of the Manuscripts in the Library of Corpus Christi College Cambridge*. 2 vols., Vol. 2. Cambridge: Cambridge University Press, 1912.

——. *The Bestiary: Being a Reproduction in Full of MS. Il 4. 26 in the University Library, Cambridge, with Supplementary Plates from Other Manuscripts of English Origin, and a Preliminary Study of the Latin Bestiary as Current in England*. Oxford: Oxford University Press, 1928.

Jarrett, Jonathan Andrew. *Rulers and Ruled in Frontier Catalonia, 880–1010: Pathways of Power*. Woodbridge: Boydell Press, 2010.

Jebe, Johanna. "Regeln, Schrift, Correctio – Karolingerzeitliche Entwürfe von Mönchtum im Spiegel der Schriftproduktion aus St. Gallen und Fulda." PhD thesis, Tübingen University, 2022.

Jessen, Tyler D., Natalie C. Ban, Nicholas XEMŦOLTW̱ Claxton, and Chris T. Darimont. "Contributions of Indigenous Knowledge to Ecological and Evolutionary Understanding." *Frontiers in Ecology and the Environment* 20 (2022): 93–101.

Jones, Charles W. *The Saint Nicholas Liturgy and its Literary Relationships (Ninth to Twelfth Centuries)*. Los Angeles: University of California Press, 1963.

Jones, Richard. *The Medieval Natural World*. Harlow: Pearson, 2013.

Jong, Mayke de. "The Empire That was Always Decaying: The Carolingians (800–888)." *Medieval Worlds* 2 (2015): 6–25.

Juste, David. "Comput et divination chez Abbon de Fleury." *Abbon de Fleury: Philosophie, science et comput autour de l'an mil: Actes des journées organisées par le Centre d'histoire des sciences et des philosophies arabes et médiévales, edited by* Barbara Obrist, 95–127. Oriens-Occidens: Sciences, Mathématiques et Philosophie de l'Antiquité à l'Âge Classique 6. Paris-Villejuif: CNRS, 2004.

Karkov, Catherine E. "Writing and Having Written: Word and Image in the Eadwig Gospels." In *Writing and Texts in Anglo-Saxon England*, edited by Alexander R. Rumble, 44–61. Publications of the Manchester Center for Anglo-Saxon Studies 5. Cambridge: D. S. Brewer, 2006.

Kavanagh, Anne K. 'The Ps-Jerome's Expositio IV euangeliorum'. In *The Scriptures and Early Medieval Ireland: Proceedings of the 1993 Conference of the Society for Hiberno-Latin Studies on Early Irish Exegesis and Homiletics*, edited by Thomas O'Loughlin, 125–31. Instrumenta Patristica et Mediaevalia 31. Turnhout: Brepols, 1999.

Keane, Catherine. *Figuring Genre in Roman Satire*. American Classical Studies 50. Oxford: Oxford University Press, 2006.

Keefe, Susan A. *A Catalogue of Works Pertaining to the Explanation of the Creed in Carolingian Manuscripts*. Instrumenta Patristica et Mediaevalia 63. Turnhout: Brepols, 2012.

——. *Water and the Word: Baptism and the Education of the Clergy in the Carolingian Empire*. 2 vols. Notre Dame: University of Notre Dame Press, 2002.

Keil, Heinrich, ed. "Diomedes: Ars grammatica." In *Grammatici Latini*, 1:299–529. Leipzig: Teubner, 1857.

Kelly, Joseph F. "A Catalogue of Early Medieval Hiberno-Latin Biblical Commentaries (II)." *Traditio* 45 (1990): 393–434.

Kelly, Mary, and Charles Doherty, eds. *Music and the Stars: Mathematics in Medieval Ireland*. Dublin: Four Courts Press, 2013.

Ker, Neil R. *Catalogue of Manuscripts Containing Anglo-Saxon*. Oxford: Clarendon Press, 1957 (reprinted Oxford: Oxford University Press, 1990.)

——. *Medieval Manuscripts in British Libraries*. 5 vols. Oxford: Oxford University Press, 1969-2002.

Kershaw, Paul J. E. "Eberhard of Friuli, a Carolingian Lay Intellectual." In *Lay Intellectuals in the Carolingian World*, edited by Patrick Wormald and Janet L. Nelson, 77–105. Cambridge: Cambridge University Press, 2007.

Kindschi Garský, Zbynek, and Rainer Hirsch-Luipold, eds. *Christus in natura. Quellen, Hermeneutik und Rezeption des Physiologus*. Studies of the Bible and its Reception 11. Berlin: De Gruyter, 2020.

King, Paul D., ed. *Charlemagne: Translated Sources*. Lambrigg: P. D. King, 1987.

King, Peter. "Augustine on Language." In *The Cambridge Companion to Augustine*, 2nd ed., edited by David Vincent Meconi and Eleonore Stump, 292–310. Cambridge: Cambridge University Press, 2014.

Koehler, Wilhelm, and Florentine Mütherich. *Die Karolingischen Miniaturen: Die Schule von Reims. Von der Mitte bis zum Ende des 9. Jahrhunderts*. Vol. 6.2. Denkmäler Deutscher Kunst. Berlin: Deutscher Verlag für Kunstwissenschaft, 1999.

Kramer, Rutger. *Rethinking Authority in the Carolingian Empire: Ideals and Expectations during the Reign of Louis the Pious*. Amsterdam: Amsterdam University Press, 2019.

Krämer, Sigrid. *Handschriftenerbe des deutschen Mittelalters*, 2 vols., Vol. 1.2. Munich: Beck, 1989.

Krapp, George Philip, ed. *The Exeter Book*. The Anglo-Saxon Poetic Records 3. New York: Columbia University Press, 2004.

Kreiner, Jamie. *Legions of Pigs in the Early Medieval West*. New Haven: Yale University Press, 2020.

Krusch, Bruno. "Das älteste fränkische Lehrbuch der dionysianischen Zeitrechnung." In *Mélanges offerts à M. Émile Chatelain*, 232–42. Paris: Champion, 1910.

——. *Studien zur christlich-mittelalterlichen Chronologie. Die Entstehung unserer heutigen Zeitrechnung*. Vol. 8. Abhandlungen der Preußischen Akademie der Wissenschaften. Berlin: Verlag der Akademie der Wissenschaften, 1938.

Kuhry, Emmanuelle. "Zoological Inconsistency and Confusion in the Physiologus Latinus." In *Fragmented Nature: Medieval Latinate Reasoning on the Natural World and Its Order*, edited by Mattia Cipriani and Nicola Polloni. Abingdon: Routledge, 2022.

Küster, Hansjörg. *Geschichte der Landschaft in Mitteleuropa: von der Eiszeit bis zur Gegenwart*. Munich: Beck, 1995.

La Rocca, Christina, and Luigi Provero. "The Dead and Their Gifts: The Will of Eberhard, Count of Friuli, and His Wife Gisela, Daughter of Louis the Pious (863–864)." In *Rituals of Power from Late Antiquity to the Early Middle Ages*, edited by Frans Thews and Janet L. Nelson. The Transformation of the Roman World 8. Leiden: Brill, 2000.

Lapidge, Michael. "Beowulf, Aldhelm, the Liber Monstrorum and Wessex." In *Anglo-Latin Literature, 600–899*, edited by Michael Lapidge, 271–312, 507. London: Hambledon Press, 1996.

——. "Surviving Booklists from Anglo-Saxon England." In *Learning and Literature in Anglo-Saxon England: Studies Presented to Peter Clemoes on the Occasion of His Sixty-Fifth Birthday*, edited by Peter Clemoes, Michael Lapidge, and Helmut Gneuss, 33–90. Cambridge: Cambridge University Press, 2010.

——. *The Anglo-Saxon Library*. Oxford: Oxford University Press, 2006.

——. "The Career of Aldhelm." *Anglo-Saxon England* 36 (2007): 15–69.

——. "The Stoic Inheritance." In *A History of Twelfth-Century Western Philosophy*, edited by Peter Dronke, 81–112. Cambridge: Cambridge University Press, 1988.

Lauchert, Friedrich. *Geschichte des Physiologus*. Strassburg: K. J. Trübner, 1889. DOI: 10.1515/9783111498324.

Lauwers, Michel. "Le glaive et la parole. Charlemagne, Alcuin et le modèle du rex praedicator: notes d'ecclésiologie carolingienne." In *Alcuin, de York à Tours: écriture, pouvoir et réseaux dans l'Europe du haut Moyen Âge*, edited by Philippe Depreux and Bruno Judic, 221–44. Annales de Bretagne et des Pays de l'Ouest 111/3. Rennes: Presses universitaires, 2004. DOI: 10.4000/abpo.1243.

Law, Vivien. *The Insular Latin Grammarians*. Studies in Celtic History 3. Woodbridge: Boydell, 1982.

Lazaris, Stavros. *Le Physiologus grec*. Vol. 1, *La réécriture de l'histoire naturelle antique*. Micrologus Library 77/1. Florence: Sismel, 2016.

——. *Le Physiologus grec*, Vol. 2, *Donner à voir la nature*. Micrologus Library 107. Florence: Sismel, Edizioni del Galluzzo, 2021.

Lazic, Milorad, and Ljubomir Kotarcic. *Fisiolog; Srednjovekovni Medicinski Spisi*. Stara Srpska Knjizevnost u 24 Knjige 24. Belgrade: Prosveta: Srpska knjizevna zadruga, 1989.

Le Goff, Jacques. "The Wilderness in the Medieval West." In *The Medieval Imagination*, by Jacques Le Goff, 47–59. Chicago: University of Chicago Press, 1992.

Le Roy Ladurie, Emmanuel. *Histoire du climat depuis l'an mil*. Paris: Flammarion, 1967.

Leclercq, Jacqueline. "De l'art antique à l'art médiéval. À propos des sources du bestiaire carolingien et de ses survivances à l'époque romane." *Gazette des Beaux-Arts* 113 (1989): 61–66.

Leclercq-Marx, Jacqueline. *Iconographie médiévale entre antiquité et art roman: d'acanthes et d'écailles. Recueil d'articles de Jacqueline Leclercq-Marx*, edited by Brigitte D'Hainaut-Zveny, Alain Dierkens, and Constantin Pion. Les Études du RILMA 9. Turnhout: Brepols, 2018.

——. *La sirène dans la pensée et dans l'art de l'antiquité et du Moyen Âge: Du mythe païen au symbole chrétien*. Mémoires de la classe des Beaux-Arts. 3rd ser., 2. Brussels: Académie royale de Belgique, 1997.

——. 'La sirène et l'(ono)centaure dans le Physiologus grec et latin." In *Bestiaires médiévaux: nouvelles perspectives sur les manuscrits et les traditions textuelles. Communications présentées au XVe colloque de la société internationale Renardienne, Louvain-La-Neuve, 19–22 août 2003*, edited by Baudouin Van den Abeele, 169–82. Louvain-la-Neuve: Brepols, 2005.

——. 'L'illustration du Physiologus grec et latin, entre littéralité et réinterprétation de l'allégorie textuelle. Le cas des manuscrits Bruxellensis 10066–77 et Smyrneus B.8 (détruit)'. In *L'allégorie dans l'art du Moyen Âge. Formes et fonctions. Héritages, créations, mutations*, edited by Christian Heck, 141–55. Les Études du RILMA 2. Turnhout: Brepols, 2011. DOI: 10.1484/M.ETRILMA-EB.3.5067.

Legendre, Paul. *Études tironiennes. Commentaire sur la VIe églogue de Virgile tiré d'un manuscrit de Chartres*. La Bibliothèque de l'École des Hautes Études 165. Paris: Champion, 1907.

Lendinara, Patrizia. *Anglo-Saxon Glosses and Glossaries*. Variorum Collected Studies Series. Aldershot: Ashgate, 1999.

——. "The *Versus sibyllae de die iudicii* in Anglo-Saxon England." In *Apocryphal Texts and Traditions in Anglo-Saxon England*, edited by Kathryn Powell and D. G. Scragg, 85–101. Cambridge: D.S. Brewer, 2003.

Leonardi, Claudio. "Intellectual Life." In *New Cambridge Medieval History: c. 900–c. 1024*, edited by Rosamond McKitterick, 3:186–211. Cambridge: Cambridge Univ Press, 1999.

Lesne, Émile. *Histoire de la propriété ecclésiastique en France: Les livres, scriptoria', et bibliothèques du commencement du VIIIe à la fin du XIe siècle*. Vol. IV. VI vols. Lille, 1938.

Letson, Douglas R. "The Old English *Physiologus* and the Homiletic Tradition." *Florilegium. Canadian Society of Medievalists* 1 (1979): 15–41. DOI: 10.3138/flor.1.003.

Lindsay, Wallace M. *Palaeographia Latina*. 6 vols. Vol. 1. St Andrews: Humphrey Milford, 1922.

Liuzza, Roy, ed. *Anglo-Saxon Prognostics: An Edition and Study of Texts from London, British Library, MS Cotton Tiberius A. Iii*. Anglo-Saxon Texts 8. Cambridge: Cambridge University Press, 2010.

Lozovsky, Natalia. "Carolingian Geographical Tradition: Was It Geography?" *Early Medieval Europe* 5 (2007): 25–43. DOI: 10.1111/j.1468-0254.1996.tb00046.x.

——. *The Earth Is Our Book: Geographical Knowledge in the Latin West ca. 400–1000*. Recentiores. Ann Arbor: University of Michigan Press, 2000.

Lubac, Henri de. *Exégèse médiévale. Les quatre sens de l'écriture*. 3 vols. Ressourcement: Retrieval and Renewal in Catholic Thought. Paris: Aubier-Montaigne, 1959.

——. *Histoire et esprit: l'intelligence de l'écriture d'après Origène. Cinquième section, écriture et eucharistie*. Oeuvres complètes 16. Paris: Aubier-Montaigne, 1950.

Lugt, Maaike van der. *Le ver, le démon et la vierge: Les théories médiévales de la génération extraordinaire. Une étude sur les rapports entre théologie, philosophie naturelle et médecine*. L'Âne d'or 20. Paris: Belles lettres, 2004.

MacCarron, Máirín. *Bede and Time: Computus, Theology and History in the Early Medieval World*. London: Routledge, 2020.

Macé, Caroline, and Jost Gippert, eds. *The Multilingual Physiologus. Studies in the Oldest Greek Recension and its Translations*. Turnhout: Brepols, 2021.

MacLean, Simon. *History and Politics in Late Carolingian and Ottonian Europe: The Chronicle of Regino of Prüm and Adalbert of Magdeburg*. Manchester: Manchester University Press, 2009.

Mähl, Sibylle. *Quadriga Virtutum: Die Kardinaltugenden in der Geistesgeschichte der Karolingerzeit*. Beihefte zum Archiv für Kulturgeschichte 9. Cologne: Böhlau, 1969.

Maniaci, "The Medieval Codex as a Complex Container: The Greek and Latin Traditions." In *One-Volume Libraries: Composite and Multiple-Text Manuscripts*, edited by Michael Friedrich and Cosima Schwarke, 27–46. Studies in Manuscript Cultures 9. Berlin: De Gruyter, 2016.

Manitius, Max. *Geschichte der lateinischen Literatur des Mittelalters: Von Justinian bis zur Mitte des 10. Jahrhunderts*. 3 vols., Vol. 1. Handbuch der Altertumswissenschaft 9. Munich: Beck, 1911.

——. *Handschriften Antiker Autoren in Mittelalterlichen Bibliothekskatalogen*. Beihefte zum Zentralblatt für Bibliothekswesen 67. Leipzig: Harrassowitz, 1935.

Mantello, Frank A. C., and Arthur G. Rigg, ed. *Medieval Latin: An Introduction and Bibliographical Guide*. Washington, DC: Catholic University of America Press, 1996.

Marchal, François J. F. *Inventaire des Manuscrits de la Bibliothèque royale des Ducs de Bourgogne*. 3 vols. Brussels: Impr. Vandooren, 1839.

Marenbon, John. *Early Medieval Philosophy (480–1150): An Introduction*. Rev. 2nd ed. London: Routledge, 1988.

——. *From the Circle of Alcuin to the School of Auxerre: Logic, Theology and Philosophy in the Early Middle Ages*. Cambridge: Cambridge University Press, 1981.

Mareschal, Valérie. "Un témoin latin illustré du Xe siècle du Physiologus: le Bruxellensis 10066–77." MA thesis, Université Catholique de Louvain, 1997.

Martindale, Charles, ed. *The Cambridge Companion to Virgil*. Cambridge: Cambridge University Press, 1997.

Masai, François. "Le manuscrit à miniatures." In *Art mosan et arts anciens du pays de Liège*, edited by Joseph De Borchgrave d'Altena, 63–80. Liège: Le Grand, 1951.

Mayeski, Marie A. "Early Medieval Exegesis: Gregory I to the Twelfth Century." In *A History of Biblical Interpretation 2: The Medieval through the Reformation Periods*, edited by Alan J. Hauser and Duane F. Watson, 86–112. Grand Rapids: Eerdmans, 2009.

McCluskey, Stephen C. "Boethius's Astronomy and Cosmology." In *A Companion to Boethius in the Middle Ages*, edited by Noel Harold Kaylor and Philip Edward Phillips, 47–74. Brill's Companions to the Christian Tradition 30. Leiden: Brill, 2012.

McCracken, Peggy. *In the Skin of a Beast: Sovereignty and Animality in Medieval France*. Chicago: The University of Chicago Press, 2017.

McCulloch, Florence. *Mediaeval Latin and French Bestiaries*. Chapel Hill: University of North Carolina Press, 1962.

McKitterick, Rosamond. "Glossaries and Other Innovations in Carolingian Book Production." In *Turning Over a New Leaf: Change and Development in the Medieval Manuscript*, edited by Erik Kwakkel, 21–78. Amsterdam: Amsterdam University Press, 2012.

——. "Roman Texts and Roman History in the Early Middle Ages." In *Rome Across Time and Space. Cultural Transmission and the Exchange of Ideas, c. 500–1400*, edited by Claudia Bolgia, 19–34. Cambridge: Cambridge University Press, 2014.

——. *Rome and the Invention of the Papacy*. Cambridge: Cambridge University Press, 2020.

——. *The Carolingians and the Written Word*. Cambridge: Cambridge University Press, 1989.

——. *The Frankish Church and the Carolingian Reforms, 789–895*. Royal Historical Society Studies in History 2. London: Royal Historical Society, 1977.

McNeill, John T. and Gamer, Helena M. *The Medieval Handbooks of Penance: A Translation of the Principal Libri Poenitentiales and Selections from Related Documents*. Records of Civilization, Sources and Studies 29. New York: Columbia University Press, 1938.

Michaud, Jean-Paul. "Des quatre sens de l'écriture. Histoire, théorie, théologie, herméneutique." *Église et théologie* 30 (1999): 165–97.

Milchsack, Gustav. *Die Handschriften der Herzoglichen Bibliothek zu Wolfenbüttel. Abth. 4: Die Gudischen Handschriften*. Wolfenbüttel: Zwissler, 1913.

Milde, Wolfgang. *Der Bibliothekskatalog des Klosters Murbach aus dem 9. Jahrhundert. Ausgabe und Untersuchung von Beziehungen zu Cassiodors "Institutiones."* Beihefte zum Euphorion, Zeitschrift für Literaturgeschichte 4. Heidelberg: Winter, 1968.

Milovanović-Barham, Čelica. "Aldhelm's Enigmata and Byzantine Riddles." *Anglo-Saxon England* 22 (1993): 51–64. DOI: 10.1017/S0263675100004300.

Minio-Paluello, Lorenzo, ed. *Aristoteles latinus. I. 1-5, Categoriae vel praedicamenta: translatio Boethii, editio composita, translatio Guillelmi de Moerbeka Lemmata e Simplicii commentario decerpta, Pseudo-Augustini paraphrasis Themistiana.* Repr. ed. Paris: Desclée de Brouwer, 1961.

Mittenhuber, Florian. "Die Berner Physiologus-Handschriften: Drei Bücher, drei Geschichten." In *Christus in Natura: Quellen, Hermeneutik und Rezeption des Physiologus*, edited by Zbyněk Kindschi Garský and Rainer Hirsch-Luipold, 177–88. Berlin: De Gruyter, 2019.

Moore, Philip S. "The Authorship of the *Allegoriae super vetus et novum testamentum*." *New Scholasticism* 9 (1935): 209–25.

Moran, Dermot. "John Scottus Eriugena." In *The Stanford Encyclopedia of Philosophy (Fall 2008 Edition)*, edited by Edward N. Zalta, n.d. https://plato.stanford.edu/archives/fall2008/entries/scottus-eriugena/.

——. *The Philosophy of John Scottus Eriugena: A Study of Idealism in the Middle Ages.* Cambridge: Cambridge University Press, 2004.

Mordek, Hubert. "Bischofsabsetzungen in Spätmerowingischer Zeit. Justelliana, Bernensis und das Konzil von Mâlay (677)." In *Papsttum, Kirche und Recht im Mittelalter*, edited by Hubert Mordek, 31–53. Tübingen: Niemeyer, 1981.

——. *Kirchenrecht und Reform im Frankenreich: Die Collectio Vetus Gallica, die älteste systematische Kanonessammlung des fränkischen Gallien. Studien und Edition.* Beiträge zur Geschiche und Quellenkunde des Mittelalters 1. Berlin: De Gruyter, 1975.

Moriarty, Christopher. "The Early Naturalists." In *Nature in Ireland: A Scientific and Cultural History*, edited by John Wilson Foster and Helena C. G. Chesney, 71–90. Montreal: McGill-Queen's University Press, 1998.

Mostert, Marco. *The Library of Fleury: A Provisional List of Manuscripts.* Middeleeuwse Studies En Bronnen 3. Hilversum: Verloren, 1989.

Munk Olsen, Birger. *L'étude des auteurs classiques latins aux XIe et XIIe siècles.* 5 vols. Paris: CNRS, 1982–2020.

Muntané, Maricarmen G. "Del 'Iudicii signum' al Canto de la Sibila: primeros testimonios." In *Hispania Vetus. Manuscritos litúrgico-musicales: de los orígenes visigóticos a la transición francorromana (siglos X–XII)*, edited by Susana Zapke, 159–73. Madrid: Fundación BBVA, 2007.

Muradyan, Gohar. *Physiologus: The Greek and Armenian Versions with a Study of Translation Technique.* Hebrew University Armenian Studies 6. Leuven: Peeters, 2005.

Muratova, Xenia. "La production des manuscrits du Physiologue grecs enluminés en Italie aux XVe–XVIe siècles et leur place dans l'histoire de la tradition de l'illustration du Physiologue." *Jahrbuch der Österreichischen Byzantinistik* 32 (1982): 330–34, 340.

——. "Problèmes de l'origine et des sources des cycles d'illustrations des MSS des bestiaires." In *Épopée animale, fable et fabliau: actes du colloque de la Société Internationale Renardienne qui s'est tenu à Amsterdam du 21 au 24 octobre 1977*, edited by Gabriel Bianciotto and Michel Salvat, 383–408. Cahiers d'études médiévales 2.3. Publications de l'Université de Rouen 83. Rouen: Presse universitaire, 1983.

Mütherich, Florentine. "Carolingian Manuscript Illumination in Rheims." In *The Utrecht Psalter in Medieval Art: Picturing the Psalms of David*, edited by Koert van der Horst, William Noel, and Wilhelmina C. M. Wüstefeld, 104–19 and 168–69. MS't Goy: HES Publishers/London: Harvey Miller, 1996.

——. *Studies in Carolingian Manuscript Illumination*. London: Prindar Press, 2004.

Nasmith, James. *Catalogus librorum manuscriptorum quos collegio corporis Christi et B. Mariæ virginis in vcademia Cantabrigiensi legauit reverendissimus Matthæus Parker, archiepiscopus Cantuariensis*. Cambridge: J. Woodyer, 1777.

Nebbiai-Dalla Guarda, Donatella. *La bibliothèque de l'abbaye de Saint-Denis en France du IXe au XVIIIe siècle*. Documents, études, répertoires 32. Paris: CNRS, 1985.

Nees, Lawrence. *Frankish Manuscripts: The Seventh to the Tenth Century*. 2 vols. A Survey of Manuscripts Illuminated in France 2. London: Harvey Miller, 2022.

——. "The Illustrated Manuscript of the Visio Baronti [Revelatio Baronti] in St. Petersburg (Russian National Library, Cod. Lat. Oct. v.I.5)." In *Court Culture in the Early Middle Ages*, edited by Catherine Cubitt, 91–128. Turnhout: Brepols, 2003.

Nelson, Anne Browning. "The Classroom of Didymus the Blind." PhD thesis, University of Michigan, 1995.

Nelson, Janet L. "On the Limits of the Carolingian Renaissance." *Studies in Church History* 14 (1977): 51–67.

Netzer, Nancy. *Cultural Interplay in the Eighth Century: The Trier Gospels and the Makings of a Scriptorium at Echternach*. Cambridge Studies in Palaeography and Codicology 3. Cambridge: Cambridge University Press, 1994.

——. "The Design and Decoration of Insular Gospel-Books and Other Liturgical Manuscripts, ca. 600–ca. 900." In *The Cambridge History of the Book in Britain*, edited by Richard Gameson, 225–43. Cambridge: Cambridge University Press, 2011.

Newton, Francis. *The Scriptorium and Library at Monte Cassino, 1058–1105*. Cambridge Studies in Palaeography and Codicology 7. Cambridge: Cambridge University Press, 1999.

Nichols, Stephen G., and Siegfried Wenzel, eds. *The Whole Book: Cultural Perspectives on the Medieval Miscellany*. Arbor: University of Michigan Press, 1996.

Nilgen, Ursula. "Der Codex Douce 292 der Bodleian Library zu Oxford: ein ottonisches Evangeliar." PhD thesis, University of Bonn, 1967.

Noble, Thomas F.X. "Carolingian Era." Oxford Bibliographies Online. Oxford: Oxford University Press, 2010. DOI: 10.1093/obo/9780195396584-0012.

Nutton, Vivian. "Early-Medieval Medicine and Natural Science." In *The Cambridge History of Science*, edited by David C. Lindberg and Michael H. Shank, 323–40. Cambridge: Cambridge University Press, 2013.

O'Brien, Conor. "Hwaetberht, Sicgfrith and the Reforming of Wearmouth and Jarrow." *Early Medieval Europe* 25 (2017): 301–19.

Obrist, Barbara. "Wind Diagrams and Medieval Cosmology." *Speculum* 72 (1997): 33–84.

Odermatt, Ambros. *Ein Rituale in Beneventanischer Schrift: Roma, Bibliotheca Vallicelliana, Cod. C 32, Ende des 11. Jahrhunderts*. Spicilegium Friburgense 26. Freiburg: Universitäts-Verlag, 1980.

O'Meara, Dominic J., ed. *Neoplatonism and Christian Thought*. Studies in Neoplatonism 3. Norfolk, VA: International Society for Neoplatonic Studies, 1981.

Oosthuizen, Susan. "Anglo-Saxon Fields." In *The Oxford Handbook of Anglo-Saxon Archaeology*, edited by Helena Hamerow, David Alban Hinton, and Sally Crawford, 377–401. Oxford: Oxford University Press, 2011.

——. "Recognizing and Moving on from a Failed Paradigm: The Case of Agricultural Landscapes in Anglo-Saxon England c. AD 400–800." *Journal of Archaeological Research* 24 (2016): 179–227.

——. *Tradition and Transformation in Anglo-Saxon England: Archaeology, Common Rights and Landscape*. London: Bloomsbury, 2013.

Orchard, Andy. *A Critical Companion to Beowulf*. Cambridge: D. S. Brewer, 2007.

——. *Pride and Prodigies: Studies in the Monsters of the Beowulf-Manuscript*. Toronto: University of Toronto Press, 2003.

Orlandi, Giovanni. "La tradizione del Physiologus e i prodromi del bestiario latino." In *L'uomo di fronte al mondo animale nell'alto medioevo*, 2:1057–1106. Settimane di studio del Centro Italiano di studi sull'alto medioevo 31. Spoleto: CISAM, 1985.

O'Sullivan, Sinéad. *Early Medieval Glosses on Prudentius' Psychomachia: The Weitz Tradition*. Mittellateinische Studien und Texte 31. Leiden: Brill, 2004.

Otten, Willemien. "Overshadowing or Foreshadowing Return: The Role of Demons in Eriugena's Periphyseon." In *Demons and the Devil in Ancient and Medieval Christianity*, edited by Nienke M. Vos and Willemien Otten, 211–30. Vigiliae Christianae Supplement 108. Leiden: Brill, 2011.

Ozment, Kate, and Lisa Maruca, ed. *New Approaches to Critical Bibliography and the Material Text Criticism*, special issue of *Criticism* 64, nos 3–4 (2022).

Pächt, Otto, and Jonathan J. G. Alexander. *Illuminated Manuscripts in the Bodleian Library*. 3 vols., Vol. 1. Oxford: Clarendon Press, 1966.

Pakis, Valentine A. "A Note in Defense of 'The Partridge' (Exeter Book 97v)." *Neophilologus* 92 (2008): 729–34.

——. "Contextual Duplicity and Textual Variation: The Siren and Onocentaur in the Physiologus Tradition." *Mediävistik* 23 (2010): 115–86.

——. "Sirens and their Victims in Version B of the Latin Physiologus." *Mittellateinisches Jahrbuch* 44 (2009): 367–75.

Palmer, James. "Climates of Crisis: Apocalypse, Nature, and Rhetoric in the Early Medieval World." *Viator* 48 (2018): 1–20.

Pardessus, Jean-Marie. "Notice sur les manuscrits de formules relatives au droit observé dans l'empire des Francs, suivis de quatorze formules inédites." *Bibliothèque de l'École des Chartes* 4 (1843): 1–22.

Parkes, Malcolm B. *Pause and Effect: An Introduction to the History of Punctuation in the West*. Aldershot: Scolar Press, 1992.

Parmentier, Elisabeth. *L'Écriture vive: interprétations chrétiennes de la Bible*. Geneva: Labor et Fides, 2004.

Parmentier, Martin. "Trying to Unravel Jacobi's Unknown Creed." *Bijdragen Tijdschrift voor Filosofie en Theologie* 52, no. 4 (1991): 354–78.

Pascual, Llúcia Martín. "La tradición animalística en Italia: el bestiario toscano." *Cultura Neolatina* 72 (2012): 145–79.

Patzold, Steffen. *Das Lehnswesen*. Beck'sche Reihe C. H. Beck Wissen 2745. Munich: Beck, 2012.

——. *Presbyter: Moral, Mobilität und die Kirchenorganisation im Karolingerreich*. Monographien zur Geschichte des Mittelalters 68. Stuttgart: Anton Hiersemann, 2020.

Patzold, Steffen, and Carine van Rhijn, eds. *Men in the Middle: Local Priests in Early Medieval Europe*. Berlin: De Gruyter, 2016.

Pelle, Stephen. "Source Studies in the Lambeth Homilies." *The Journal of English and Germanic Philology* 113 (2014): 34–72.

Pellegrin, Élisabeth. "Membra disiecta Floriacensia." *Bibliothèque de l'École des Chartes* 117 (1959): 5–56 (reprinted in *Bibliothèques retrouvées. Manuscrits, bibliothèques et bibliophiles du Moyen Âge et de la Renaissance*, edited by Élisabeth Pellegrin, 159–210. Paris: CNRS, 1988).

——. "Membra disiecta Floriacensia (II)." In *Miscellanea codicologica F. Masai dicata*, edited by Pierre Cockshaw, Monique-Cécile Garand and Pierre Jodogne, 83–103 (reprinted in *Bibliothèques retrouvées. Manuscrits, bibliothèques et bibliophiles du Moyen Âge et de la Renaissance*, edited by Élisabeth Pellegrin, 285–97. Paris: CNRS, 1988).

Pellegrin, Élisabeth, and Jean-Paul Bouhot. *Catalogue des manuscrits médiévaux de la bibliothèque municipale d'Orléans*. Documents, études et répertoires. Institut de recherche et d'histoire des textes 78. Paris: CNRS, 2010.

Péricard, Jacques. *Ecclesia Bituricensis: le diocèse de Bourges des origines à la réforme Grégorienne*. Collection des thèses 7. Clermont-Ferrand: Fondation Varenne, 2006.

Perry, Ben E. *Babrius and Phaedrus*. Cambridge, MA: Harvard University Press, 1965.

——. "Physiologus." In *Paulys Realencyclopädie des classischen Alterturmswissenschaft* 20, edited by Wilhelm K. Mittelhaus, cols. 1074–129. Stuttgart: Alfred Druckenmüller,1941.

——. *Studies in the Text History of the Life and Fables of Aesop*. Haverford: American Philological Association, 1936.

Petrov, Valery V. "Karolingskiye schkolniye teksty: glossy iz kruga Ioanna Skotta i Remigiya iz Oserra" [Carolingian School Texts: Glosses from the Circle of John Scottus and Remigius of Auxerre]. In *Filosofiya prirody v antichnosti i v srednie veka*, edited by Piama P. Gaĭdenko and Valery V. Petrov, 530–91. Moscow: Progress-Traditsiia, 2000.

Petrucci, Armando. "From Unitary Book to Miscellany." In *Writers and Readers in Medieval Italy: Studies in the History of Written Culture*, by Armando Petrucci, 15–42. Edited and translated by Charles M. Radding. New Haven: Yale University Press, 1995.

Pohl, Walter. "History in Fragments: Montecassino's Politics of Memory." *Early Medieval Europe* 10 (2001): 343–74.

Porcier, Stéphanie, Salima Ikram, and Stéphane Pasquali, ed. *Creatures of Earth, Water and Sky: Essays on Animals in Ancient Egypt and Nubia*. Leiden: Sidestone Press, 2019.

Quain, Eduino A., and Julian G. Plante. "Catalogue of manuscripts in the Bibliothek der phil.-theol.-Hochschule der Diözese Linz." *Traditio* 32 (1976): 427–74.

Raaijmakers, Janneke. *The Making of the Monastic Community of Fulda, c. 744–c. 900*. Cambridge: Cambridge University Press, 2012.

Ramírez-Weaver, Eric M. *A Saving Science: Capturing the Heavens in Carolingian Manuscripts*. University Park: The Pennsylvania State University Press, 2017.

——. "Carolingian Innovation and Observation in the Paintings and Star Catalogs of Madrid, Biblioteca Nacional, Ms. 3307." PhD thesis, New York University, 2008.

Rand, Edward K. "Is Donatus's Commentary on Vergil Lost?" *Classical Quarterly* 10 (1916): 158–64.

Rankin, Susan. "Ego Itaque Notker Scripsi." *Révue Bénédictine* 101 (1991): 268–98.

Reifferscheid, August. *Bibliotheca patrum latinorum Italica III: Die Ambrosianische Bibliothek in Mailand*. Vol. 2.1. Sitzungsberichte der Kaiserlichen Akademie der Wissenschaften, philosophisch historische classe. Vienna: Gerold in Komm., 1871.

Reimitz, Helmut. *History, Frankish Identity and the Framing of Western Ethnicity*. Cambridge: Cambridge University Press, 2015.

Reuter, Timothy. "'Kirchenreform' und 'Kirchenpolitik' im Zeitalter Karl Martells: Begriffe und Wirklichkeit." In *Karl Martell in Seiner Zeit*, edited by Jörg Jarnut, Ulrich Nonn, and Michael Richter, 35–59. Beihefte der Francia: Forschungen zur Westeuropäischen Geschichte 37. Sigmaringen: Thorbecke, 1994.

Reynolds, Roger. *The Ordinals of Christ from Their Origins to the Twelfth Century*. Beiträge zur Geschichte und Quellenkunde des Mittelalters 7. Berlin: De Gruyter, 1978.

Riché, Pierre, and Georges Tate, eds. *Textes et documents d'histoire du Moyen-Âge, Ve-Xe siècles. Tome II: Milieu VIIIe siècle–Xe siècle*. Regards sur l'histoire 17. Paris: Société d'enseignement supérieur, 1974.

Rio, Alice. *Legal Practice and the Written Word in the Early Middle Ages: Frankish Formulae, c. 500–1000*. Cambridge Studies in Medieval Life and Thought: Fourth Series. Cambridge: Cambridge University Press, 2009.

Robertson, Duncan. *Lectio Divina: The Medieval Experience of Reading*. Cistercian Studies 238. Trappist: Cistercian Publications, 2011.

Robinson, Pamela R. "The 'Booklet': A Self-Contained Unit in Composite Manuscripts." *Codicologica* 3 (1980): 46–69; repr. in *The History of the Book in the West: 400 AD–1455*, edited by Pamela R. Robinson and Jane Roberts, 159–182. The History of the Book in the West: A Library of Critical Essays 1. Farnham: Ashgate, 2010 and Abingdon: Routledge, 2016.

Rossi-Reder, Andrea. "The Physiologus and Beast Lore in Anglo-Saxon England." PhD thesis, University of Connecticut, 1992.

Rowland, Beryl. "The Relationship of St. Basil's Hexameron to the Physiologus." In *Épopée animale, fable et fabliau: actes du colloque de la Société Internationale Renardienne qui s'est tenu à Evreux, 7–11 Septembre 1981*, edited by Gabriel Bianciotto, 489–98. Cahiers d'études médiévales 2–3. Paris: Presses universitaires de France, 1984.

Ryan, Michael. "Some Aspects of Sequence and Style in the Metalwork of Eighth- and Ninth-Century Ireland." In *Ireland and Insular Art, A.D. 500–1200: Proceedings of a Conference at University College Cork, 31 October–3 November 1985*, edited by Michael Ryan, 66–74. Dublin: Royal Irish Academy, 1987.

——. "The Menagerie of the Derrynaflan Chalice." In *The Age of Migrating Ideas: Early Medieval Art in Northern Britain and Ireland*, edited by Michael Spearman and John Higgitt, 151–61. Edinburgh: National Museums of Scotland, 1993.

Salisbury, Joyce E. *The Beast Within: Animals in the Middle Ages*. New York: Routledge, 2010.

Salvador-Bello, Mercedes. "Evidence of the Use of the Physiologus as a Source in Aldhelm's Enigmata." *The Review of English Studies*, 2020.

Salvat, Michel. "Notes sur les bestiaires catalans." In *Épopée Animale, Fable, Fabliau: Actes du IVe colloque de la Société Internationale Renardienne, Évreux, 7–11 Septembre 1981*, edited by Gabriel Bianciotto and Michel Salvat, 499–508. Paris: Presses Universitaires de France, 1984.

Sargan, James D. "What Could a Trans Book History Look Like? Toward Trans Codicology." *Criticism* 64 (2022): 571–86.

Sbordone, Francesco. "La tradizione manoscritta del Physiologus latino." *Athenaeum: Studi periodici di letteratura e storia dell'antichità* 27 (1949): 246–80.

——. *Physiologus*. Milan: Società Anonima Editrice "Dante Alighieri," 1936.

Scafi, Alessandro. *Mapping Paradise: A History of Heaven on Earth*. Chicago: University of Chicago Press, 2006.

Schannat, Johann Friedrich. *Historia Fuldensis in Tres Partes Divisa*. Frankfurt am Main: John Benjamin Andrae and Henry Hort, 1729.

Scheibelreiter, Georg. "Gegenwart und Vergangenheit in der Sicht Fredegars." In *The Medieval Chronicle II: Proceedings of the 2nd International Conference on the Medieval Chronicle, Driebergen/Utrecht, 16–21 July 1999*, edited by Erik Kooper, 212–22. Costerus New Series 144. Amsterdam: Rodopi, 2002.

Schenkluhn, Wolfgang. "Ottonische Renaissance." In *Auf den Spuren der Ottonen I: Protokoll des Wissenschaftlichen Kolloquiums anläßlich des 1000. Todestages der Reichsäbtissin Mathilde von Quedlinburg am 06.02.1999 in Quedlinburg*, edited by Cornelia Kessler, 89–98. Beiträge zur Regional- und Landeskultur Sachsen-Anhalts 13. Halle: Landesheimatbund Sachsen-Anhalt, 1999.

Scherrer, Gustav. *Verzeichniss der Handschriften der Stiftsbibliothek von St. Gallen*. Halle: Georg Olms, 1875.

Schmid, Joseph. *Die Osterfestberechnung in der abendländischen Kirche: Vom I. allgemeinen Konzil zu Nicäa bis zum Ende des VIII. Jahrhunderts*. Strassburger theologische Studien 9.1 Freiburg im Breisgau: Herder, 1907.

Schmitz, Wilhelm. "Notenschriftliches aus der Berner Handschrift 611." In *Commentationes Woelfflinianae: Eduard Woelfflino natalicia sexagesima gratulantur qui colligendo thesauro linguae latinae eo auctore per VII annos operam dederunt*, 7–13, Leipzig: Teubner, 1891.

——. "Schnellschriftliches aus der Berner Handschrift 611." *Deutsche Stenographenzeitung* 3 (1888): 360–67.

——. "Tironianum." In *Mélanges Julien Havet: Recueil de travaux d'érudition dédiés à la mémoire de Julien Havet (1853–1893)*, 77–80. Paris: E. Leroux, 1895.

——. "Tironische Noten aus der Pariser lateinischen Handschrift 10756." In *Franz Xaver Gabelsberger und seine Kunst: Festschrift Herausgegeben bei Gelegenheit der Enthüllung des Gabelsberger Denkmals am 10. August 1890 vom Hauptfestausschusse*, by Franz X. Gabelsberger, 116–20. Munich: G. Franz'scher, 1890.

Schönberger, Rolf, Andrés Quero Sánchez, Lu Jiang, and Brigitte Berges, eds. *Repertorium edierter Texte des Mittelalters aus dem Bereich der Philosophie und angrenzender Gebiete*. 2nd ed. Berlin: De Gruyter, 2012.

Schramm, Percy E., "Karl der Große: Denkart und Grundauffassungen. Die von ihm bewirkte correctio." *Historische Zeitschrift* 198 (1964): 306–45.

Schramm, Percy E., and Florentine Mütherich. *Denkmale der deutschen Könige und Kaiser*, Vol. 1, *Ein Beitrag zur Herrschergeschichte von Karl dem Großen bis Friedrich II. 768–1250*. Veröffentlichungen des Zentralinstituts für Kunstgeschichte in München 2. Munich: Prestel, 1964.

Schubel, Friedrich. *Englische Literaturgeschichte*, Vol. 1, *Die alt- und mittelenglische Periode*. Sammlung Göschen 1114. Berlin: De Gruyter, 1954.

Schulz, Monika. *Beschwörungen im Mittelalter: Einführung und Überblick*. Beiträge zur älteren Literaturgeschichte. Heidelberg: C. Winter, 2003.

Scott, Alan. "The Date of the Physiologus." *Vigiliae Christianae* 52 (1998): 430–41.

Scragg, Donald G. "Exeter Book." In *The Wiley-Blackwell Encyclopedia of Anglo-Saxon England*, edited by Michael Lapidge. Chichester: Wiley Blackwell, 2014.

Sears, Elizabeth. "Word and Image in Carolingian Carmina Figurata." In *World Art: Themes of Unity and Diversity. Acts of the XXVIth International Congress of the History of Art*, edited by Irving Lavin, 2:341–45. University Park: Pennsylvania State University Press, 1989.

Sebastián, Santiago. *El Fisiólogo atribuido a San Epifanio, seguido de el bestiario toscano*. Madrid: Ediciones Tuero, 1986.

Seel, Otto. *Der Physiologus: Tiere und ihre Symbolik*. 5th ed. Zurich: Artemis & Winkler, 1987.

Shailor, Barbara A. "A Cataloger's View." In *The Whole Book: Cultural Perspectives on the Medieval Miscellany*, edited by Stephen G. Nichols and Siegfried Wenzel, 153–67. Ann Arbor: University of Michigan Press, 1996.

Siegmund, Albert. *Die Überlieferung der griechischen christlichen Literatur in der lateinischen Kirche bis zum zwölften Jahrhundert.* Abhandlungen der bayerischen Benediktiner-Akademie 5. Munich: Filser, 1949.

Siewers, Alfred K. *Strange Beauty: Ecocritical Approaches to Early Medieval Landscape.* Basingstoke: Palgrave Macmillan, 2009.

Silvestre, Hubert. "À propos du 'Bruxellensis' 10066–77 et de son noyau primitif." In *Miscellanea codicologica F. Masai dicata 1*, edited by Pierre Cockshaw, Monique-Cécile Garand, and Pierre Jodogne, 131–56. Les publications de Scriptorium 8. Ghent: Story-Scientia, 1979 (reprinted in *Scriptorium* 34 (1980): 131–50).

———. "Antiennes de matines d'un ancien office de l'Assumption ou de la purification." *Ephemerides Liturgicae* 67 (1953): 138–46.

———. "Aperçu sur les commentaires carolingiens de Prudence." *Sacris Erudiri* 9 (1957): 50–74 and 396.

———. "Un cliché peu étudié: 'fortis in armis'." *Archivum latinitatis medii aevi* 32 (1962): 255–57.

———. "Une adaptation du commentaire de Macrobe sur le songe de Scipion dans un manuscrit de Bruxelles." *Archives d'histoire doctrinale et littéraire du Moyen-Âge* 29 (1961): 93–101.

———. "Une copie du Xe siècle non utilisée du premier glossaire grec-latin Abscida lucida (fragment)." *Archivum latinitatis medii aevi* 21 (1951): 159–70.

Sinner, Jean R. *Catalogus Codicus MSS Bibliothecae Bernesis.* Bern: Brunner & Haller, 1760.

Smith, Julia Mary Howard. "'Emending Evil Ways and Praising God's Omnipotence': Einhard and the Uses of Roman Martyrs." In *Conversion in Late Antiquity and the Middle Ages: Seeing and Believing*, edited by Kenneth Mills and Anthony Grafton, 189–223. Rochester: University of Rochester Press, 2003.

Smyth, Marina B. "The Seventh-Century Hiberno-Latin Treatise *Liber de Ordine Creaturarum*: A Translation." *The Journal of Medieval Latin* 21 (2011): 137–222.

———. "The Word of God and Early Medieval Irish Cosmology: Scripture and the Creating Word." In *Celtic Cosmology: Perspectives from Ireland and Scotland*, edited by Jacqueline Borsje, Ann Dooley, Séamus Mac Mathúna, and Gregory Toner, 112–43. Papers in Mediaeval Studies 26. Toronto: Pontifical Institute of Mediaeval Studies, 2014.

Squatriti, Paolo. *Landscape and Change in Early Medieval Italy: Chestnuts, Economy, and Culture.* Cambridge: Cambridge University Press, 2013.

———. *Water and Society in Early Medieval Italy, AD 400–1000.* Cambridge: Cambridge University Press, 2002.

———. *Weeds and the Carolingians: Empire, Culture, and Nature in Frankish Europe, AD 750–900.* Cambridge: Cambridge University Press, 2022.

Stancliffe, Clare. "The Thirteen Sermons Attributed to Columbanus and the Question of Their Authorship." In *Columbanus: Studies on the Latin Writings*, edited by Michael Lapidge, 93–202. Studies in Celtic History 17. Woodbridge: Boydell Press, 1997.

Stansbury, Mark. "*Sammelhandschriften* and the *Breuiarium librorum* in Sankt Gallen 728." In *The Art of Compilation: Manuscripts and Networks in the Early Medieval Latin West*, edited by Anna Dorofeeva and Michael Kelly. Binghamton: Gracchi Books, forthcoming in 2023.

Stefaniw, Blossom. *Christian Reading: Language, Ethics, and the Order of Things.* Oakland: University of California Press, 2019.

Steiger, Christoph von, and Otto Homburger. *Physiologus Bernensis. Voll-Faksimile-Ausgabe des Codex Bongarsianus 318 der Burgerbibliothek Bern.* Basel: Alkuin, 1964.

Steinmeyer, Elias von. *Die kleineren althochdeutschen Sprachdenkmäler*. Berlin: Weidmann, 1916.

Steinmeyer, Elias von, and Eduard Sievers, ed. *Die althochdeutschen Glossen*. 5 vols. Berlin: Weidmann, 1879–1922 (reprinted 1968–1969).

Steinová, Evina. *Notam superponere studui: The Use of Annotation Symbols in the Early Middle Ages*. Bibliologia 52. Turnhout: Brepols, 2019.

——. "*Psalmos, notas, cantus*: On the Meanings of *Nota* in the Carolingian Period." *Speculum* 90 (2015): 424–57.

Stettiner, Richard. *Die illustrierten Prudentius-Handschriften*. 2 vols., Vol. 1. Berlin: Grote, 1895–1905.

Stewart, Patricia. "The Mediaeval Bestiary and its Textual Tradition." PhD thesis, University of St Andrews, 2012.

Stock, Brian. *The Implications of Literacy: Written Language and Models of Interpretation in the Eleventh and Twelfth Centuries*. Princeton: Princeton University Press, 1983.

Stork, Nancy P. "A Spanish Bishop Remembers the Future: Oral Traditions and Purgatory in Julian of Toledo." *Oral Tradition* 23 (2008): 43–70.

Story, Joanna. "Aldhelm and Old St Peter's, Rome." *Anglo-Saxon England* 39 (2010): 7–20.

Suchier, Walther. *Das mittellateinische Gespräch Adrian und Epictitus nebst verwandten Texten (Joca monachorum)*. Tübingen: Niemeyer, 1955.

Suciu, Alin. "Quotations from the Physiologus in a Homily of the Coptic Holy Week Lectionary." In *Synaxis Katholikē: Beiträge zu Gottesdienst und Geschichte der fünf Altkirchlichen Patriarchate für Heinzgerd Brakmann zum 70. Geburtstag*, edited by Diliana Atanassova, Tinatin Chronz, and Heinzgerd Brakmann, 677–89. Orientalia, Patristica, Oecumenica, 6.1–2. Vienna: Lit, 2014.

Sullivan, Richard E. *The Gentle Voices of Teachers: Aspects of Learning in the Carolingian Age*. Columbus: Ohio State University Press, 1995.

Swarzenski, Georg. "Die karolingische Malerei und Plastik in Reims." *Jahrbuch der Königlich-Preußischen Kunstsammlungen* 23 (1902): 81–100.

——. *Die Regensburger Buchmalerei des X. und XI. Jahrhunderts. Studien zur Geschichte der deutschen Malerei des frühen Mittelalters*. Denkmäler der süddeutschen Malerei des frühen Mittelalters 1. Leipzig: Hiersemann, 1901.

Swarzenski, Hanns. *Monuments of Romanesque Art: The Art of Church Treasures in North-Western Europe*. London: Faber, 1954.

Tahkokallio, Jaakko. "Manuscripts as Evidence for the Use of Classics in Education, *c.* 800–1200: Estimating the Randomness of Survival." *Interfaces: A Journal of Medieval European Literatures* (2017): 28–45.

Teeuwen, Mariken. "Glossing in Close Co-Operation: Examples from Ninth-Century Martianus Capella Manuscripts." In *Practice in Learning: The Transfer of Encyclopaedic Knowledge in the Early Middle Ages*, edited by Rolf H. Bremmer and Cornelis Dekker, 85–100. Mediaevalia Groningana 16. Paris: Peeters, 2010.

——. *The Vocabulary of Intellectual Life in the Middle Ages*. Études sur le vocabulaire intellectuel du Moyen Age 10. Turnhout: Brepols, 2003.

Teeuwen, Mariken, and Sinéad O'Sullivan, eds. *Carolingian Scholarship and Martianus Capella: Ninth-Century Commentary Traditions on De Nuptiis in Context*. Cultural Encounters in Late Antiquity and the Middle Ages 12. Turnhout: Brepols, 2011.

Thielo, Karl. *De glossario codicis Monac. 14388*. Commentationes philologae Ienenses 11. Leipzig: Teubner, 1914.

Traube, Ludwig. *Vorlesungen und Abhandlungen*, Vol. 2, *Einleitung in die lateinische Philologie des Mittelalters*, edited by Franz Boll and Paul Lehmann. Munich: Beck, 1911.

Travis, William J. "Of Sirens and Onocentaurs: A Romanesque Apocalypse at Montceaux-l'Étoile." *Artibus et Historiae* 23 (2002): 29–62.

Treharne, Elaine. *Perceptions of Medieval Manuscripts: The Phenomenal Book.* Oxford: Oxford University Press, 2021.

Treu, Ursula. "Zur biblischen Überlieferung im Physiologus." In *Überlieferungsgeschichtliche Untersuchungen*, edited by Franz Paschke, 549–52. Berlin: Akademie-Verlag, 1981.

——. "Zur Datierung des Physiologus." *Zeitschrift für die neutestamentliche Wissenschaft* 57 (1966): 101–4.

Tselos, Dimitri. "A Greco-Italian School of Illuminators and Fresco Painters: Its Relation to the Principal Reims Manuscripts and to the Greek Frescoes in Rome and Castelseprio." *The Art Bulletin* 38 (1956): 1–30.

Turcan-Verkerk, Anne-Marie. "Langue et littérature latines du Moyen Âge." *Annuaire de l'École Pratique des Hautes Études (EPHE), Section des Sciences Historiques et Philologiques* 145 (2014): 138–54.

Ullmann, Walter. *The Carolingian Renaissance and the Idea of Kingship.* London: Methuen, 1969.

Vaccari, Alberto. "Una fonte del 'De ortu et obitu patrum' di S. Isidoro." In *Scritti di erudizione e di filologia*, Vol. 2, *Per la storia del testo e dell'esegesi biblica*, edited by Alberto Vaccari, 271–81. Storia e Letteratura. Raccolta di studi e testi 67. Rome: Edizioni di Storia e Letteratura, 1958.

Van den Broek, Roelof. *The Myth of the Phoenix, According to Classical and Early Christian Traditions*. Études préliminaires aux religions orientales dans l'empire romain 24. Leiden: Brill, 1972.

van Rhijn, Carine. "Introduction." In *Rethinking the Carolingian Reform*, edited by Carine van Rhijn, Ingrid Rembold, and Arthur Westwell. Manchester: Manchester University Press, 2023.

——. *Leading the Way to Heaven. Pastoral Care and Salvation in the Carolingian Period.* Abingdon: Routledge, 2022.

Vanderputten, Steven. *Dark Age Nunneries: The Ambiguous Identity of Female Monasticism, 800–1050*. Ithaca: Cornell University Press, 2018.

Vidal, José A. V., Pilar D. Álvarez, and Javier M. Osende. "La versión C del Fisiólogo latino: El Codex Bongarsianus 318 de Berna." *Medievalismo* 10 (2000): 27–67.

Vieillard-Troiekouroff, May. "Sirènes-poissons carolingiennes." *Cahiers archéologiques* 19 (1969): 61–83.

Villar Vidal, José A., and Pilar D. Álvarez. "El Fisiólogo latino: versión B. 1. Introducción y texto latino." *Revista de Literatura Medieval* 15 (2003): 9–52.

Vocino, Georgia. "Migrant Masters and Their Books. Italian Scholars and Knowledge Transfer in Post-Carolingian Europe." In *Using and Not Using the Past After the Carolingian Empire, c. 900–c. 1050*, edited by Sarah Greer, Alice Hicklin, and Stefan Esders, 241–61. Abingdon: Routledge, 2020.

Waagmeester, Bastiaan. "Pastoral Works. Priests, Books and Compilative Practices in the Carolingian Period." PhD thesis, Tübingen University, 2021.

Wallach, Liutpold. *Alcuin and Charlemagne. Studies in Carolingian History and Literature.* Cornell Studies in Classical Philology 32. New York: Cornell University Press, 1959.

Walz, Dorothea. *Die historischen und philosophischen Handschriften der Codices Palatini Latini in der Vatikanischen Bibliothek (Cod. Pal. Lat. 921–1078)*, edited by Veit Probst and Karin Zimmermann. Kataloge der Universitätsbibliothek Heidelberg 3. Wiesbaden: Ludwig Reichert, 1999.

Warntjes, Immo. *The Munich Computus: Text and Translation. Irish Computistics Between Isidore of Seville and the Venerable Bede and Its Reception in Carolingian Times*. Stuttgart: Steiner, 2010.

Wegera, Klaus-Peter. "Zur Rezeption des Physiologus im Spätmittelalter und in der frühen Neuzeit." *Germanistik. Publications du Centre universitaire de Luxembourg* 8 (1996): 73–86.

Wellmann, Max. *Der Physiologos: Eine religionsgeschichtlich-naturwissenschaftliche Untersuchung*. Philologus, Supplementband 22. Leipzig: Dietrich'sche Verlagsbuchhandlung, 1930.

West, Charles. *Reframing the Feudal Revolution: Political and Social Transformation Between Marne and Moselle, c. 800–c. 1100*. Cambridge: Cambridge University Press, 2013.

Westwell, Arthur. "The Content and the Ideological Construction of the Early Pontifical Manuscripts." *Mélanges de l'École Française de Rome - Moyen Âge* 132, no. 1 (2020). DOI: 10.4000/mefrm.7681.

White, Hayden. "The Forms of Wildness: The Archaeology of an Idea." In *Tropics of Discourse: Essays in Cultural Criticism*, by Hayden White, 150–82. Baltimore: Johns Hopkins University Press, 1978.

White, Lynn T. "The Historical Roots of Our Ecologic Crisis." *Science* 155, no. 3767 (1967): 1203–7 (reprinted in Lynn T. White, *Dynamo and Virgin Reconsidered: Essays in the Dynamism of Western Culture*. Cambridge, MA: MIT Press, 1968, 75–94).

——. *Medieval Religion and Technology: Collected Essays*. Center for Medieval and Renaissance Studies, UCLA 13. Berkeley: University of California Press, 1978.

Whited, Tamara L., Jens I. Engels, Richard C. Hoffmann, Hilde Ibsen, and Wybren Verstegen, eds. *Northern Europe: An Environmental History*. Santa Barbara: ABC-CLIO, 2005.

Whitney, Elspeth. *Medieval Science and Technology*. Westport: Greenwood Press, 2004.

——. "Paradise Restored: The Mechanical Arts from Antiquity Through the Thirteenth Century." *Transactions of the American Philosophical Society* 80 (1990: 1–169.

Wildberg, Christian. "Neoplatonism." In *The Stanford Encyclopedia of Philosophy*, edited by Edward N. Zalta, n.d. https://plato.stanford.edu/archives/spr2016/entries/neoplatonism/.

Williamson, Tom. *Environment, Society and Landscape in Early Medieval England: Time and Topography*. Rochester: Boydell Press, 2013.

Willjung, Harald. "Zur Überlieferung der Epistola de processione Spiritus sancti Smaragds von Saint-Mihiel." *Deutsches Archiv für Erforschung des Mittelalters* 47 (1991): 161–66.

Wilmanns, Wilhelm. "Ein Fragebüchlein aus dem neunten Jahrhundert." *Zeitschrift für Deutsches Altertum* 15 (1872): 166–80.

Wittkower, Rudolph. "'Physiologus' in Beatus Manuscripts." *Journal of the Warburg Institute* 1 (1938): 253–54.

Wolfram, Herwig, Andreas Kusternig, and Herbert Haupt, eds. *Die vier Bücher der Chroniken des sogenannten Fredegar (Buch 2, Kapitel 53 bis Buch 4, unwesentlich gekürzt). Die Fortsetzungen der Chroniken des sogenannten Fredegar*. Ausgewählte Quellen zur deutschen Geschichte des Mittelalters 4a. Darmstadt: Wissenschaftliche Buchgesellschaft, 1982.

Wood, Ian N. "Fredegar's Fables." In *Historiographie im frühen Mittelalter*, edited by Anton Scharer and Georg Scheibelreiter, 359–66. Veröffentlichungen des Instituts für Österreichische Geschichtsforschung 32. Vienna: Oldenbourg, 1994.

——. "The Problem of Late Merovingian Culture." In *Exzerpieren – Kompilieren – Tradieren*, edited by Stephan Dusil, Gerald Schwedler, and Raphael Schwitter, 199–222. Berlin: De Gruyter, 2016.

Woodruff, Helen. "The Illustrated Manuscripts of Prudentius." *Art Studies* 7 (1929): 33–79.

——. "The *Physiologus* of Bern: A Survival of Alexandrian Style in a Ninth Century Manuscript." *The Art Bulletin* 12 (1930): 226–53.

Wormald, Patrick, and Janet L. Nelson, eds. *Lay Intellectuals in the Carolingian World*. Cambridge: Cambridge University Press, 2011.

Wright, Charles D. "Bischoff's Theory of Irish Exegesis and the Genesis Commentary in Munich Clm 6302: A Critique of a Critique." *Journal of Medieval Latin* 10 (2000): 136–37.

——. "From Monks' Jokes to Sages' Wisdom: The *Joca Monachorum* Tradition and the Irish *Immacallam in Dá Thúarad*." In *Spoken and Written Language: Relations Between Latin and the Vernacular Languages in the Earlier Middle Ages: Papers from the Second Utrecht Symposium on Medieval Literacy, Utrecht, 24–26 June 1999*, edited by Mary Garrison, Arpad P. Orbán, and Marco Mostert, 199–225. Turnhout: Brepols, 2003.

——. *The Irish Tradition in Old English Literature*. Cambridge Studies in Anglo-Saxon England 6. Cambridge: Cambridge University Press, 1993.

Wright, Roger. *Late Latin and Early Romance in Spain and Carolingian France*. ARCA: Classical and Medieval Texts, Papers, and Monographs 8. Liverpool: F. Cairns, 1982.

Wüstefeld, Wilhelmina C. "Catalogue. The School of Rheims 9: Bern Physiologus." In *The Utrecht Psalter in Medieval Art: Picturing the Psalms of David*, edited by Koert van der Horst, William Noel, and Wilhelmina C. M. Wüstefeld, 190–91. Utrecht: HES, 1996.

Yapp, Brunsdon W. "A New Look at English Bestiaries." *Medium Aevum* 54 (1985): 1–19.

Yavuz, N. Kıvılcım. "Transmission and Adaptation of the Trojan Narrative in Frankish History between the Sixth and Tenth Centuries." PhD thesis, University of Leeds, 2015.

Young, Arthur M. *Troy and Her Legend*. Pittsburgh: University of Pittsburgh Press, 1948.

# GENERAL INDEX

Page numbers in italics refer to plates.

Printed and bound by CPI Group (UK) Ltd, Croydon, CR0 4YY

13/09/2026

14961066-0002